Lecture Notes
in Business Information Processing

534

AF147905

Series Editors

Wil van der Aalst , *RWTH Aachen University, Aachen, Germany*
Sudha Ram , *University of Arizona, Tucson, AZ, USA*
Michael Rosemann , *Queensland University of Technology, Brisbane, QLD, Australia*
Clemens Szyperski, *Microsoft Research, Redmond, WA, USA*
Giancarlo Guizzardi , *University of Twente, Enschede, The Netherlands*

LNBIP reports state-of-the-art results in areas related to business information systems and industrial application software development – timely, at a high level, and in both printed and electronic form.

The type of material published includes

- Proceedings (published in time for the respective event)
- Postproceedings (consisting of thoroughly revised and/or extended final papers)
- Other edited monographs (such as, for example, project reports or invited volumes)
- Tutorials (coherently integrated collections of lectures given at advanced courses, seminars, schools, etc.)
- Award-winning or exceptional theses

LNBIP is abstracted/indexed in DBLP, EI and Scopus. LNBIP volumes are also submitted for the inclusion in ISI Proceedings.

Katarzyna Gdowska ·
María Teresa Gómez-López ·
Jana-Rebecca Rehse
Editors

Business Process Management Workshops

BPM 2024 International Workshops
Krakow, Poland, September 1–6, 2024
Revised Selected Papers

 Springer

Editors
Katarzyna Gdowska Ⓘ
AGH University of Krakow
Kraków, Poland

María Teresa Gómez-López Ⓘ
Universidad de Sevilla
Seville, Spain

Jana-Rebecca Rehse Ⓘ
University of Mannheim
Mannheim, Germany

ISSN 1865-1348   ISSN 1865-1356 (electronic)
Lecture Notes in Business Information Processing
ISBN 978-3-031-78665-5   ISBN 978-3-031-78666-2 (eBook)
https://doi.org/10.1007/978-3-031-78666-2

This Springer imprint is published by the registered company Springer Nature Switzerland AG
The registered company address is: Gewerbestrasse 11, 6330 Cham, Switzerland

If disposing of this product, please recycle the paper.

Preface

The International Conference on Business Process Management (BPM) was established 22 years ago as a venue where people from academia and industry meet to discuss the latest developments in the area of business process management. In 2024, the conference took place in Krakow, Poland, from September 1st to 6th. The BPM conference is accompanied by a workshop program, offering a platform for workshops focused on specialized themes, interdisciplinary issues and emerging trends and paradigms in all aspects of BPM. This volume collects the proceedings of the BPM 2024 workshops, held prior to the main conference on September 2nd, 2024.

BPM 2024 invited proposals for one-day or half-day workshops. During the review and acceptance process, the goal was to curate a workshop program that covered the full spectrum of BPM research. Also, priority was given to workshops that not only covered engaging topics but also promised an innovative format aimed at sparking active discussions and generating new ideas. Examples of such formats included panels to bring together practitioners and researchers or sessions for young researchers to present research-in-progress papers. From 14 submitted proposals, eleven were originally selected. These included ten established workshops and one new one, focused on process innovation and value creation. In the end, nine workshops were held in conjunction with BPM 2024:

- *2nd International Workshop on Formal Methods for Business Process Management* (FM-BPM) organized by Claudio Di Ciccio, Alessandro Gianola and Andrey Rivkin. FM-BPM centered on the application of Formal Methods for Business Process Management (BPM), addressing the challenges of ensuring process correctness amidst the complexity of contemporary business operations. Formal Methods, which leverage rigorous mathematical tools and techniques, provide a robust framework for specifying, developing and analyzing business processes to prevent design errors that could lead to unintended outcomes or violations of objectives. Although there were no formal submissions, the event, had strong participation, with 20 to 25 attendees actively engaging in discussions on various relevant topics. A panel of experts also explored the current state of Formal Methods in BPM and identified key challenges to be addressed in the future, making the workshop a significant contribution to the BPM field[1].
- *8th International Workshop on Artificial Intelligence for Business Process Management* (AI4BPM) organized by Chiara Di Francescomarino, Fabrizio Maria Maggi, Andrea Marrella, Arik Senderovich and Emilio Sulis. AI4BPM provided a forum for researchers and professionals to explore, envision, and discuss the challenges and opportunities associated with transitioning from traditional programmatic approaches to emerging AI-enabled BPM solutions.

---

[1] Papers presented at this workshop are not included in this volume.

- *3rd International Workshop on Data-Driven Business Process Optimization* (BPO) organized by Remco Dijkman and Arik Senderovich. BPO brought together researchers from BPM, Operations Research, and other related disciplines, aiming to create techniques to optimize business processes in organizations using models based on real-world data.
- *8th International Workshop on Business Processes Meet the Internet-of-Things* (BP-Meet-IoT) organized by Agnes Koschmider, Francesco Leotta, Massimo Mecella, Estefanía Serral and Victoria Torres. BP-Meet-IoT discussed the current state of ongoing research, industry needs, future trends and practical experiences in integrating the IoT and BPM fields.
- *17th International Workshop on Social and Human Aspects of Business Process Management* (BPMS2) organized by Rainer Schmidt and Selmin Nurcan. BPMS2 focused on the social and human aspects of BPM, promoting a human-centric approach that prioritizes people in process design. It leverages AI assistants for personalized service delivery and highlights the importance of social interactions and platforms. Additionally, the workshop highlighted the role of organizational culture in adopting new processes, as well as the ethical principles, diversity and inclusion essential to BPM, advocating for a more participatory, collaborative and equitable BPM practice.
- *3rd International Workshop on Natural Language Processing for Business Process Management* (NLP4BPM) organised by Manuel Resinas, Han van der Aa, Adela del-Río-Ortega and Henrik Leopold. NLP4BPM served as a platform for researchers and practitioners to present, discuss, and evaluate how natural language processing (NLP) can be used to create new methods or improve existing techniques, tools and process-aware systems that aid in the different phases of the BPM lifecycle.
- *2nd International Workshop on Object-centric processes from A to Z* (OBJECTS) organised by Marco Montali, Andrey Rivkin and Jan Martijn van der Werf. OBJECTS welcomed contributions that present innovative ideas and advancements in object-centric processes, aiming to stimulate discussions on the conceptual and technical challenges present in the research domain of object-centric processes.
- *2nd International Workshop on Change, Drift, and Dynamics of Organizational Processes* (ProDy) organised by Bastian Wurm, Waldemar Kremser and Jan Mendling. ProDy called for conceptual, empirical and algorithm engineering papers that explore the dynamic aspects of business processes and organizational routines.
- *1st International Workshop on Managing Process Innovation and Value Creation in the Era of Digital Transformation* (Innov8BPM) organized by Banu Aysolmaz, Amy Van Looy, Oktay Turetken, Marta Indulska, Flavia Santoro and Panagiotis Keramidis. Innov8BPM focused on the essential advancements needed in the BPM discipline to effectively manage process innovation and generate business value in today's rapidly changing environment. Contributions examine the managerial implications of emerging BPM capabilities in the context of digital transformation and the creation of business value through BPM.

A total of 64 submissions were received across all workshops. Each workshop had its own independent Program Committee responsible for selecting papers for publication. Every submission underwent at least three single-blind reviews, where 30 papers were chosen for presentation at the workshops, resulting in an acceptance rate of 47%. Additionally, a full paper from the BPO competition was invited.

We extend our gratitude to all the workshop proposers and organizers, authors, reviewers, keynote speakers, presenters, and attendees of the BPM 2024 workshops for their invaluable contributions to the creation and dissemination of knowledge in the field of business process management. Our thanks also go to the organizers and volunteers of the BPM 2024 conference and to AGH University of Krakow for being an excellent host and ensuring a successful and enjoyable event. For the preparation of these proceedings, we are particularly indebted to the proceedings chair Simone Agostinelli, who invested an incredible amount of time and patience to facilitate the process. In addition, we also want to thank Springer for their support in publishing the proceedings.

September 2024

Katarzyna Gdowska
María Teresa Gómez-López
Jana-Rebecca Rehse

Contents

8th International Workshop on Artificial Intelligence for Business Process Management (AI4BPM 2024)

The Role of Trust in AI-Augmented Business Process Management Systems . . .   5
Giacomo Acitelli, Simone Agostinelli, Angelo Casciani, and Andrea Marrella

Knowledge Graphs: A Key Technology for Explainable Knowledge-Aware
Process Automation?   18
Leon Bein and Luise Pufahl

Graph Neural Networks for PPM: Review and Benchmark for Next
Activity Predictions   31
Sebastiano Dissegna and Chiara Di Francescomarino

Bridging Domain Knowledge and Process Discovery Using Large
Language Models   44
Ali Norouzifar, Humam Kourani, Marcus Dees, and Wil M. P. van der Aalst

3rd International Workshop on Data-Driven Business Process Optimization (BPO 2024)

The Business Process Optimization Competition   61
Felix Schumann, Matthias Ehrendorfer, Michel Kunkler, Kiran Busch, Henrik Leopold, Leon Urny, Martin Schmauch, Olga Rodzik, Stefanie Lanz, Efe Tıraş, Amgad Al-Zamkan, Jana El Kari, and Remco Dijkman

Ad-Hoc Subprocesses – The Missing Link Between Scheduling
and Business Process Modelling   73
Asvin Goel

Applying Process Mining on Scientific Workflows: A Case Study on High
Performance Computing Data   84
Zahra Sadeghibogar, Alessandro Berti, Marco Pegoraro, and Wil M. P. van der Aalst

8th International Workshop on Business Processes Meet Internet-of-Things (BP-Meet-IoT 2024)

Check My Flow: Distributed Conformance Checking at the Source 101
  Julia Andersen, Patrick Rathje, and Olaf Landsiedel

From IoT Event Logs to Human Routines via Community Detection
Algorithms 113
  Massimo Callisto De Donato, Fabrizio Fornari,
  and Abel Armas-Cervantes

Machinery Activity Recognition in the Industry Based on Heterogeneous
Data 125
  Marta Podobińska-Staniec, Marek Kęsek, and Edyta Brzychczy

LLM-Based Event Abstraction and Integration for IoT-Sourced Logs 138
  Mohsen Shirali, Mohammadreza Fani Sani, Zahra Ahmadi,
  and Estefanía Serral

17th International Workshop on Social and Human Aspects of Business Process Management (BPMS2 2024)

Non-visual Process Models: How Do Blind and Low-Vision Users Model
Business Processes? 155
  Lisa Baumann, Anjo Seidel, and Mathias Weske

Diversity and Inclusion in HR Processes in Financial Sector 168
  Sylwia Białas and Piotr Wróbel

Mining for Well-Being: The Potential of Process Mining for Evaluating
Employee Well-Being 180
  Mari A. J. Braakman, Jos Zuijderwijk, Iris Beerepoot, Sven Lugtigheid,
  Thomas Martens, Maria Peeters, Eva Knies, and Hajo A. Reijers

3rd International Workshop on Natural Language Processing for Business Process Management (NLP4BPM 2024)

Straight Outta Logs: Can Large Language Models Overcome
Preprocessing in Next Event Prediction? 197
  Katharina Brennig, Sascha Kaltenpoth, and Oliver Müller

Enhancement of Low-Level Event Abstraction with Large Language
Models (LLMs) 209
  Edyta Brzychczy, Krzysztof Kluza, and Leszek Szała

ProcessLLM: A Large Language Model Specialized in the Interpretation,
Analysis, and Optimization of Business Processes 221
    Alina Buss, Wolfgang Kratsch, Sebastian Johannes Schmid,
    and Hongyang Wang

Towards a Benchmark for Causal Business Process Reasoning with LLMs 233
    Fabiana Fournier, Lior Limonad, and Inna Skarbovsky

Using Large Language Models to Generate Process Knowledge
from Enterprise Content 247
    Sandro Franzoi, Maxime Delwaulle, Julian Dyong, Jan Schaffner,
    Mara Burger, and Jan vom Brocke

Efficient LLM-Based Conversational Process Modeling 259
    Julius Köpke and Aya Safan

Leveraging Generative Vision Models for Extracting Process Models
from Documents 271
    Marvin Voelter, Raheleh Hadian, Timotheus Kampik,
    Marius Breitmayer, and Manfred Reichert

2nd International Workshop on Object-centric Processes from A to Z (OBJECTS 2024)

Coordination Process Verification for Object-Centric Business Processes 287
    Lisa Arnold and Manfred Reichert

Transforming Object-Centric Event Logs to Temporal Event Knowledge
Graphs 300
    Shahrzad Khayatbashi, Olaf Hartig, and Amin Jalali

2nd International Workshop on Change, Drift, and Dynamics of Organizational Processes (ProDy 2024)

Forging the LongSWORD: Exaptation and Enhancement of the SWORD
Framework for Workaround Detection 319
    Bernd Löhr, Christian Bartelheimer, Frank Köhne, Sina Nordlohne,
    Daniel Alile, and Andrees Latten

First Insights into the Impact of Concept Drift on Process Complexity 332
    Maxim Vidgof

Visualizing Routine Dynamics in Outpatient Medical Clinics
with Topological Data Analysis 338
    Li Zhang, Julie Ryan Wolf, Alice P. Pentland, and Brian T. Pentland

**1st International Workshop on Managing Process Innovation and
Value Creation in the Era of Digital Transformation (Innov8BPM 2024)**

Process Mining Pipelines with Controlled Sharing of Data and Algorithms 357
    Andrea Burattin, Ekkart Kindler, Nicholas Dyhre, Sebastian Vestrup,
    Francesca Zerbato, and Barbara Weber

How Can We Develop Inclusive Business Process Management? 370
    E. R. Mahendrawathi

Business Value of Process Mining: A Contingency Perspective 376
    Astria Hijriani and Marco Comuzzi

Towards a Unified Approach: Developing a Reference Model for Digital
Twins and Business Process Management in Clinical Trials 382
    Gerald Kremer, Luiz Ricardo Brito Ribeiro, Till Blüher,
    Silvia Dallavalle, and Rainer Stark

Business in the Age of Platform Economics: Managing Decentralised
Business Processes Beyond Blockchain 394
    Fabian Stiehle, Finn Klessascheck, Martin Kjäer, and Ingo Weber

Author Index 401

8th International Workshop on Artificial Intelligence for Business Process Management (AI4BPM 2024)

8th International Workshop on Artificial Intelligence for Business Process Management (AI4BPM 2024)

With the growing importance of Artificial Intelligence (AI), numerous novel methodologies and techniques are emerging and being applied in a number of different areas. The application of AI methods to Business Process Management (BPM) is one of these areas which is recently attracting the attention of both industry and academia. The use of AI in BPM has been discussed as one of the emerging technologies that will touch upon almost all business process activities performed by humans. In some cases, AI will dramatically simplify human interaction with processes, while in other cases it will enable full automation of tasks that have traditionally required manual labor. We believe that, in the future, AI may have an important impact in all the phases of the BPM lifecycle: modeling, analysis, automation, implementation and monitoring. Future AI-augmented BPM technologies will enable constant improvement and adaptation based on continuous experiential learning, thus supporting humans in a number of tasks, such as analysis and decision making.

The goal of the AI4BPM workshop is to establish a forum for researchers and professionals interested in understanding, envisioning and discussing challenges and opportunities of moving from current, largely programmatic approaches for BPM to emerging forms of AI-driven BPM.

This year, a keynote speech by Marco Comuzzi opened the program of the workshop with a talk on challenges and future directions of the Predictive Process Monitoring research field. In addition, the workshop attracted 8 submissions on different topics including the usage of Large Language Models for different purposes, Predictive Process Monitoring, AI-augmented Business Process Management and Process Automation. All submissions were reviewed by at least 3 program committee members (or their sub-reviewers) and eventually 4 papers were accepted.

Acitelli et al. delved into the importance and the role of trust in AI-augmented Business Process Management Systems, proposing and testing a classification framework that identifies the trustworthy AI principles each AI-augmented Business Process Management stage should conform to. Norouzifar et al. proposed an approach leveraging Large Language Models to integrate domain knowledge in business process discovery. Bein et al. investigated the potential of knowledge graphs for explainable knowledge-aware process automation. Finally, Dissegna et al. reviewed and benchmarked existing Predictive Process Monitoring approaches using Graph Neural Networks.

The presentation of each paper stimulated questions and an interesting discussion in the audience. The workshop confirmed itself as a reference venue for researchers working on the application of AI techniques in the BPM field.

Organization

Organizing Committee

Chiara Di Francescomarino	University of Trento, Italy
Fabrizio Maria Maggi	Free University of Bozen-Bolzano, Italy
Andrea Marrella	Sapienza Università di Roma, Italy
Arik Senderovich	York University, Canada
Emilio Sulis	University of Turin, Italy

Program Committee

Han van der Aa	University of Mannheim, Germany
Anti Alman	University of Tartu, Estonia
Matteo Baldoni	University of Turin, Italy
Patrizio Bellan	FBK, Italy
Andrei Buliga	FBK, Italy
Marco Comuzzi	Ulsan Institute, Republic of Korea
Francesco Corcoglioniti	Free University of Bozen-Bolzano, Italy
Claudio Di Ciccio	Sapienza University of Rome, Italy
Ivan Donadello	Free University of Bozen-Bolzano, Italy
Joerg Evermann	Memorial University of Newfoundland, Canada
Stephan Fahrenkrog-Petersen	Humboldt University of Berlin, Germany
Peter Fettke	DFKI, Germany
Francesco Folino	CNR, Italy
Fabiana Fournier	IBM Research AI, Israel
Krzysztof Kluza	AGH University of Science and Technology, Poland
Henrik Leopold	Kühne Logistics University, Germany
Lior Limonad	IBM Research AI, Israel
Elisa Marengo	University of Turin, Italy
Francesca Meneghello	FBK and Sapienza University, Italy
Roberto Micalizio	University of Turin, Italy
Roberto Nai	University of Turin, Italy
Andrey Rivkin	Technical University of Denmark, Denmark
Massimiliano Ronzani	Fondazione Bruno Kessler, Italy
Giulia Ruffini	University of Turin, Italy
Tijs Slaats	University of Copenhagen, Denmark

Heiner Stuckenschmidt   University of Mannheim, Germany
Stefano Tedeschi   Università della Valle d'Aosta, Italy
Daniele Theseider Dupré   University of Eastern Piedmont, Italy
Matteo Zavatteri   University of Padua, Italy

The Role of Trust in AI-Augmented Business Process Management Systems

Giacomo Acitelli, Simone Agostinelli[(✉)], Angelo Casciani,
and Andrea Marrella

Sapienza Universitá di Roma, Rome, Italy
{acitelli,agostinelli,casciani,marrella}@diag.uniroma1.it

Abstract. The significant impact of Artificial Intelligence (AI) on academia, industry, and government has led to a strong focus on the realization of trustworthy AI systems. Among them, there is an emerging class of process-aware information systems infused with AI, called AI-Augmented Business Process Management Systems (ABPMSs), which autonomously unfold and adapt the execution flow of business processes (BPs) through continuous conversation with their human principals, who oversee the system decision. While much research on trustworthy AI has been conducted on devising general-purpose trust recommendations, in this paper we take a first step toward exploring the role of trust to develop trustworthy ABPMSs. Specifically, we assess a relevant subset of trustworthy AI principles against the lifecycle stages of an ABPMS, thus providing a classification framework that identifies to which principles the ABPMS stages should conform. Then, we test the applicability of our framework on a real-world healthcare BP, and we evaluate its reliability through a user study involving 15 academics at the intersection of AI and BPM. The results show a promising consensus that our framework reasonably aligns trustworthy AI principles with the ABPMS stages.

1 Introduction

In the expanding landscape of Artificial Intelligence (AI), ensuring that AI systems are designed and developed to be *trustworthy* throughout their lifecycle is paramount [6,15]. A trustworthy AI system fosters human trust by achieving its goals while complying with legal regulations and ethical principles [16,19]. Trust is a central component of the interaction between people and AI, in that incorrect levels of trust may cause misuse, abuse or disuse of the AI technology [10]. Over the past few years, research institutes, private organizations, and government agencies have proposed various guidelines to make AI systems trustworthy [7,13,19], including the EU's recent AI Act [14].

Although much research has been conducted on devising trust recommendations for developing general-purpose trustworthy AI systems [11,17], these are often high-level statements that are difficult to translate into concrete implementation strategies for an emerging class of process-aware information systems infused with AI, called *AI-augmented Business Process Management Systems* (ABPMSs) [9]. ABPMSs are capable of dynamically unfolding and adapting business process (BP) execution flows by enabling improvement opportunities in autonomy and through continuous conversation with their human principals, who oversee the system decisions. In this context, improving trust between human users and ABPMSs is crucial for the practical adoption of these solutions.

In this paper, we take a first step toward exploring the principle needed to build trustworthy ABPMSs. Specifically, we explore the role played by the well-known principles for trustworthy AI in [15] to support the development of trustworthy ABPMSs by addressing two fundamental research questions, namely:

- **RQ1:** *Which trustworthy AI principles are necessary to achieve trust in an ABPMS?*
- **RQ2:** *How reliable are the selected principles in evaluating the trustworthiness of an ABPMS?*

To answer RQ1, we build a classification framework to connect the trustworthy AI principles in [15] with the lifecycle stages of an ABPMS, and we show its applicability on a real-world healthcare BP. Then, to tackle RQ2, we present the results of a preliminary user study conducted with 15 academics at the intersection of AI and BPM to gain insights on the reliability of our classification framework in identifying to which principles the ABPMS should conform.

The rest of the paper is organized as follows. Section 2 provides the background on ABPMSs and an overview of the literature on trustworthy AI. Section 3 introduces our classification framework, supported by a running example detailed in Sect. 4. Subsequently, Sect. 5 is dedicated to evaluating the reliability of our framework through a user study. Finally, Sect. 6 concludes the paper by reporting on the limitation of our study and tracing future work.

2 Preliminaries

The objective of this section is to give the reader the main ingredients for understanding the rest of the paper. Specifically, Sect. 2.1 introduces the fundamental concepts of an ABPMS and the stages involved during its lifecycle. Then, Sect. 2.2 reports the related works concerning trustworthy AI systems, discussing different points of view for fostering human trust in AI systems.

2.1 Background

Introduced by [9], ABPMSs represent a new generation of process-aware information systems, empowered by trustworthy AI technology, where the BP execution flows are not pre-determined, adaptations do not require explicit changes to software applications, and improvement opportunities are autonomously discovered, validated, and enabled on-the-fly.

Unlike conventional BPMSs where each task or decision in a BP is driven either by a human agent or by a software application according to pre-determined logic, an ABPMS can reason autonomously about the current state of the BP to determine the course of actions that improves its performance. Additionally, ABPMSs support BPs with unknown or incomplete structures at design time, which can emerge during runtime through reasoning and learning from past executions or due to exogenous events. Therefore, ABPMSs do not merely operate as traditional BPMSs integrating AI for performing isolated tasks; instead, they adopt a comprehensive approach wherein AI technology is employed to carry out and optimize the entire BP. The lifecycle of an ABPMS, as illustrated in Fig. 1, extends that of a traditional BPMS in two directions [9]:

- augmenting with AI capabilities the traditional lifecycle stages of a BPMS (*frame, perceive, reason, enact*);
- introducing advanced stages (i.e., *explain, adapt, improve*) that are specific to ABPMSs and only feasible when AI is an integral part of the system.

We now proceed to examine in detail each stage of the ABPMS lifecycle.

- *Frame*: This initial stage of the lifecycle involves an agent (an autonomous entity entitled to use the ABPMS, either a human or a digital entity) framing the ABPMS by providing it with initial constraints and goals. These constraints may incorporate different types of knowledge, such as predefined procedures, norms, regulatory constraints, common-sense rules, and best practices. The frame defines the boundaries wherein the ABPMS should operate. Upon completion of this initial stage, the lifecycle proceeds to its central phase, namely, the process-aware execution, which comprises three main stages: *perceive, reason,* and *enact.*
- *Perceive*: First, an ABPMS needs to perceive data concerning the BP execution produced by the working environment (e.g., collected through IoT sensors). Such data may be heterogeneous, varying in granularity and level of certainty. For example, the ABPMS may acquire event data directly from legacy systems, or from unstructured textual and visual sources [12].
- *Reason*: Subsequently, an ABPMS is expected to reason on the collected data, converting them into valuable information. The purpose is to equip the ABPMS with the ability to independently and continuously reason about BP enactment outcomes, most often retrospectively. This includes ongoing capturing of key conditions (e.g., historical framing reflecting timely assumptions and beliefs) and the ability to draw inferential associations between such conditions and intermediate BP execution results.
- *Enact*: Following this, the ABPMS leverages its actuators to interact (either directly or indirectly) with the environment and execute the BP.

At any stage throughout the lifecycle, the ABPMS may decide to perform one of the previously introduced advanced stages:

- *Explain*: An ABPMS can provide explanations regarding the past, current, and anticipated states of the system. Ensuring the explainability of an ABPMS is crucial to fostering trust and the cooperation of human agents.
- *Adapt*: BPs are frequently inserted within dynamic contexts marked by uncertainty and unpredictable changes [3]. Consequently, an ABPMS needs the

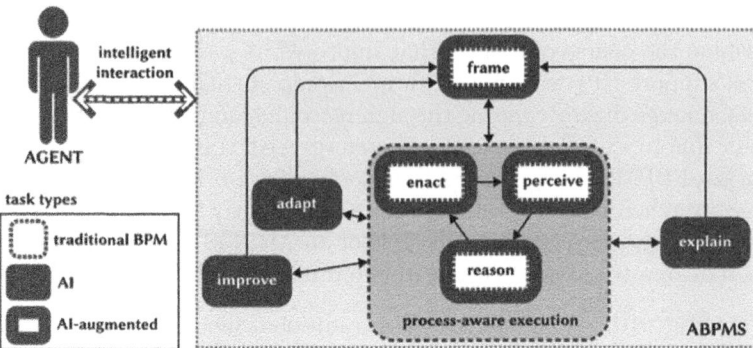

Fig. 1. The ABPMS lifecycle [9].

ability to detect alterations within its dynamic and uncertain environment and react properly, adapting itself to the evolving BP context in real-time.
– *Improve*: Additionally, an ABPMS should proactively anticipate changes and constantly improve itself by optimizing its execution against its goals, the available resources, and the framing constraints.

It is worth noticing that the execution of these advanced stages may prompt the ABPMS to revise and update its internal knowledge base. Therefore, the ABPMS may autonomously reframe itself based on the newly acquired knowledge.

2.2 Related Work

In light of the ever-increasing interest originating not only from academia but also from industry and government agencies, the construction of trustworthy AI systems has emerged as a central theme across numerous research domains. Various publications (a recent literature review discussing the most up-to-date research advancements in trustworthy AI is presented in [11]) have tackled this prominent subject, striving to formally define both the concept of trustworthy AI and the underlying principles that entail it, while also designing evaluation methodologies to rigorously assess the trustworthiness of AI systems.

For instance, one of the first pioneering papers to discuss and formalize the notion of trust in AI is [17]. The author delineates the potential societal benefits that AI can yield while also underscoring the necessity to address pertinent concerns for its responsible and transparent realization within an interdisciplinary and collaborative environment.

Similarly, in [10], a study formalizing trust-related concepts is conducted, investigating the factors determining human trust and exposing the need for contracts to guide the design and evaluation of trustworthy AI systems. Additionally, the study reported in [6] examines the role of human trust concerning AI voice assistants, undertaking a quantitative assessment of the degree of trust in this technology. It also proposes a Technology Acceptance Model (TAM) featuring a multidimensional trust measure. Conversely, in [20], the authors introduce a conceptual framework for trustworthiness, backed by legal theory and case law and validated through legal precedent. Another application is reported in [16], where a reference architecture is proposed for the implementation of trustworthy AI systems. The paper analyzes its practical application and associated challenges within a real-world financial scenario.

Differently from previous works that are targeted at exploring the nature of human trust and presenting applications of AI systems to assess their trustworthiness in various ways, in [15], the authors systematically outline a complete collection of trustworthy AI principles derived from human rights, ethical norms, and legal precedents. For this reason, in the next section, we primarily refer to [15] as it offers the most overarching overview of the trustworthy AI principles that can contribute to the trustworthiness of ABPMSs.

3   Classification Framework

The main objective of this work is to propose a classification framework (cf. Table 1) aimed at assessing the trustworthiness of an ABPMS, a critical endeavor to anticipate the degree of trust perceived by human workers in its real utilization. To this end, we draw upon a relevant subset of the trustworthy AI principles outlined in [15] and we align them with each stage of the ABPMS introduced and comprehensively examined in Sect. 2.1, thus answering **RQ1**.

Our contribution diverges from the existing literature on trustworthy AI by extending this concept to the domain of ABPMS, thereby establishing the foundation for future research in developing AI systems within BPM.

Before proceeding with our main contribution, it is worth briefly reviewing the definitions of the trustworthy principles for a generic system empowered by AI technologies as provided in [15]:

- *Autonomy and control* allows the AI systems to perform actions independently, without the direct intervention of the human agent.
- *Explainability* enables the AI system to justify the rationale behind its decisions in a human-understandable manner.
- *Functionality and performance* are the attributes of an AI system that enable it to fulfill its objectives.
- *Privacy* characterizes the ability of the AI system to maintain confidentiality of the data it handles.
- *Robustness and reliability* empower the AI system to maintain a good level of performance even in operations regarding atypical data.
- *Security* implements the AI system's capability of maintaining integrity, authenticity, and reliability.
- *Transparency* allows the AI system to ensure that stakeholders possess all relevant information about its operations.
- *Verifiability* is the principle enabling the verification of the AI system's behavior, ensuring it meets the specified requirements.

Table 1. The ABPMS stages and the corresponding trustworthy AI principles.

Trustworthy AI Principles	ABPMS Stages						
	Frame	Perceive	Reason	Enact	Explain	Adapt	Improve
Autonomy and control		✓	✓	✓		✓	✓
Explainability					✓		
Functionality and performance		✓	✓	✓		✓	✓
Privacy		✓	✓	✓			
Robustness and reliability		✓	✓	✓			
Security		✓	✓	✓			
Transparency	✓						
Verifiability	✓						

In the following, we discuss how the aforementioned trustworthy AI principles can be satisfied by each stage of an ABPMS. The results are summarized in Table 1, where each row highlights the relationship between the ABPMS stages and the corresponding principles necessary to achieve human trust.

To ensure the trustworthiness of the *Frame* stage, we posit that it should adhere to the following subset of trustworthy AI principles:

- *Transparency*: this property ensures that relevant stakeholders receive appropriate information about the BP under analysis.
- *Verifiability*: by relying on a specific BP representation, the ABPMS can conduct checks to verify that it will be executed effectively and meet the specified requirements.

From a trustworthiness perspective, the stages of *Perceive*, *Reason*, and *Enact* can be collectively considered as constituting a unified *execution* phase. The trustworthiness of the execution phase could be accomplished through the fulfillment of the following trustworthy AI principles:

- *Autonomy and control*: the ABPMS should support the automatic execution of BPs, without direct human intervention.
- *Functionality and performance*: this property allows ABPMSs to meet their specified objectives according to the BP under analysis.
- *Privacy*: the ABPMS should maintain the confidentiality of BP's data and attributes.
- *Robustness and reliability*: this principle entails the ABPMS's ability to perform consistently across various scenarios, including atypical ones.
- *Security*: it ensures the integrity, authenticity, and reliability of the ABPMS.

To establish the trustworthiness of the *Explain* stage, it is essential to comply with the corresponding trustworthy AI principle, namely *Explainability*. This attribute embodies the ABPMS's capacity to clarify essential aspects influencing its decisions in a manner understandable to humans. The *Explain* stage is inherently linked to the *intelligent interaction* that the ABPMS should support.

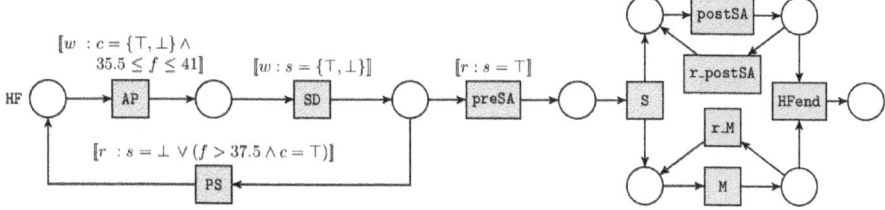

Fig. 2. The data Petri net representing the clinical BP model introduced by [2].

While this interaction is not directly mapped to a specific trustworthy property, it holds fundamental importance as it contributes to fostering trust in the ABPMS. When users understand the rationale behind the system's actions, they perceive a sense of control and are more inclined to trust it [9].

Proceeding in our effort to align the trustworthy AI principles with the ABPMS's stages, we postulate that the *Adapt* stage should be realized in conformance to:

- *Autonomy and control*: the ABPMS should be designed with autonomous capabilities to dynamically adjust itself in response to overcome unforeseen circumstances as they arise.
- *Functionality and performance*: the ABPMS must adapt its behavior to consistently achieve its objectives.

Ultimately, the last stage of the ABPMS that must be covered by the classification framework is *Improve*. The ABPMS can reach trustworthiness in this phase by fulfilling the following properties:

- *Autonomy and control*: this property prescribes that the ABPMS should autonomously optimize its execution to align with its objectives, available resources, and framing constraints.
- *Functionality and performance*: in the pursuit of optimization, the ABPMS should enhance its effectiveness to consistently achieve its objectives.

4   Running Example

To demonstrate how our classification framework can be applied to ensure the trustworthiness of an ABPMS, we use as a running example the real-world scenario outlined in [2]. The scenario is described by the BP model in Fig. 2, represented with a data Petri net. It illustrates a healthcare procedure concerning the treatment process for patients suffering from hip fractures.

In this example, the *Frame* stage is realized by framing the BP through a combination of a data Petri net and a bundle of Declare constraints capturing some basic medical knowledge rules that must not be violated. For example, if the doctor suspects the patients is suffering from a chest infection, a chest X-ray must be taken on the patient before the surgery decision (SD). Additionally, the frame can be dynamically adjusted throughout the subsequent stages of the ABPMS. According to the classification framework presented in Sect. 3, the trustworthiness of this stage is achieved if the following principles hold:

- *Transparency*: the frame stage achieves transparency by providing a clear and multi-model representation of the underlying BP. Consequently, the stakeholders (i.e., both clinicians and patients) can indirectly benefit from this transparency through proper interfaces towards the ABPMS, enabling them to comprehend essential information about the clinical procedure.
- *Verifiability*: by incorporating the data Petri net alongside multi-perspective Declare constraints, the ABPMS can easily verify the conformance of the process to the specified constraints and particular requirements, such as preserving the patient's health in every assisted operation.

Regarding the *execution* phase (i.e., *Perceive*, *Reason*, and *Enact*), a detailed examination of the treatment process for hip fractures illustrated in Fig. 2 is necessary to better contextualize how the *sense-think-act* cycle [18] is put in place within this scenario by the ABPMS. The clinical procedure begins with an initial assessment of the patient's condition (AP) to determine whether the surgery is safe or not (SD). Surgical intervention is delayed when the assessment yields negative results, indicating symptoms such as high fever or cough and necessitating a reassessment after resolving these issues (PS). Conversely, a positive evaluation leads to the patient undergoing pre-surgery anesthesia in a designated room (preSA), followed by the surgery (S) in an operating room. Following the intervention, the patient is prescribed post-surgery analgesia (postSA) for pain relief and physiotherapy (M) to mobilize the hip. These activities may be repeated until the patient's condition improves, signifying the conclusion of treatment (HFend). According to the proposed classification framework, the execution phase of an ABPMS is expected to adhere to these principles:

- *Autonomy and control*: the ABPMS should be able to autonomously decide whether to proceed with surgery or wait based on the patient's clinical information.
- *Functionality and performance*: during the execution, the ABPMS should consistently aim for the patient's optimal post-surgical recovery and safe rehabilitation from the hip fracture.
- *Privacy*: the ABPMS is required to provide strict confidentiality of the patient's conditions and the assigned medical prescriptions.
- *Robustness and reliability*: the ABPMS should operate effectively in both complete and partial executions of the clinical process, such as instances where surgery is not performed due to unforeseen conditions. An example of this situation could involve the premature termination of the process following the assessment of the patient's condition due to a previous pathology preventing him from receiving anesthesia.
- *Security*: the ABPMS should defend against external threats that could compromise the integrity of the entire process, such as unauthorized modifications to a patient's health data to allow surgical intervention under unsafe conditions, causing unexpected complications and menacing his survival.

Guided by the view exposed in [9], the previous stages can be extended also with the advanced stages unique to ABPMSs, as explained as follows.

In the context of the running example, the *Explain* stage can be enacted whenever explanations regarding the system's states are needed within the BP execution. In fulfilling the trustworthy *Explainability* property, the ABPMS should possess the capacity to clarify the rationale behind its decision-making regarding whether to proceed with surgery for the patient under evaluation in a comprehensible way. Additionally, an *intelligent interaction* with the ABPMS during the execution of the clinical procedure could also be enacted. The intelligent interaction can be achieved by leveraging a conversational interface in natural language [5] for informing clinicians about the issues encountered during the execution. This is crucial to aid them in assessing the patient's conditions, evaluating alternative treatment solutions for more informed decision-making [4], and providing useful feedback to the ABPMS.

Another stage presented by the ABPMS is *Adapt*, wherein the system continuously monitors the patient's vitals, medical history, and environmental factors, to modify its operations in response to any unexpected change. In this case, the trustworthy AI principles to adhere to are:

- *Autonomy and control*: for instance, if the patient presents symptoms of an allergic reaction after the anesthesia, the ABPMS should autonomously trigger additional diagnostic tests before proceeding with the surgery.
- *Functionality and performance*: if the designated expert for the intervention is unavailable, the system should either reschedule the surgery or assign it to another qualified doctor to ensure prompt treatment of the patient's hip fracture while minimizing additional health complications.

In the *Improve* stage, the ABPMS managing the hip fracture process should incorporate a predictive model to recommend the most suitable timing for surgery based on the historical data regarding patients' recovery times, treatment outcomes, and individual familiarity risks. Additionally, continuous feedback and suggestions gathered through interactions with clinicians can be leveraged to enhance the decision-making process. The ABPMS can reach trustworthiness in this stage by fulfilling the following properties:

- *Autonomy and control*: the ABPMS should evolve in response to emerging medical best practices and research to stay aligned with the latest standards. For example, it may suggest and implement new anesthesia or post-surgery treatments to facilitate faster and more effective patient recovery.
- *Functionality and performance*: as it improves, the ABPMS should increasingly optimize its performance to achieve its target, namely the optimal recovery outcome for the patient, within the specified frame.

5   Evaluation

To answer **RQ2**, we conducted a preliminary evaluation with the goal of assessing the reliability of the proposed classification framework in connecting the trustworthy AI principles with the lifecycle stages of ABPMSs.

Table 2. The results of the administered survey in percentage. The values for the classification framework pairs are reported in bold.

Trustworthy AI Principles	ABPMS Stages						
	Frame	Perceive	Reason	Enact	Explain	Adapt	Improve
Autonomy and control	0.07	**0.53**	**0.60**	**0.93**	0.07	**0.87**	**0.73**
Explainability	0.20	0.00	0.53	0.13	**0.93**	0.27	0.20
Functionality and performance	0.20	**0.60**	**0.93**	0.73	0.07	**0.67**	**0.80**
Privacy	0.27	**0.87**	0.53	0.40	0.13	0.13	0.00
Robustness and reliability	0.13	**0.60**	**0.80**	**0.60**	0.07	0.47	0.33
Security	0.47	**0.87**	**0.80**	**0.73**	0.20	0.40	0.33
Transparency	**0.53**	0.07	0.07	0.20	0.47	0.00	0.13
Verifiability	**0.73**	0.27	0.33	0.40	0.20	0.07	0.40

We developed a comprehensive survey providing first relevant background information about ABPMSs, the selected trustworthy AI principles to assess, and the example scenario presented in Sect. 4. Then, we asked participants to fill in the empty classification framework table, aligning the trustworthy AI principles with the corresponding stages of the ABPMS. We administered the survey to 15 academics at Sapienza University of Rome, including 12 doctoral students, 2 post-doctoral researchers, and 1 professor. We selected only participants who declared themselves highly familiar with BPM-related topics and the use of AI systems. Ultimately, we collected the responses from the participants while ensuring their anonymity.

After receiving all responses, we assessed the reliability of our classification framework by calculating the percentage of participants who selected each combination of trustworthy AI principle and ABPMS stage. This gave us an overview of the presence or lack of consensus regarding specific matches. Table 2 reports the outcomes of the survey, emphasizing in bold the associations envisioned by the proposed classification framework in Sect. 3.

We considered as accepted all the pairs ⟨*trustworthy AI principle, ABPMS stage*⟩ that reached an agreement level of the relative majority, namely 51%, while rejecting the others. Specifically, based on the outcomes, we reserved the possibility to adapt the classification framework accordingly, incorporating matches not initially included but reaching a consensus exceeding the predetermined 51%, and discarding those included in the framework but failing to meet the previously established threshold.

To interpret the survey's results and estimate their confidence intervals, we can leverage the empirical rules proposed in [8], based on simplified forms of formulas for the variance of a Binomial distribution. In particular, for a small proportion of the sample selecting a particular option, we can compute the confidence interval as twice the square root of the number of users who chose that option. For instance, for the pair ⟨*Verifiability* principle, *Adapt* step⟩, we can be reasonably confident in the 7% within a margin of ±13%. Conversely, for a larger majority (e.g., 93% for the pair ⟨*Explainability* principle, *Explain* step⟩), we use the same formula but we consider the number of respondents who did not choose this option (i.e., within ±13%). Moreover, for percentages in the middle range, we estimate the confidence interval by taking 1.5 times the square root of the number of participants selecting this option.

From analyzing the survey's outcomes, we found that all predicted associations met the minimum level of consensus, with a few exceptions. The first discrepancy concerns the pair ⟨*Privacy* principle, *Enact* stage⟩, which did not reach the established threshold for acceptance, achieving a score of 40%. This indicates that there might be differing opinions on the importance or applicability of privacy during the enactment phase of ABPMS. The second exception is that the academic experts involved in the study agreed on the pair ⟨*Explainability* principle, *Reason* stage⟩ that was not initially included in the framework, achieving a score of 53%. This could be due to the possibility that participants misunderstood the distinction between the *Reason* stage and the *Explain* stage.

Therefore, based on this preliminary evaluation, we can conclude that our classification framework effectively aligns with the mental models of participants concerning the relationship between trustworthy AI principles and the stages of the ABPMS, demonstrating a promising degree of reliability.

6   Concluding Remarks

In this paper we assessed a relevant subset of trustworthy AI principles against the stages of an ABPMS, thus providing a classification framework that links specific trust principles to each stage. Our framework has been designed to ensure *transparency, verifiability, autonomy and control, functionality and performance, privacy, robustness and reliability, security* and *explainability* principles throughout the lifecycle of ABPMSs, thus answering **RQ1**.

Instead, to address **RQ2**, the proposed classification framework was showcased through a real-world healthcare use case scenario and validated through a preliminary study involving researchers at the intersection of AI and BPM fields. The obtained results emphasized a high degree of consensus among participants, confirming the alignment of our framework with their mental models regarding the association between trustworthy AI principles and ABPMS stages. Despite some discrepancies, such as the failure to accept the *Privacy* principle for the *Enact* stage and the acknowledgement of the *Explainability* principle for the *Reason* stage, our framework has proven to be valid.

In conclusion, this work contributes to the existing literature on ABPMSs, by extending trustworthy AI principles to this domain and providing a convincing foundation for future research. The adoption of our framework can promote the development of more transparent, reliable, and accountable systems, capable of managing the complexities of modern BPs. However, our work presents some limitations. Although the classification framework is general, in this study, we limit its applicability to the healthcare domain [1]. Therefore, as future work, a possible extension could include validating the framework in different contexts to confirm its broad applicability. In addition, the selection of trustworthy AI principles was based on a subset identified in the current literature, and there may be other relevant principles that were not considered. Further studies could explore the integration of additional principles to enrich the classification framework proposed in this paper.

Acknowledgments. This work is supported by the H2020 project DataCloud (Grant 101016835), the Sapienza project FOND-AIBPM, the PRIN 2022 project MOTOWN and the PNRR MUR project PE0000013-FAIR. The work of Angelo Casciani is in the range of the Italian National Doctorate on AI run by Sapienza.

References

1. Agostinelli, S., Covino, F., D'Agnese, G., De Crea, C., Leotta, F., Marrella, A.: Supporting governance in healthcare through process mining: a case study. IEEE Access **8**, 186012–186025 (2020)
2. Alman, A., Maggi, F.M., Montali, M., Patrizi, F., Rivkin, A.: Multi-model monitoring framework for hybrid process specifications. In: Advanced Information Systems Engineering - 34th International Conference, CAiSE 2022, pp. 319–335. Springer (2022)
3. Baiyere, A., Salmela, H., Tapanainen, T.: Digital transformation and the new logics of business process management. EU J. Inf. Syst. **29**(3), 238–259 (2020)
4. Bernardi, M.L., Casciani, A., Cimitile, M., Marrella, A.: Conversing with business process-aware large language models: the BPLLM framework. Preprint Research Square (2024). https://doi.org/10.21203/rs.3.rs-4125790/v1
5. Casciani, A., Bernardi, M.L., Cimitile, M., Marrella, A.: Conversational systems for AI-augmented business process management. In: RCIS 2024 (2024)
6. Choung, H., David, P., Ross, A.: Trust in AI and its role in the acceptance of AI technologies. Int. J. Hum.-Comput. Interact. **39**(9), 1727–1739 (2023)
7. Díaz-Rodríguez, N., Del Ser, J., et al.: Connecting the dots in trustworthy artificial intelligence: from AI principles, ethics, and key requirements to responsible AI systems and regulation. Inf. Fusion **99**, 101896 (2023)
8. Dix, A.J.: Statistics for HCI: Making Sense of Quantitative Data. Synthesis Lectures on Human-Centered Informatics. Morgan & Claypool Publishers (2020)
9. Dumas, M., Fournier, F., Limonad, L., Marrella, A., et al.: AI-augmented business process management systems: a research manifesto. ACM TMIS **14**(1) (2023)
10. Jacovi, A., Marasović, A., Miller, T., Goldberg, Y.: Formalizing trust in artificial intelligence: prerequisites, causes and goals of human trust in AI. In: 2021 Conference on Fairness, Accountability, and Transparency, FAccT 2021, pp. 624–635. ACM (2021)
11. Kaur, D., et al.: Trustworthy artificial intelligence: a review. **55**(2) (2022)
12. Knoch, S., Ponpathirkoottam, S., Schwartz, T.: Video-to-model: unsupervised trace extraction from videos for process discovery and conformance checking in manual assembly. In: Business Process Management, pp. 291–308. Springer (2020)
13. Li, B., et al.: Trustworthy AI: from principles to practices. ACM Comput. Surv. **55**(9) (2023)
14. Madiega, T.: AI Act. EU Parliament Research Service, EU Parliamentary (2021)
15. Mariani, R., Rossi, F., et al.: Trustworthy AI - part 1. Computer **56**(2), 14–18 (2023)

16. Mazumder, S., Dhar, S., Asthana, A.: A framework for trustworthy AI in credit risk management: perspectives and practices. Computer **56**(5), 28–40 (2023)
17. Rossi, F.: Building trust in AI. J. Int. Affairs **72**(1), 127–134 (2018)
18. Russell, S., Norvig, P.: Artificial Intelligence: A Modern Approach. Pearson (2020)
19. Schelble, B.G., Lopez, J., et al.: Towards ethical AI: empirically investigating dimensions of AI ethics, trust repair, and performance in human-AI teaming. Hum. Fact. **66**(4) (2024)
20. Singh, A.M., Singh, M.P.: Wasabi: a conceptual model for trustworthy artificial intelligence. Computer **56**(2), 20–28 (2023)

Knowledge Graphs: A Key Technology for Explainable Knowledge-Aware Process Automation?

Leon Bein[✉][iD] and Luise Pufahl[iD]

School of Computation, Information, and Technology,
Technical University of Munich, Heilbronn, Germany
{leon.bein,luise.pufahl}@tum.de

Abstract. Process automation is a key subfield of business process management. Recent advances in AI research promise to yield a new type of *intelligent* process automation that can support high-variability, flexible, knowledge-intensive processes previously hard to enhance with process automation. However, primarily proposed, subsymbolic deep learning approaches fail to reliably consider the complex knowledge inherent to these processes and provide adequate explanations for their decisions. Neuro-symbolic reasoning approaches based on *knowledge graphs* promise to address these challenges by allowing to holistically encode complex domain knowledge and to perform explainable reasoning thereupon. In this vision paper, we investigate the potential of knowledge graphs for intelligent process automation. Using tangible examples, we show how they can be used to enable explainable, knowledge-aware process automation, integrating a wide range of process knowledge. We show that such knowledge-aware process automation can contribute to addressing two current challenges of the BPM community: the automation of knowledge-intensive processes and the design of AI-augmented business process management systems. Finally, we discuss avenues for future research.

Keywords: Business Process Automation · Knowledge Graphs · Knowledge-intensive Processes · Neuro-symbolic AI

1 Introduction

The progressing digitization of organizations and recent rapid advances in Artificial Intelligence (AI) capabilities are promising to revolutionize the *Process Automation* domain. This subfield of *Business Process Management (BPM)* is concerned with supporting the execution of business processes using *Business Process Management Systems (BPMS)*, software systems that orchestrate process execution by determining, prioritizing, and assigning tasks, managing handovers between them, and detecting and escalating business errors [10,23]. Visions have been formulated, of *intelligent* or *AI-augmented* process automation

K. Gdowska et al. (Eds.): BPM 2024 Workshops, LNBIP 534, pp. 18–30, 2025.
https://doi.org/10.1007/978-3-031-78666-2_2

systems that integrate AI technologies to enable an autonomous and adaptive orchestration of business processes and extend software support to previously hard-to-support high-variability, flexible processes that are driven by expert decisions based on domain knowledge [9,23].

To actually support these *Knowledge-intensive Processes (KiP)*, several requirements must be considered [8,18]. In particular, it is imperative to consider the vast knowledge inherent to KiPs to inform system recommendations and decisions [3,8]. Consequently, KiPs are considered hard to support with traditional BPMSs [8,20]. Further, explainability has been identified as an important aspect of intelligent process automation systems [9]. It is also particularly necessary for the support of knowledge-intensive processes, where a huge responsibility is attached to decisions due to their high impact on process executions [8]. However, existing approaches often fail to fulfill both requirements, either only considering shallow flow-based process models and execution data in the form of event logs or failing to produce interpretable explanations for their decisions, rather relying on black-box subsymbolic AI approaches [5,21].

Neuro-symbolic AI approaches, which can provide explainability while retaining learning power by combining symbolic and subsymbolic AI techniques, appear promising to address this problem. Specifically, we argue that *knowledge graph reasoning* techniques that apply machine learning approaches to *knowledge graphs* appear appealing to the BPM field as they allow a holistic integration of process knowledge and reasoning upon it in an explainable way.

Existing work touches on how certain kinds of process knowledge can be encoded into knowledge graphs [1,3,12]. Further, their usage as part of an intelligent process automation system has been proposed in [4] and [9]. However, the existing works on process knowledge encoding do not consider the full width of process knowledge that must be encoded, especially for the support of knowledge-intensive processes. The works that propose knowledge graphs as part of intelligent process automation systems do not yet provide pointers on how the reasoning upon such graphs could look like in the context of business process automation. Further, an extensive overview of research opportunities on knowledge graphs for process automation has not yet been provided.

Consequently, in this paper, we discuss the potential of knowledge graphs for intelligent process automation. Using tangible examples, we show how they can be used to integrate a broad width of process knowledge and how, this way, they enable *explainable* and *knowledge-aware* process automation. Further, we highlight how these technological capabilities can be used to advance the design of intelligent process automation systems, and, in particular, support the execution of knowledge-intensive processes. Lastly, we carve out, motivate, and give an overview of avenues for future research on knowledge graphs for explainable knowledge-aware process automation.

The remainder of this paper is structured as follows: In Sect. 2, we introduce the necessary background on knowledge graph reasoning. Section 3 introduces an example process and outlines the potentials of knowledge graphs for process knowledge encoding and reasoning. In Sect. 4, we show how knowledge graph reasoning can contribute to two BPM challenges. Finally, in Sects. 5 and 6, we illustrate avenues for future research and provide concluding remarks.

2   Knowledge Graph Reasoning

Knowledge graph (KG), notoriously, has been a hard-to-define term [11,22]. For this paper, we adhere to the definition of Hogan et al. [11], according to whom a knowledge graph is "a graph of data intended to accumulate and convey knowledge of the real world, whose nodes represent entities of interest and whose edges represent relations between these entities." From a technical perspective, knowledge graphs additionally allow the definition of context for the graph data in the form of, i.a., ontologies and taxonomies, data constraints, and rules to infer further information [11].

Due to their generality, knowledge graphs allow encoding a wide range of knowledge. Organizations then use this knowledge, e.g., for product recommendations using the graph to encode product and user relations [11,24], for self-service chatbots using the graph as source for question answering [11] or for personalized medicine using the graph to encode biochemical interactions [19].

As the definition suggests, the main units of data for KGs are entity nodes and relations. Entities are defined by a unique id. Relations are defined by a source node (*subject*), an edge label (*predicate*), and a target node (*object*). More complex graph traits, such as node attributes, n-ary relationships, and meta-information, can be translated to the entity-relation abstraction using attribute relations, literal nodes, and relational nodes [11]. Formally, we can express relations as logical facts using the notation $predicate(subject, object)$ if a relationship is present in the graph, and $\neg\ predicate(subject, object)$, if not. Further, consider Fig. 1 for a small example visualization as used through the remainder of this paper. For simplification, we use node labels in the form "id:typename" which imply the relation $type(id, typename)$.

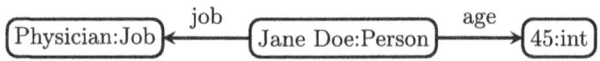

Fig. 1. Knowledge graph example. Node labels include type names for simplification.

In addition to entities and relationships, common knowledge graph technology stacks allow defining ontological schemata and inference rules. Schemata can be used to define information about the domain, e.g., types of entities and properties of relations. Relation properties include source or target type limitations, symmetry and transitivity, or attribute ranges [11]. Inference rules use general logical statements on the graph data to infer new relations, e.g., $parent(?a, ?b) \land parent(?b, ?c) \Rightarrow grandparent(?a, ?c)$ to infer grandparentship from two consecutive links of parentship.

Knowledge graph reasoning (KGR) describes this task of inferring new knowledge from a given graph [6]. As KGs are usually incomplete, reasoning is often applied to *graph completion* problems. These often take the form of *link prediction*, i.e., determining whether a certain relation should exist between two entities [6,7].

Using logical rules to infer new relations is a simple symbolic method to achieve this. However, advances in subsymbolic artificial neural nets have also yielded new, more powerful approaches to achieve graph reasoning. These approaches integrate – to varying degrees – the symbolic graph structure and logic-based reasoning techniques into subsymbolic neural-net-based ones. They thus largely fall into the category of neuro-symbolic AI [7].

Multiple families of such approaches exist; for a more extensive overview of categories, consider [7] or [11]. As for many neuro-symbolic techniques, a major tradeoff for the KGR approaches is between the neural component's power and the symbolic component's interpretability. The subsymbolic power usually yields better generalizability, efficiency, and often simply better results. However, better interpretability allows for KGR techniques to provide understandable explanations of how they arrived at their result, placing them as *explainable AI (XAI)* methods [11,22,24]. For instance, one set of KGR approaches learns and considers weighted path rules for link prediction, providing the deciding rules as fairly interpretable output along with the edge [6,7,11].

3   Technological Potentials for Knowledge Graphs in BPM

We argue that knowledge graphs and applied reasoning may prove to be a valuable tool for the BPM domain and more specifically for the intelligent automation of knowledge-intensive processes. In this section, we first introduce a running example process. We then highlight two technological potentials of knowledge graphs for BPM, showing how explainable knowledge-aware process automation can be made possible. Finally, we provide some technical considerations.

3.1   Running Example

The *SEPSIS* healthcare process is based on the observed process of handling patients with sepsis in a Dutch regional hospital, which was captured in [14] and for which there is an event log publicly available under [13].

The process tracks patients from arrival to release. It contains activities related to registration and triage, measurement of clinical parameters, inserting infusions, admission and transfer to normal and intensive care units, and different types of discharge. As the process is driven by the expert decisions of the attending medical staff, it is characterized by a high amount of variability, evidenced by 846 variants in 1050 traces. While there is little strongly regular process flow, there are descriptive rules, e.g., a medical guideline stating that sepsis patients should get an antibiotics infusion within one hour after triage [14].

The process is a good showcase for this work for several reasons. It is a prime example of a knowledge-intensive process, and has shown to be comparatively difficult to process with existing methods [14, 21]. Yet, there are regularities that can explain some of its behavior, like the medical guidelines. Further, there is an event log for future engineering work readily available.

3.2   Rich Integrated Process Encoding

Through the generality of knowledge graphs, it becomes possible to holistically integrate the various process knowledge perspectives into one, uniform data structure, a *process knowledge graph*, even reusing many formalizations investigated in previous works. In the following, we list some of the process aspects (cp. Fig. 2 for an overview) that can be encoded, show how they relate, and give (non-exhaustive) pointers on which existing formalization might be reused.

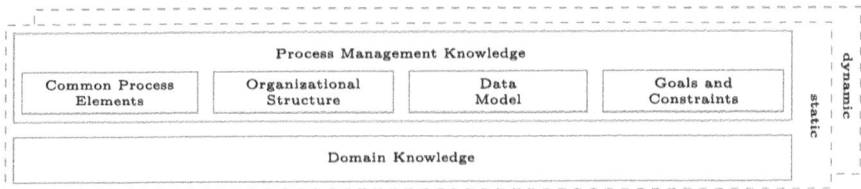

Fig. 2. Overview of selected process knowledge, adapted from [8, 10, 25]

First, *common process elements*, such as activities, events, choices, and data artifacts, and their dependencies can be integrated. This constitutes the knowledge inherent to traditional flow-based process modeling languages. For this, an encoding translation from XML to knowledge graph already exists for the de-facto standard process modeling language BPMN (see [1, 3]).

However, knowledge graphs allow to encode, and notably integrate, further aspects of process knowledge. *Organizational structure*, i.e., concrete resources, their roles, and their associations, can be modeled, and related to the aforementioned activities by means of permissions and requirements. For a taxonomy of resource aspects, e.g., [2] could be considered. Detailed *data models*, including types, attributes, relations, cardinality constraints, etc., can be encoded into the graph ontology, extending the data artifacts of classical process models. Structure paralleling UML class diagrams could be used here. Process *goals and constraints*, as often modeled with declarative process modeling languages, can be integrated using graph rules, directly referencing the above-described entities. E.g., logical formulae could be adapted from [20].

Further, it is possible to integrate highly *domain-specific knowledge* graphs with the process-oriented one. For many domains, including healthcare, public knowledge graphs have been published [11, 19].

Lastly, knowledge graphs allow to integrate knowledge about *process executions*, historic and ongoing, with the process as modeled. This dynamic knowledge represents an orthogonal dimension to the other perspectives, relating instances to static concepts. Taking the notion of events logs, for instance, events can be connected to the encoded activities, resources, and other entities. Further, encoded process constraints and goals can be directly checked against past (and present) process executions. Existing work to make such an integration has been performed in [12], and could be extended by domain-specific and data aspects.

Notably, it is also possible to encode multiple processes into one graph, which will then be connected via shared entities, such as resources and domain concepts.

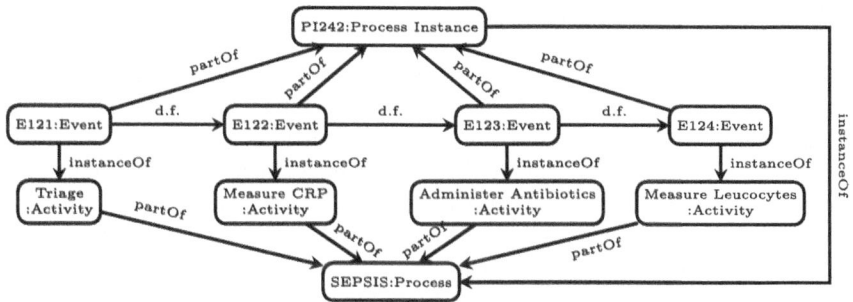

Fig. 3. Example process knowledge graph snippet with basic event log information. "d.f." abbreviates "directlyFollowedBy".

Examples. Figure 3, exemplifies how process knowledge of modeled activities and instances can be integrated for the SEPSIS process. The snippet displays four of the activities, but without relations between them, and one process instance with four events, in a strict *directlyFollowedBy* order.

As a more complex example, Fig. 4 shows an excerpt of how domain-specific knowledge can be integrated with knowledge about activities and process executions. It shows an activity instance that produced a certain measurement, how the measured clinical parameter and the case patient relate to medical conditions, and how those lead back to required actions, i.e., activities. Note that both examples show different excerpts from the same graph, so nodes with the same ID are identical. This highlights how the different process knowledge aspects can be integrated.

3.3 Explainable Knowledge-Aware Process Automation

Having all process knowledge in this regular structure allows applying explainable AI reasoning techniques that operate on the notion of entities, relations, and rules. This allows for building Business Process Management Systems that are able to (a) intelligently reason on the process knowledge and to (b) communicate their reasoning, i.e., they work *knowledge-aware* and *explainable*.

Example and Vision. Consider the following motivating example from the SEPSIS process: During the treatment of the patient, the physician might have to decide to have the patient transferred from the normal care unit to the intensive care unit (ICU). This decision is based on the years of experience the physician has, as well as on certain clinical guidelines. Both are, in turn, informed by the general scientific knowledge of how the human body works.

A BPMS powered by knowledge graph reasoning can consider this knowledge. The physician's knowledge can be implicitly encoded into historical process executions and the clinical guidelines and general medical knowledge can be explicitly encoded in the same KG using domain-specific ontologies. After a recent measurement of clinical parameters, the system might propose actions *based on that stored knowledge* and *providing interpretable reasoning*, such as: "The patient should be transferred to the ICU, as the just measured out-of-range leucocytes and earlier measured c-reactive proteins are early indicators of an organ failure, to which the patient is especially vulnerable due to their high age and diabetes". The physician can then consider said reasoning for their decision but also take into account their direct assessment of the patient, based on facts potentially not recorded in the system. Figure 4 visualizes (a simplified subset of) graph facts that could lead to such reasoning, with two paths leading from the current case to the ICU transfer activity via the organ failure medical condition.

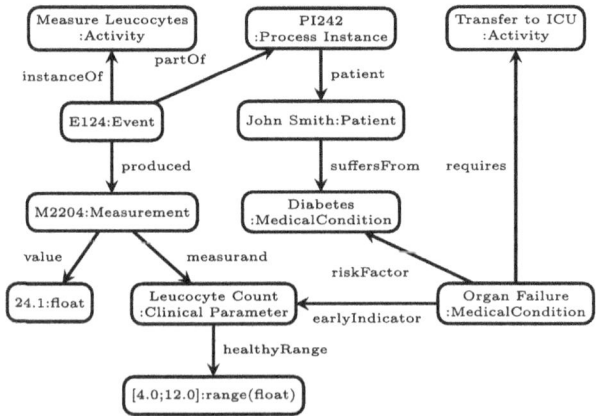

Fig. 4. Example process knowledge graph snippet with reasoning knowledge for ICU transfer.

To extend the vision, the system might even proactively consider an increased risk and schedule future observatory measurements to decide to send a patient to the ICU. The system could also ask the physician why a specific decision was arrived at and learn new knowledge from that, to be taken into account for future decisions. Especially when multiple courses of action are possible, the system could also give multiple recommendations with different reasonings and confidence measures. Finally, especially in other domains, the system might take over decision-making completely instead of giving recommendations.

3.4 Technical Considerations

As promising as the vision and the examples are, some technical challenges need to be considered. Many graph reasoning techniques assume a certain level of regularity. While this might be given for domain-independent entities and relations, such as activities and their dependencies, domain-specific information, such as medical interrelationships, often are rather heterogenous in addition to differing from domain to domain [7]. Rules in a process might be based not only on path-like structures, i.e., sequences of specific relations. As seen in the examples above, more complex structures might need to be considered. These structures might also include node literals (such as specific activities), which, according to our preliminary survey, not all explainable reasoning techniques support. Further, reasoning based on numerical relations, such as "blood pressure greater than 130", or based on negations or temporal relations, such as "there has not been a measurement of lactic acid in this case in the last 2 h", must be able to be detected. Handling both the domain-specificness and this heterogeneity requires reasoning techniques to show greater generalizability, which, in turn, often negatively correlates with explainability [7]. Potentially, a tradeoff between graph expressiveness and explainability must be made.

Consequently, the applicability of existing knowledge graph reasoning techniques for BPM has to be evaluated, and, if necessary, dedicated techniques should be developed.

4   Addressed BPM Challenges

The explainable knowledge-aware process automation as envisioned in Sect. 3 contributes to addressing two current (strongly related) challenges of the BPM community, the automation of knowledge-intensive processes and the design of AI-augmented Business Process Management Systems.

4.1 Automation of Knowledge-Intensive Processes

Knowledge-intensive Processes (KiP) constitute a category of processes that are inherently driven by decisions based on domain experts' knowledge [8]. Consequently, KiPs are characterized by a high degree of variability and unpredictability, with actual flow emerging at run time based on decisions driven by the process state and complex knowledge [8]. This makes KiPs hard to support with traditional BPMSs, whose modeling languages require explicit control flow definitions at design time, but cannot map complex domain knowledge [8]. Dedicated, so-called case management approaches to modeling KiPs, such as CMMN

[17], can provide basic decision support, often based on data constraints. However, even they usually consider only parts of the knowledge involved in KiPs, instead retaining the human-readability of created models [8]. In contrast, in our envisioned knowledge-aware process automation, where process flow is driven by the full width of encoded process and domain knowledge, which seems an ideal fit to support KiPs.

Requirements for systems supporting KiPs have been formalized in previous works [8,18]. One major group of these requirements is concerned with the process aspects that the system needs to model, including process data, activities and their data constraints, rules and constraints, process goals, resources and roles, and business events. Further, systems should record and learn from event logs [8]. As described in Sect. 3.2, the abstraction of knowledge graph allows all of this, even in one integrated structure.

Another group of requirements is concerned with process flexibility. Systems should enforce process flow only loosely, allowing process participants to decide which activities to execute at any time, and even to add new activities, constraints, goals, etc., at runtime [8,18]. In our proposed approach, while constraints would be checked and considered for decision-making, any process execution can be logged in the graph using event nodes and respective attributes. Further, new knowledge can be added to the graph at any time. Adding, e.g., new goals at runtime might, however, necessitate at least partial retraining of the applied reasoning techniques. Systems are further required to be able to adapt to unanticipated process exceptions [8,18]. We envision reasoning toward goals at any time during the process. While this does not explicitly entail specification on how to recover constraint violations, the system should, in theory, be able to do so. Reasoning techniques have in the past been used to achieve process adaptation [15].

We also want to highlight that many knowledge-intensive domains, such as healthcare, entail high responsibility for knowledge workers, due to the high impact their decisions have [8]. Consequently, it is essential for supporting systems to be highly transparent and capable of explaining their decisions, as does our envisioned approach.

We thus surmise that process automation based on explainable knowledge graph reasoning promises to be a good fit to support knowledge-intensive processes, although technical challenges have to be investigated in future research.

4.2 Process Framing for AI-Augmented Business Process Management Systems

Recently, the vision of *AI-augmented Business Process Management Systems (ABPMS)* has been proposed [9]. ABPMSs are defined as software systems that utilize AI technologies to autonomously orchestrate process execution inside a given *frame* of goals and constraints. In contrast to traditional BPMSs, where control flow is explicitly modeled, this process frame is intended to define valid execution paths much less explicitly. Instead, actual process flow emerges at runtime based on the system reasoning on the frame and the current process state. Besides autonomous orchestrations, ABPMSs promise capabilities such as autonomous error detection, adaptation, and escalation, self-optimization, and proactive reporting and process explanation.

Our vision extends the ABPMS one, giving more concrete pointers how such a system could be designed. BPMSs that follow the knowledge-aware approach for autonomous execution are, per the given definition, ABPMSs. Knowledge graphs hereby realize the process frame, defining the data and constraints that drive the process orchestration. Graph reasoning techniques concretize the AI technology used to attain process goals within this frame. While knowledge graphs were named as technology for process frame in the original ABPMS vision, such questions of detail for ABPMS design were left open for future research.

Consequently, our proposal strongly contributes to addressing this gap. First, while previous work has pointed out the value of existing constraint-based declarative languages for process framing, they have also pointed out that an integration of these (purely symbolic) rules with more sophisticated AI techniques is necessary [16]. We provide details for the proposed usage of knowledge graphs for process framing, utilizing neuro-symbolic reasoning approaches.

Second, the ABPMS vision for situation-aware explainability, i.e., the system being able to on the fly explain its reasoning for a decision. The more responsibility a system gets, the more important trust in it becomes, which can be built from explanations. In ABPMSs, users additionally need to understand process and system dynamics to be able to reframe the system. Given the possible explainability of knowledge-graph reasoning, highlighted in Sect. 3.3, our proposal also addresses this requirement.

5   Avenues for Further Research

We discovered multiple avenues for future research throughout our work. Knowledge extraction, i.e., collecting the facts to be put into the process knowledge graph, has been largely bracketed out in this paper. The body of work on the translation of models and event logs needs to be extended with declarative models and process knowledge aspects such as resources, data, or goals. Further, it is to be investigated how well existing knowledge extraction methods on unstructured data can be applied for process knowledge graph creation based on, e.g., regulations, guidelines, policy documents, etc.

Related, more research on the encoding of the process knowledge graph appears valuable. While we already provide naïve ways to encode most of the width of process knowledge, we are confident that more efficient ways can be found, which, in the end, improve the performance of applied reasoning techniques. For instance, the current encoding of previous process executions leads to an explosion of entities, which may or may not impact performance negatively – to be investigated. Similarly to be improved is the naïve approach to encode running process instances using incomplete traces. Further, the translation of declarative process aspects, i.e., rules and constraints, into graph schemata and rules has to be investigated further. Finally, the open question of how open-ended the graph schema should be might be investigated.

Extensive investigations of applying reasoning techniques on process knowledge graphs should be conducted in future. Requirements have to be formalized, and existing techniques have to be evaluated. If existing techniques prove inadequate, or potentials for improvement are identified, adaptations or new methods should be developed. We especially want to highlight the explainability aspect to be considered. Investigating the use of Large Language Models to translate explanations provided by existing techniques into natural language or similar better human-interpretable formats seems valuable in that direction. Further, we mostly focus on logical rules for explainable knowledge graph reasoning in our examples. Other XAI methods on knowledge graphs might be feasible for BPM use cases and should be investigated. Lastly, while we show the theoretical feasibility of our approach, further proof is needed. Hence, we call for a proof of concept that actually encodes and reasons upon one example, for instance the SEPSIS process.

Additionally, a bridge to existing symbolic reasoning techniques of the BPM field (e.g., [15]) should be thrown, both to transfer them onto the knowledge graph abstraction, but also then to integrate them into neuro-symbolic approaches based thereupon.

Finally, we want to highlight that we assume the two demonstrated BPM use cases not to be the only ones where knowledge graph reasoning might be useful. Consequently, future work should investigate their applicability and usefulness to other BPM challenges.

6   Concluding Remarks

In this paper, we investigated how knowledge graph reasoning can be used to enable explainable process automation based on the width of process knowledge. Using the well-known SEPSIS process as running example, we showed how various dimensions of process knowledge can be integrated into knowledge graphs, and how said knowledge can be used to support the execution of processes in a knowledge-aware and explainable fashion. We generalized these technological potentials to show how the BPM challenges of autonomous process framing and knowledge-intensive process support can be addressed. Finally, we motivated and highlighted avenues for future research. We thus conclude that knowledge graphs are indeed a key technology to be further researched for explainable and knowledge-aware process automation.

References

1. Annane, A., Aussenac-Gilles, N., Kamel, M.: BBO: BPMN 2.0 based ontology for business process representation. In: 20th European Conference on Knowledge Management (ECKM 2019), vol. 1, pp. 49–59 (2019)
2. Arias, M., Munoz-Gama, J., Sepúlveda, M.: Towards a taxonomy of human resource allocation criteria. In: Business Process Management Workshops, vol. 308, pp. 475–483. Springer, Cham (2018)

3. Bachhofner, S., Kiesling, E., Revoredo, K., Waibel, P., Polleres, A.: Automated process knowledge graph construction from BPMN models. In: Database and Expert Systems Applications, vol. 13426, pp. 32–47. Springer, Cham (2022)
4. Beheshti, A., et al.: ProcessGPT: transforming business process management with generative artificial intelligence. In: 2023 IEEE International Conference on Web Services (ICWS), pp. 731–739. IEEE, Chicago (2023)
5. Buliga, A., Di Francescomarino, C., Ghidini, C., Maggi, F.M.: Counterfactuals and ways to build them: evaluating approaches in predictive process monitoring. In: Advanced Information Systems Engineering. Lecture Notes in Computer Science, pp. 558–574. Springer, Cham (2023)
6. Chen, Y., et al.: An overview of knowledge graph reasoning: key technologies and applications. J. Sens. Actuat. Netw. **11**(4), 78 (2022)
7. DeLong, L.N., Mir, R.F., Fleuriot, J.D.: Neurosymbolic AI for reasoning over knowledge graphs: a survey. arXiv preprint arXiv:2302.07200 (2024)
8. Di Ciccio, C., Marrella, A., Russo, A.: Knowledge-intensive processes: characteristics, requirements and analysis of contemporary approaches. J. Data Semant. **4**(1), 29–57 (2015)
9. Dumas, M., et al.: AI-augmented business process management systems: a research manifesto. ACM Trans. Manag. Inf. Syst. **14**(1), 1–19 (2023)
10. Dumas, M., La Rosa, M., Mendling, J., Reijers, H.A.: Fundamentals of Business Process Management. Springer, Heidelberg (2018)
11. Hogan, A., et al.: Knowledge graphs. ACM Comput. Surv. **54**(4), 1–37 (2022)
12. Krause, F., Kurniawan, K., Kiesling, E., Paulheim, H., Polleres, A.: On the representation of Dynamic BPMN process executions in knowledge graphs. In: Knowledge Graphs and Semantic Web, vol. 14382, pp. 97–105. Springer, Cham (2023)
13. Mannhardt, F.: Sepsis cases - event log (2016). https://data.4tu.nl/articles/dataset/Sepsis_Cases_-_Event_Log/12707639/1
14. Mannhardt, F., Blinde, D.: Analyzing the trajectories of patients with sepsis using process mining. RADAR+EMISA 2017, Essen, Germany, 12–13 June 2017, pp. 72–80 (2017)
15. Marrella, A., Mecella, M., Sardina, S.: Intelligent process adaptation in the SmartPM system. ACM Trans. Intell. Syst. Technol. **8**(2), 25:1–25:43 (2016)
16. Montali, M.: Constraints for process framing in AI-augmented BPM. In: Business Process Management Workshops. Lecture Notes in Business Information Processing, pp. 5–12. Springer, Cham (2023)
17. Object Management Group (OMG): Case management model and notation version 1.0. Standard (2014). http://www.omg.org/spec/CMMN/1.0/
18. Reichert, M., Weber, B.: Enabling Flexibility in Process-Aware Information Systems. Springer, Heidelberg (2012)
19. Ruiz, C., Zitnik, M., Leskovec, J.: Identification of disease treatment mechanisms through the multiscale interactome. Nat. Commun. **12**(1), 1796 (2021)
20. Seidel, A., Haarmann, S., Weske, M.: Model-based decision support for knowledge-intensive processes. J. Intell. Inf. Syst. **61**(1), 143–165 (2023)
21. Tama, B.A., Comuzzi, M.: An empirical comparison of classification techniques for next event prediction using business process event logs. Expert Syst. Appl. **129**, 233–245 (2019)
22. Tiddi, I., Schlobach, S.: Knowledge graphs as tools for explainable machine learning: a survey. Artif. Intell. **302**, 103627 (2022)
23. Vu, H., Leopold, H., van der Aa, H.: What is business process automation anyway? In: 56th Hawaii International Conference on System Sciences (HICSS 2023), pp. 5462–5471. ScholarSpace (2023)

24. Wang, X., Wang, D., Xu, C., He, X., Cao, Y., Chua, T.S.: Explainable Reasoning over Knowledge Graphs for Recommendation. In: Proceedings of the AAAI Conference on Artificial Intelligence, vol. 33, no. 01, pp. 5329–5336 (2019)
25. Weske, M.: Business Process Management: Concepts, Languages, Architectures, 3rd edn. Springer, Heidelberg (2019)

Graph Neural Networks for PPM: Review and Benchmark for Next Activity Predictions

Sebastiano Dissegna[✉] and Chiara Di Francescomarino

Department of Computer Science Engineering, University of Trento, Trento, Italy
{sebastiano.dissegna,c.difrancescomarino}@unitn.it

Abstract. Predictive Process Monitoring (PPM) is a subfield of Process Mining, which focuses on using Machine Learning (ML) methods to predict the future of an ongoing execution based on its early stages. Predictions provided by PPM approaches include the time remaining until the end of the process execution, the next event activity to be performed, or the overall outcome of the execution trace. While different machine and deep learning methods have been leveraged in the literature of PPM, in recent years, Graph Neural Networks (GNNs) have emerged as a new type of neural network and have been leveraged by few approaches in PPM. GNN models offer the advantage of working with a more natural representation of a trace prefix, allowing more expressive and semantically rich encodings. In this paper, we review three GNNs-based approaches from the PPM literature and we compare them with state-of-the-art approaches on the task of next activity prediction. The results show that GNNs are able to achieve an accuracy gain of more than 10% with respect to traditional approaches for some datasets, thus making them a promising solution in PPM.

Keywords: Graph Neural Network · Predictive Process Mining · Next activity predictions

1 Introduction

Predictive Process Monitoring (PPM) is a sub-field of Process Mining that leverages Machine Learning (ML) techniques to predict the future of an ongoing execution based on its early stages. Predictions provided by PPM approaches include the time remaining until the end of the process execution, the next event activity to be performed, or the overall outcome of the execution trace. Various ML and Deep Learning (DL) methods, e.g., techniques like decision trees, support vector machines, and recurrent neural networks, have been explored in the PPM literature, however in recent years, a new class of neural networks called Graph Neural Networks (GNNs) has gained attention for its ability to handle graph-structured data. GNNs offer a significant advantage in PPM by allowing a

© The Author(s), under exclusive license to Springer Nature Switzerland AG 2025
K. Gdowska et al. (Eds.): BPM 2024 Workshops, LNBIP 534, pp. 31–43, 2025.
https://doi.org/10.1007/978-3-031-78666-2_3

more natural and expressive representation of trace prefixes. This enhanced representation enables the encoding of richer semantic information, potentially leading to more accurate predictions. Despite their potential, GNN-based approaches are still quite unexplored in the PPM domain.

This paper aims to review and benchmark the existing recent literature on GNN-based approaches for PPM. In detail we compare the approaches on the task of next activity prediction. Our findings reveal that GNNs can achieve an accuracy gain of more than 10 with respect to state-of-the-art PPM approaches on certain datasets, proving their effectiveness at the cost of a much higher computational power and longer training times.

The paper is structured as follows: Sect. 2 introduces the concepts needed for understanding the paper, Sect. 3 describe the GNNs approaches and, finally Sect. 4 and Sect. 6 report about the benchmarking evaluation and conclude the paper, respectively.

2   Background

In this section we provide a brief introduction to the main background concepts used in the paper on Process Mining, Predictive Process Monitoring and Graph Neural Networks.

2.1   Process Mining

Process Mining uses techniques to analyze event data from an operational process to achieve various objectives. These objectives may include discovering the process model, identifying bottlenecks and deviations, and gaining insights from the data. The starting point of any Process Mining technique resides in the creation of an event log from the data.

Definition 1 (Event Log). *An event log $\mathcal{E}$ is a collection of events that correspond to the selected process. Each event $e \in \mathcal{E}$ has a series of attributes $a \in \mathcal{AT}$, where $\mathcal{AT}$ is the set of all possible attributes. The three main attributes are as follows:*

- *CaseID (e_{id}): this identifies the unique case that a specific event belongs to;*
- *activity ($e_{act} \in \mathcal{A}$): this is the activity (in the set of activities $\mathcal{A}$) performed in that event;*
- *timestamp (e_{ts}): This indicates the time at which the event occurred.*

Other examples of attributes belonging to an event include the amount of a certain resource utilized during the process or information about how much time has elapsed since the start of the process or since the previous event occurred.

Definition 2 (Trace). *A trace $\sigma \models \langle e_1, .., e_{|\sigma|} \rangle$ is a non-empty sequence of events $e \in \mathcal{E}^* \backslash \{\emptyset\}$, such that $\forall e_i, e_j \in \sigma, 1 \leq i \leq j \leq |\sigma| : e_{i_{id}} = e_{j_{id}} \wedge e_{i_{ts}} \leq e_{j_{ts}}$, i.e., they all have the same CaseID and are ordered according to the timestamps.*

Definition 3 (Trace Prefix). *Given a trace* $\sigma \models \langle e_1, .., e_{|\sigma|} \rangle$, *the trace prefix of length* k, *where* $k \leq |\sigma|$, *is the sub-trace* $tp(\sigma, k) \models \langle e_1, .., e_k \rangle$.

Starting from event logs it is possible to automatically discover process models. Process models can be described using different representations, ranging from more informal ones - as the Directly Follows Graph (DFG) – up to more formal ones - as Petri Nets. Differently from DFGs Petri nets offer a more formal representation of the control flow and allow for representing concurrency.

Definition 4 (Petri Net). *A Petri Net* $\mathcal{PN} = (\mathcal{P}, \mathcal{T}, \mathcal{FL})$ *is a triplet where* $\mathcal{P}$ *is a finite set of places,* $\mathcal{T}$ *is a finite set of transitions and* $\mathcal{FL} \subseteq (\mathcal{P} \times \mathcal{T}) \cup (\mathcal{T} \times \mathcal{P})$ *is a set of directed arcs representing the flow relation. A marked Petri Net is a pair* $(\mathcal{PN}, \mathcal{M})$ *where* $\mathcal{M}$ *is the marking of the net (marking places with tokens), that is a multi-set over* $\mathcal{P}$ *that represents a state of the Petri net.*

In Process Mining, typically labelled Petri nets are used, that is Petri nets where transitions are labeled to refer to activities in the event log [1].

Definition 5 (Labeled Petri Net). *We can define a Labeled Petri Net as a tuple* $\mathcal{LPN} = (\mathcal{P}, \mathcal{T}, \mathcal{FL}, \mathcal{L})$, *where* $(\mathcal{P}, \mathcal{T}, \mathcal{FL})$ *is a Petri net, and* $\mathcal{L} : \mathcal{T} \rightarrow \mathcal{A}$ *is a labeling function that assigns to a transition* $t \in \mathcal{T}$ *a label belonging to the set of activities* $\mathcal{A}$.

2.2 Predictive Process Monitoring

Predictive Process Monitoring (PPM) is a branch of Process Mining that specifically deals with forecasting the future of a process execution. This usually entails building a model using a machine-learning algorithm to learn from historical traces. The model takes a trace prefix as input and returns a predicted key performance indicator (KPI) as output. KPIs may include the sequence of upcoming activities, the remaining time until the process is finished, or the outcome of the trace.

One challenge is that we cannot directly input a prefix into a machine-learning model; we need to encode it first. There are various methods for encoding a prefix. The simplest approach treats the prefix as a basic symbolic sequence, disregarding data and data flow information. Two examples of this type of encoding are the *boolean* and the *frequency-based* methods. In the *boolean* method, we create a feature vector denoted as v with boolean values, where each element indicates the presence or absence of an event in the prefix. The *frequency-based* method uses the frequency of each event class activity types instead of boolean values. These encodings are typically used with machine-learning models like Random Forest or SVM.

When dealing with the prediction of next activities, deep learning architectures are usually leveraged. One of the state-of-the-art approaches for trace suffix prediction is the work by Tax et al. [15], which utilizes an LSTM model. When working with an LSTM, a more complex encoding needs to be created. In this scenario, each event has a corresponding vector that is a concatenation of its different attributes. Numerical attributes can be used as they are, while categorical ones, such as *activities*, are converted into numerical attributes using a One-Hot Encoder. This means that a prefix will be represented by a matrix where each row corresponds to the event's vector. Additionally, LSTMs require working with inputs of the same length, but because we are dealing with prefixes of varying lengths, zero padding is necessary.

There has been little focus on using Graph Neural Networks for this type of predictive analysis. However, these networks offer several advantages. For instance, they can handle prefixes of different sizes without requiring zero padding. Additionally, they can represent connections between different events in a prefix, which is beneficial for representing events of different types, such as when working with Object Centric Event Log (*OCEL*). One drawback of GNNs is that they typically require more computational power, leading to longer training times compared to simpler machine-learning approaches.

2.3   Graph Neural Networks

Graph Neural Networks (GNNs) are a type of Neural Network created to work with data structured as graphs, where graphs are mathematical structures that model pairwise relationships between objects, consisting of nodes and edges connecting them. This type of network can work with two different types of graphs, the homogeneous ones, and the heterogeneous ones. In the first case, all nodes of a graph represent the same type of entity, while in the second case, we can have nodes of different types in the same graph. For this work, only the homogeneous case is considered. Compared to classical approaches like Node2Vec [8] and Nodesketch [20], which return a feature vector from a graph via some statistical analysis on the graph, with a GNN the graph representation is learned. Graph Neural Networks (GNNs) can be used for a range of tasks such as node classification and link prediction. In this paper, we use GNNs for graph classification. In this context, a feature vector is obtained from a graph and then fed into a Multi-layer Perceptron (MLP) for regression or discriminant analysis. The simplest GNN is the Graph Convolutional Neural Network (GCNN). We describe in the following how to use GCNN for graph classification and we then sketch more advanced GNN architectures.

GNN for Graph Classification. The training of a Graph Neural Network consists of 4 phases. In the first phase, both the node and edge features are initialized based on the available input data. Since we are dealing with a homogeneous case, all node feature vectors contain the same type of information. After the initialization, we need to propagate the information from each node to its neighboring nodes (second phase). In this phase, we perform a convolution step called message passing, which differs from a standard convolution operation in CNN due to the variable size of the neighborhood. We aggregate the information from neighboring nodes using Eq. 1, where i is the index of the current layer,

$h_u^{(i-1)}$ is the hidden representation of a neighboring node at the previous layer, and $N(v)$ is the set of neighbors of v.

$$h_{N(v)}^{(i)} = AGGR^{(i)}(\{h_u^{(i-1)} : u \in N(v)\}) \tag{1}$$

Different forms of aggregation exist. For instance, we can have mean (Eq. 2) and max (Eq. 3), with the second aggregation performed on a transformed representation of the nodes.

$$h_{N(v)}^{(i)} = Mean^{(i)}(\{h_u^{(i-1)} : u \in N(v)\}) \tag{2}$$

$$h_{N(v)}^{(i)} = Max^{(i)}(\{\sigma(W^{(i)}h_u^{(i-1)} + b) : u \in N(v)\}) \tag{3}$$

After obtaining information on the neighboring nodes, we can update the current node representation, via an *UPDATE* function, like in Eq. 4. In this equation, we apply a non-linear function, such as a Rectified Linear Unit (ReLU), to the linear mapping between the learned weights $W^{(i)}$ and the concatenation of the previous node representation $h_v^{(i-1)}$ and the neighborhood information $h_{N(v)}^{(i)}$.

$$h_v^{(i)} = \sigma(W^{(i)}[h_v^{(i-1)}; h_{N(v)}^{(i)}]) \tag{4}$$

In the third phase, multiple layers of message passing can be stacked to further propagate information across the graph, allowing to capture and model more complex dependencies between nodes. Finally, in the fourth phase, we perform a final aggregation of all nodes, called *READOUT*, by introducing a fake node connected to all nodes in the graph, whose representation will be the output feature vector of our GNN. This feature vector can then be fed into an MLP to perform the classification or regression step.

More Advanced GNN Architectures. In the literature different more sophisticated architectures than GCNN are available, each with its own unique characteristics. In this paper we will consider two of them. The first one is the GraphSAGE [9] model, which addresses the problem of the size of the neighborhood when updating a node representation. Intuitively, as the number of neighbours of a node in a dense graph may be in the thousands, taking the full neighborhood becomes very inefficient. They solve this by adopting sampling to obtain a fixed number of neighbours for each node when updating. The second model of interest is the Graph Attention Network (GAT) [17], which introduces an attention mechanism that learns the relative weights between two connected nodes.

3   GNN-Based Approaches

In this paper we review existing GNN-based approaches used for PPM tasks. In the next subsections we first report the review procedure used and we then describe the three final selected papers.

3.1   Review Procedure

To review GNN-based approaches for PPM tasks, we queried the online repository of one of the most popular technical publisher, Scopus with the following query on TITLE-ABS-KEY "Graph" AND "Predictive" AND "Process" AND "Monitoring". The search returned 117 results and we only considered conference and journal papers. We manually inspected the remaining 80 papers and applied some exclusion criteria. From the initial list, we excluded: (i) 71 papers not addressing PPM tasks; (ii) 4 papers using graph encodings but not GNNs. After applying the exclusion criteria only 5 papers remained. Out of them, two were one the extension of the other (workshop and journal papers) by the same authors, so we only kept the journal version. The initial search was hence finally reduced to 4 papers. Three of them focus on predicting the next activity [3, 13, 19] and one on predicting the remaining cycle time of a process [6]. Since the work in [19] proposes a similar but preliminary approach of the one in [6] based on the encoding of the feature vector and Gated Graph Recurrent Neural Networks, we focus on the three main approaches, which are described in the following.

3.2   Gated Graph Recurrent Neural Network

The approach introduced by Duong et al. [6] returns predictions on the remaining time. The encoding they use is straightforward. Each feature vector of an event in the prefix corresponds to a node feature vector in the graph. As mentioned in the previous section, GNN can handle graphs of any size directly, so there is no need for zero padding as compared to LSTM. The node features used are based on the *activity* and *timestamp* attributes. The activity is transformed using a one-hot encoder, while the *timestamps* are converted to four time-based features: time from a previous event, time from the start of the case, time within a day, and the day within a week. As for the edge feature vector, they are categorized into three types: *forward* (edges from one activity to a new activity within a case), *backward* (edges from one activity to an activity that has been performed within a case), and *repeat* (edges between two events associated with the same activity).

In the GNN model, they utilize a standard message-passing network as illustrated in Eq. (5). The *UPDATE* function is replaced with a Gated Recurrent Unit (GRU) cell, depicted in Eq. (6).

$$m_i^{l+1} = \sum_{j \in N(i)} e_{ji} \times W \times h_j^l \qquad (5)$$

$$h_i^{l+1} = GRU(m_i^{l+1}, h_i^l) \qquad (6)$$

For the final layer, they employ global mean pooling, which averages node features across the node dimension, to acquire a fixed-size graph embedding. This embedding is then fed as input to a Multilayer Perceptron (MLP), to obtain the desired prediction.

3.3   Recurrent Graph Convolutional Process Predictor (TACO)

The approach proposed by Rama-Maneiro et al. [13] provides predictions on the next event. Compared to the previous approach, this one combines information from traces with information extracted directly from the process model. In the initial phase, a set of process models represented as Petri Nets is extracted. The best model is converted into a place graph, illustrating connections between places in the Petri Net. This graph is then transformed into an adjacency matrix, and its normalized Laplacian is used as input for a GRNN (Graph Recurrent Neural Network). Concurrently, the node feature matrix is computed by replaying tokens from the prefix on the initial Petri Net. These two inputs are fed into the GRNN, and a max pooling function is used to obtain the graph embedding. The obtained graph embedding is combined with an attribute feature matrix, which contains information about the entire prefix derived from the event log. For each event in the prefix, four features are extracted: the event log activity that triggers a token in the places, the encoded time elapsed since the previous event, the encoded time elapsed since the first event of the prefix, and the attributes associated with the event. All of these components are then fed in input to an LSTM with a final dense layer to predict the next activity.

3.4   Multi Building Instance Graph Deep Graph Convolutional Neural Network (Multi-BIG-DGCNN)

The third approach by Chiorrini et al. [3] also focuses on predicting the next events. Differently from the previous ones, it relies on instance graphs to encode a prefix. The notion of instance graph is based on the notion of causal relation, as described in the following.

Definition 6 (Causal Relation). *A Causal Relation (CR) is a relation on the set of activities, $CR \subseteq \mathcal{A} \times \mathcal{A}.a_1 \rightarrow_{CR} a_2$ denotes that $(a_1, a_2) \in CR$. It represents the order of execution of a pair of activities in a process.*

Definition 7 (Instance Graph). *Let σ be a trace. An Instance Graph (IG) γ_σ of σ is a directed acyclic graph (V, E) where:*

- *$V = \{e_{act} \in \sigma\}$ is the set of vertexes, corresponding to activities in the trace.*
- *$E = \{(e_i, e_j) \in V \times V \mid i < j \wedge e_{i_{act}} \rightarrow_{CR} e_{j_{act}} \wedge (\forall e_k \in V(i < k < j \Rightarrow e_{i_{act}} \nrightarrow_{CR} e_{k_{act}})) \vee \forall e_w \in V(i < w < j \Rightarrow e_{w_{act}} \nrightarrow_{CR} e_{j_{act}}))\}$ is the set of arcs, which defines a partial order over V*

In order to extract a graph suitable for a graph neural network from an event log, three steps are required. First, each trace is represented with the corresponding IG. Second, each IG is enriched with additional event attributes, if available, and with additional perspectives regarding the sequential execution. Finally, all of them can be fed to a graph neural network along with their label.

The GNN model utilizes a Deep Graph Convolutional Neural Network (DGCNN) consisting of three main layers. The first layer comprises a series of convolutional layers, which aim to extract features from the local substructure of nodes and establish a node ordering. This ordering is then processed in a SortPoolingLayer, where the top nodes are selected to standardize the input dimension. Finally, a one-dimensional convolutional layer is followed by an additional dense layer and a softmax layer to obtain the representation used to make the desired prediction.

4   Benchmarking Evaluation

In this section we aim at benchmarking the three GNN-based state-of-the-art approaches for PPM from the literature on a specific PPM task: the next event prediction. We benchmarked two out of the three approaches, that is the Gated Graph Recurrent Neural Network by Doung et al. [6] and the Recurrent Graph Convolutional Process Predictor (TACO) by Rama-Maneiro et al. [13] since they made their code available.

Specifically, we are interested to investigate the following research question:

- How do existing GNNs-based approaches perform in the task of next-activity prediction?

The research question aims at benchmarking the accuracy and the training time required by the GNN-based approaches on the next activity prediction task and at comparing them on the same task with one of the state-of-the-art approaches for next event prediction, that is the work by Tax et al. [15].

4.1   Datasets

The evaluation is carried out on the following seven datasets:

- **Helpdesk:** This dataset contains a compilation of records from a helpdesk ticketing system, documenting events related to the handling and resolving of helpdesk tickets.
- **EnvPermit:** This dataset contains event logs from an environmental permit application process, which typically captures the sequence of activities performed to process permit applications.
- **BPI-2012-A:** This dataset is part of the BPI Challenge 2012 and contains detailed logs of a loan application process
- **BPI-2012-O:** This dataset is part of the BPI Challenge 2012 and contains logs from the organization perspective of a loan application process.
- **BPI-2012-WC:** This dataset is part of the BPI Challenge 2012 and contains logs from a combined workflow and case perspective related to a loan application process.
- **BPI-2013-Closed-Problems**: This dataset is part of the BPI Challenge 2013 focusing on Volvo IT incidents and problem management and contains logs related to closed problems. It often captures the resolution process of incidents or problems reported within an organization.

- **BPI-2013-Incidents:** Another part of the BPI Challenge 2013 dataset on Volvo IT incidents and problem management, whose focus is on reported incidents, documenting the events involved in handling and resolving these incidents.

Table 1 reports some detailed statistics related to the event logs.

Table 1. Dataset information

	Helpdesk	EnvPermit	BPI-2012-A	BPI-2012-O	BPI-2012-WC	BPI-2013-CP	BPI-2013-I
n. cases	4552	1434	13087	5015	9658	1487	7554
n. activities	10	27	10	7	6	7	13
n. events	21197	8577	60849	31244	72413	6660	65533
avg. case len	4.66	5.98	4.65	6.23	7.5	4.48	8.68
max case len	15	25	8	30	74	35	123
avg. case duration	40.85	5.41	8.08	17.19	11.4	178.89	12.08
max case duration	59.99	275.8	91.5	89.59	91.08	2254.85	771.35
n. variants	207	116	17	168	2263	327	2278

4.2 Evaluation Setting

All experiments were carried out on a laptop equipped with an Intel (R) Core (TM) i7-13700H, 32 GB of RAM, and a 4070 Nvidia GPU. The experiments were carried out with the following settings. For all approaches the default hyperparameters were used, while for the trace attributes employed, we considered only the activity of each event and the four time-related features described in Sect. 3.2. The maximum number of epochs was set to 10 for TACO [13], 50 for Duong et al. [6], and 100 for the work by Tax et al. [15]. This variation in epochs is due to the longer training time for TACO and Duong et al. compared to the simpler LSTM model. However, TACO and Duong et al. also converge in fewer epochs. An early-stop technique was also utilized for all three approaches. Each dataset was divided into a training set and a test set, with an 80/20% split. Additionally, a validation set of size 1/3 was extracted from the training set.

4.3 Results

Table 2 reports for each dataset and each technique the overall accuracy on the test set. Moreover, the table reports the gain or the loss obtained by each GGN-based approach with respect to the LSTM by Tax et al.. We report in bold the highest accuracy.

The table shows that, while Duong et al. only slightly worsens the performance of the state-of-the-art approach Tax et al. for most of the datasets, TACO shows an average gain with respect to Tax et al.'s performance of 4.3% across all datasets. Tax et al. indeed sligthly outperforms TACO only for the EnvPermit and the BPI-2012-WC datasets. For all the other four datasets, and especially

Table 2. Next activity accuracy on test set of all approaches

Approach	Helpdesk	EnvPermit	BPI-2012-A	BPI-2012-O	BPI-2012-WC	BPI-2013-CP	BPI-2013-I
Tax et al. [15]	72.34	**91.91**	66.71	79.18	**79.65**	56.92	57.43
Duong et al. [6]	63.04 (−9.3)	86.97 (−4.94)	64.41 (−2.3)	71.96 (−7.22)	77.51 (−2.14)	52.52 (−4.4)	52.73 (−4.7)
TACO [13]	**78.53** (+6.19)	88.67 (−3.24)	**78.01** (+11.3)	**81.51** (+2.33)	65.51 (−14.14)	**70.34** (+13.42)	**71.66** (+14.23)

for BPI2013 and BPI-2021-A, where the gain is higher than 10%, TACO presents a quite high gain.

Furthermore, by looking at the table, we can observe that TACO outperforms the Duong et al. approach with significantly higher accuracy across all datasets, except for BPI-2012-WC. This is possibly due to the low number of activity classes and to the high number of variants characterizing the BPI-2012-WC dataset. In general, the better performance of TACO are likely due to the capability of this approach to take into account not only the trace information but also the information extracted from the event log.

However, it is worth noting that the training time for TACO and Duong et al. is longer compared to Tax et al.. TACO takes around 5 to 10 min per epoch, Duong et al. takes 1 to 2 min, while Tax et al. only takes 2–3 s per epoch. While in certain scenarios, a faster training model might be preferred due to some hardware constraints, for applications in fields such as medicine, where accuracy remains the top priority, GNN models represent an important breakthrough. We can hence answer to our research question stating that, at a higher training time cost, GNN-based approaches, and specifically the TACO approach, show promising results in terms of next activity prediction accuracy.

A First Look Into More Complex Architectures. One thing that is common to all the approaches presented above is the fact that they employ some very simple models of GNN. We performed an initial experimentation taking the implementation of the Duong et al. approach and replacing the Graph Convolutional part with more complex GNN models such as GAT [17] and GraphSage [9]. Table 3 shows the results of the preliminary experiment carried out.

Table 3. Next activity prediction accuracy obtained with different architectures of the Duong et al. approach on the dataset test sets

Architecture	Helpdesk	EnvPermit	BPI-2012-A	BPI-2012-O	BPI-2012-WC	BPI-2013-CP	BPI-2013-I
GGNN	63.04	86.97	64.41	71.96	77.51	52.52	52.73
GAT	**75.36** (+12.32)	88.34 (+1.37)	64.05 (−0.36)	66.07 (−5.69)	78.2 (+0.51)	42.56 (−9.96)	49.72 (−3.01)
GraphSage	75.08 (+12.04)	**91.47** (+4.5)	**66.7** (+2.29)	**79.52** (+7.76)	**79.56** (+2.05)	**57.3** (+4.78)	**56.9** (+4.17)

Although also for this experiment no hyperparameter optimization was carried out, we can see an increase in accuracy across all datasets. The accuracy gain in some cases is marginal, around 2–4%, but for the BPI-2012-O and Helpdesk datasets, this amounts to 7.56% and 12.32%, respectively. This is a promising development for future investigations. Using a more complex architecture and a richer representation, similar to the one employed in TACO, can lead to even better results.

5   Related Work

In Predictive Process Monitoring (PPM), the literature can be categorized into three main groups [4]. A first group aims at forecasting the outcome of a process execution [16], a second group deals with time-related predictions [18], while a third one focuses on predicting the next activity within a sequence of events. In this work we mainly focus on this latter category when benchmarking GNN-based approaches. Many existing approaches in the literature leverage deep-learning techniques [12] to build models for making such predictions. Most of these approaches [2,7,10,14,15] utilize Long Short-Term Memory (LSTM) models to predict the next activity in an ongoing sequence of events. LSTMs are preferred due to their ability to effectively handle sequential data. In addition to predicting the next activity, some approaches also focus on forecasting other attributes of an event. For instance, [15] built an LSTM model to predict the timestamp of the next event in a sequence. Apart from LSTMs, other deep learning models like Convolutional Neural Networks (CNNs) are also utilized in the literature. Mauro et al. [5] used a CNN network with inception modules, while Pasquadibisceglie et al. [11] employed a frequency-based encoding of spatial representation of event logs as input for the CNN.

6   Conclusion

This study compared two GNN-based approaches in PPM (Duong et al. and TACO) and demonstrated the superior performance of TACO in the next-activity prediction task with respect to state-of-the-art approaches. GNNs have shown great potential in improving the predictive capabilities of PPM systems, when leveraging the process model graph structure and combining it with data, as in the case of TACO. The drawback of these approaches is the high computational cost. As the model becomes more complex, it demands for a greater computational power and longer training times. However, the 10% gain in accuracy justifies the additional costs. Future work will focus on exploring more advanced GNN architectures to improve results on the next activity prediction task.

References

1. van der Aalst, W.M.P., Carmona, J. (eds.): Process Mining Handbook. LNBIP, vol. 448. Springer (2022). https://doi.org/10.1007/978-3-031-08848-3
2. Camargo, M., Dumas, M., González-Rojas, O.: Learning accurate LSTM models of business processes. In: Hildebrandt, T., van Dongen, B.F., Röglinger, M., Mendling, J. (eds.) BPM 2019. LNCS, vol. 11675, pp. 286–302. Springer, Cham (2019). https://doi.org/10.1007/978-3-030-26619-6_19

3. Chiorrini, A., Diamantini, C., Genga, L., Potena, D.: Multi-perspective enriched instance graphs for next activity prediction through graph neural network. J. Intell. Inf. Syst. **61**, 5–25 (2023). https://doi.org/10.1007/s10844-023-00777-1

4. Di Francescomarino, C., Ghidini, C.: Predictive process monitoring. In: van der Aalst, W.M.P., Carmona, J. (eds.) Process Mining Handbook. Lecture Notes in Business Information Processing, vol. 448, pp. 320–346. Springer, Cham (2022)

5. Di Mauro, N., Appice, A., Basile, T.M.A.: Activity prediction of business process instances with inception CNN models. In: Alviano, M., Greco, G., Scarcello, F. (eds.) AI*IA 2019. LNCS (LNAI), vol. 11946, pp. 348–361. Springer, Cham (2019). https://doi.org/10.1007/978-3-030-35166-3_25

6. Duong, L.T., Travé-Massuyès, L., Subias, A., Merle, C.: Remaining cycle time prediction with graph neural networks for predictive process monitoring. In: International Conference on Machine Learning Technologies, ICMLT 2023, pp. 95–101 (2023). https://doi.org/10.1145/3589883.3589897

7. Evermann, J., Rehse, J., Fettke, P.: Predicting process behaviour using deep learning. Decis. Support Syst. **100**, 129–140 (2017). https://doi.org/10.1016/J.DSS.2017.04.003

8. Grover, A., Leskovec, J.: node2vec: scalable feature learning for networks. CoRR abs/1607.00653 (2016), http://arxiv.org/abs/1607.00653

9. Hamilton, W.L., Ying, Z., Leskovec, J.: Inductive representation learning on large graphs. In: Annual Conference on Neural Information Processing Systems 2017 (NIPS 2017), pp. 1024–1034 (2017)

10. Lin, L., Wen, L., Wang, J.: MM-pred: a deep predictive model for multi-attribute event sequence. In: SIAM International Conference on Data Mining, pp. 118–126. SIAM (2019). https://doi.org/10.1137/1.9781611975673.14

11. Pasquadibisceglie, V., Appice, A., Castellano, G., Malerba, D.: Using convolutional neural networks for predictive process analytics. In: International Conference on Process Mining, ICPM 2019, pp. 129–136. IEEE (2019). https://doi.org/10.1109/ICPM.2019.00028

12. Rama-Maneiro, E., Vidal, J., Lama, M.: Deep learning for predictive business process monitoring: review and benchmark. IEEE Trans. Knowl. Data Eng. (2021)

13. Rama-Maneiro, E., Vidal, J.C., Lama, M.: Embedding graph convolutional networks in recurrent neural networks for predictive monitoring. IEEE Trans. Knowl. Data Eng. **36**(1), 137–151 (2024). https://doi.org/10.1109/TKDE.2023.3286017

14. Schönig, S., Jasinski, R., Ackermann, L., Jablonski, S.: Deep learning process prediction with discrete and continuous data features. In: 13th International Conference on Evaluation of Novel Approaches to Software Engineering, ENASE 2018, pp. 314–319. SciTePress (2018). https://doi.org/10.5220/0006772003140319

15. Tax, N., Verenich, I., La Rosa, M., Dumas, M.: Predictive business process monitoring with LSTM neural networks. In: International Conference on Advanced Information Systems Engineering, pp. 477–492. Springer (2017)

16. Teinemaa, I., Dumas, M., La Rosa, M., Maggi, F.M.: Outcome-oriented predictive process monitoring: review and benchmark. ACM Trans. Knowl. Discov. Data (TKDD) **13**(2), 1–57 (2019)

17. Velickovic, P., Cucurull, G., Casanova, A., Romero, A., Liò, P., Bengio, Y.: Graph attention networks. In: International Conference on Learning Representations. OpenReview.net (2018)

18. Verenich, I., Dumas, M., Rosa, M.L., Maggi, F.M., Teinemaa, I.: Survey and cross-benchmark comparison of remaining time prediction methods in business process monitoring. ACM Trans. Intell. Syst. Technol. **10**(4), 34:1–34:34 (2019). https://doi.org/10.1145/3331449

19. Weinzierl, S.: Exploring gated graph sequence neural networks for predicting next process activities. In: Marrella, A., Weber, B. (eds.) BPM 2021. LNBIP, vol. 436, pp. 30–42. Springer, Cham (2022). https://doi.org/10.1007/978-3-030-94343-1_3

20. Yang, D., Rosso, P., Li, B., Cudre-Mauroux, P.: NodeSketch: highly-efficient graph embeddings via recursive sketching. In: International Conference on Knowledge Discovery & Data Mining, KDD 2019, pp. 1162–1172. ACM (2019). https://doi.org/10.1145/3292500.3330951

Bridging Domain Knowledge and Process Discovery Using Large Language Models

Ali Norouzifar[1]([✉])(iD), Humam Kourani[1,2](iD), Marcus Dees[3](iD), and Wil M. P. van der Aalst[1](iD)

[1] RWTH University, Aachen, Germany
{ali.norouzifar,wvdaalst}@pads.rwth-aachen.de
[2] Fraunhofer FIT, Sankt Augustin, Germany
humam.kourani@fit.fraunhofer.de
[3] UWV Employee Insurance Agency, Amsterdam, The Netherlands
marcus.dees@uwv.nl

Abstract. Discovering good process models is essential for different process analysis tasks such as conformance checking and process improvements. Automated process discovery methods often overlook valuable domain knowledge. This knowledge, including insights from domain experts and detailed process documentation, remains largely untapped during process discovery. This paper leverages Large Language Models (LLMs) to integrate such knowledge directly into process discovery. We use rules derived from LLMs to guide model construction, ensuring alignment with both domain knowledge and actual process executions. By integrating LLMs, we create a bridge between process knowledge expressed in natural language and the discovery of robust process models, advancing process discovery methodologies significantly. To showcase the usability of our framework, we conducted a case study with the UWV employee insurance agency, demonstrating its practical benefits and effectiveness.

Keywords: Process Mining · Process Discovery · Process Knowledge · Large Language Models

1 Introduction

Recorded event data within information systems provides a rich source of information for process mining applications, enabling organizations to gain insights and improve their operational processes. In the field of process mining, various automated techniques are utilized to discover descriptive models that explain process executions. Despite the development of numerous methodologies for process discovery, the task remains inherently complex and challenging [3]. Discovering process models that do not align with domain knowledge presents significant

This research was supported by the research training group "Dataninja" (Trustworthy AI for Seamless Problem Solving: Next Generation Intelligence Joins Robust Data Analysis) funded by the German federal state of North Rhine-Westphalia.

K. Gdowska et al. (Eds.): BPM 2024 Workshops, LNBIP 534, pp. 44–56, 2025.
https://doi.org/10.1007/978-3-031-78666-2_4

challenges, particularly when these models are intended for conformance checking and process improvement.

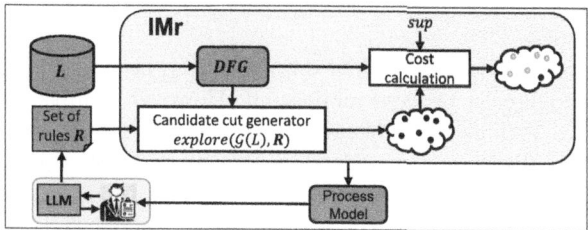

Fig. 1. Our proposed framework to integrate process knowledge in the IMr framework employing LLMs. (Color figure online)

In addition to the extracted event data from information systems, we often have access to domain experts, process documentation, and other resources collectively referred to as *domain knowledge*, which cannot be directly used for process discovery. These valuable resources typically remain untapped when aiming to discover process models. Incorporating domain knowledge into the discovery of process models poses several challenges. For instance, domain experts usually have a thorough understanding of their processes, but they can only explain them in natural language. Furthermore, textual process documents, although potentially rich in detail, also pose integration difficulties. In our paper, we address these challenges by enabling the direct involvement of such information in process discovery through the use of Large Language Models (LLMs). LLMs have demonstrated the ability to handle user conversations and comprehend human reasoning effectively.

Our framework builds upon the IMr framework proposed in [12]. IMr is an inductive mining-based framework that recursively selects the process structure that best explains the actual process. Within this framework, the algorithm encounters various possibilities for constructing the process structure. To guide this selection, rules are introduced as inputs to prune the search space and eliminate potentially suboptimal process structures. Although the concept of rules is broad, the Declare rule specification language is proposed as an example [11]. Declarative rules are advantageous due to their similarity to human reasoning and logic, supported by extensive literature. They are based on logical statements and have specific semantics, though it is unrealistic to expect users to provide these rules directly.

Our proposed framework, illustrated in Fig. 1, leverages LLMs and prompt engineering to integrate domain knowledge into process discovery. Starting with an event log, it employs process knowledge in various forms. LLMs play a crucial role by translating textual inputs into declarative rules, which IMr then integrates. This framework allows for the use of textual process descriptions prior to initiating process discovery, enables domain experts to provide feedback on the discovered models, and facilitates interactive conversations with domain experts to gather information and improve the models.

2   Related Work

In traditional process discovery, event data are often used as the primary source of information to create process models [3]. However, additional information resources, such as various forms of process knowledge, can significantly enhance the quality of the discovered models [14]. When available, this supplementary knowledge can be utilized before discovery to filter the event log [5], during the discovery phase to influence the process model structure [12], or within an interactive framework [4,15]. Despite these benefits, the direct involvement of process experts is often limited due to the complexities involved in integrating their knowledge into process discovery.

In [12], declarative rules are used as an additional input for process discovery, which can be provided by the user or generated by automated methods. However, expecting users to be proficient in declarative rule specification language is not always feasible. The proposed method in [4] requires users to engage at a low level to position transitions and places based on guiding visualizations. The approach in [15] begins with an initial model discovered from a user-selected subset of variants and incrementally allows adding more variants to update the process model. Some research focuses on repairing process models after discovery, primarily to improve the correspondence between the process models and event logs, rather than incorporating process knowledge [13]. In contrast, our paper aims to minimize the effort required from domain experts by using natural language conversations to influence process discovery.

Translating the natural language to process models using natural language processing is investigated in [1]. Anomaly detection is examined in [2] by focusing on semantic inconsistencies in event labels within event logs, utilizing natural language processing to identify anomalous behavior. Recently, LLMs have been employed for various process mining tasks. The opportunities, strategies, and challenges of using LLMs for process mining and business process management are discussed in [16]. Additionally, several studies propose the extraction of process models directly from textual inputs [6,8,9]. Unlike these approaches, our method maintains the event log as the main source of information while incorporating textual process knowledge into the discovery process.

3   Background

The blue box in Fig. 1 highlights one recursion of the IMr framework [12]. Each recursion extracts a Directly Follows Graph (DFG) from the event log, representing the set of activities Σ and their direct succession. The algorithm searches for all binary cuts that divide Σ into two disjoint sets considering a structure specification type, i.e., sequence, exclusive choice, concurrent, or loop type. IMr filters out candidate cuts that may violate any rule $r \in R$, where R is the set of rules given by the user or discovered using automated methods. While [12] incorporates declarative constraints listed in Table. 1, the framework is flexible to support other rule specification languages. Cost functions evaluate the quality of candidate cuts, based on counting the number of deviating edges and estimating the number of missing edges considering parameter $sup \in [0, 1]$. In each recursion, the algorithm selects the cut with the minimum cost, splits the event log accordingly, and recursively processes each sub-log until only base cases remain.

Table 1. Declarative templates supported by IMr [12].

Declarative Template	Description
$at-most(a)$	a occurs at most once
$existence(a)$	a occurs at least once
$response(a, b)$	If a occurs, then b occurs after a
$precedence(a, b)$	b occurs only if preceded by a
$co\text{-}existence(a, b)$	a and b occur together
$not\text{-}co\text{-}existence(a, b)$	a and b never occur together
$not\text{-}succession(a, b)$	b cannot occur after a
$responded\text{-}existence(a, b)$	If a occurs in the trace, then b occurs as well

4   Motivating Example

To motivate the research question addressed in this paper, consider the following event log extracted from a synthetic process $L = [\langle$A-created, A-canceled$\rangle^{300}$, $\langle$A-created, Doc-checked, Hist-checked, A-accepted$\rangle^{200}$, $\langle$A-created, Hist-checked, Doc-checked, A-accepted$\rangle^{50}$, $\langle$A-created, Doc-checked, Hist-checked, A-rejected$\rangle^{300}$, $\langle$A-created, Hist-checked, Doc-checked, A-rejected$\rangle^{80}$, $\langle$A-created, A-canceled, A-accepted$\rangle^{20}$, $\langle$A-created, A-canceled, A-rejected$\rangle^{15}$, $\langle$A-created, Doc-checked, Hist-checked, A-rejected, A-accepted $\rangle^{35}]$, where A stands for application, Doc for documents, and Hist for history.

Figure 2a illustrates the process model discovered using the IMf algorithm as a state-of-the-art process discovery technique [10]. The IMr framework with parameter $sup = 0.2$ and utilizing the Declare Miner [11] with $confidence = 1$ discovers the same process model. Consider that in addition to the provided event log, we have some additional process knowledge that helps us verify this model and pinpoint the possible unexpected behavior represented in the process model. In this paper, ChatGPT refers to ChatGPT-4o. We provided a text as feedback on this discovered model and asked ChatGPT to translate natural language feedback into understandable rules for the IMr framework. Here is our written feedback:

The discovered process does not fully adhere to our intuitions. Specifically, if a claim is canceled, the application cannot be either rejected or accepted. Furthermore, a claim cannot be both rejected and accepted for a single individual. Additionally, the history is always checked after the documents have been reviewed.

The following declarative rules, as explained in this paper, were extracted by ChatGPT:

not-co-existence(A-cancelled, A-accepted), not-co-existence(A-cancelled, A-rejected), not-co-existence(A-accepted, A-rejected), response(Doc-checked, Hist-checked)

Figure 2b presents the process model discovered using our proposed pipeline. In this approach, we utilized ChatGPT to interpret the textual feedback and generate declarative constraints, which are then used as input for the IMr framework.

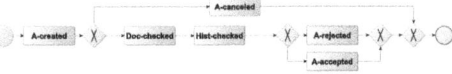

(a) Discovered model with deviations from the process knowledge. This model is discovered using IMf with f=0.2 and IMr with sup=0.2 and rules discovered employing Declare Miner [11] with $confidence$=1.

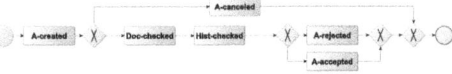

(b) The desired process model considering both the event log and process knowledge. IMr with sup=0.2 and rules extracted from process description employing ChatGPT discover this model.

Fig. 2. Discovered models from the motivating example event log using different techniques.

5   Domain-Enhanced Process Discovery with LLMs

In this section, we present our framework that leverages LLMs to integrate domain knowledge into the process discovery task. Figure 3 illustrates an overview of our proposed framework. The core idea is to utilize domain knowledge to generate a set of rules R which serves as input for the IMr framework. This can be done before starting the discovery by encoding process descriptions as rules, or after the process discovery by having a domain expert review the process model and provide feedback. Engaging in interactive conversations with

LLMs in both scenarios helps address uncertainties and improve the quality of the extracted rules. An implementation of the framework is publicly available[1].

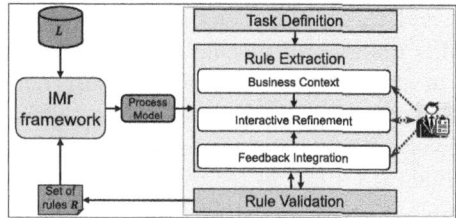

Fig. 3. Different components of the designed framework to bridge domain knowledge and process discovery using LLMs.

5.1 Task Definition

As outlined in [9], role promoting, knowledge injection, few-shot learning, and negative prompting techniques have significant potential to effectively prepare LLMs for specific process mining tasks. In our initial prompt, we define the role of the LLM as an interface between the domain expert and process discovery framework, such that LLM should encode the domain knowledge to declarative constraints as we need in IMr. Despite the similarity of declarative templates to human logic and reasoning, we observed the difficulties of LLMs in adhering to strict expectations. Therefore, we explain in our prompt the set of constraints we support, detailing both the syntax and the semantics of these constraints (cf. Table 1). We leverage the LLM's ability to derive insights from examples by providing multiple pairs of textual process descriptions and their corresponding declarative constraints. Additionally, we include instructions to avoid common issues, such as syntactic mistakes, and extend our learning pairs to include examples of undesirable constraints. The detailed written prompt is available in our GitHub repository[2].

5.2 Rule Extraction

After introducing the task, the LLM is ready to receive textual input and produce output as declarative constraints. As illustrated in Fig. 3, domain experts can contribute in three distinct ways: providing business context, offering feedback after reviewing process models, and engaging in interactive conversations with the LLM. In the following sections, we explain these contributions in detail and discuss their respective roles.

---

[1] https://github.com/aliNorouzifar/IMr-LLM.git.
[2] https://github.com/aliNorouzifar/IMr-LLM/blob/master/files/prompts.pdf.

Business Context. The domain expert can introduce the actual business process to the LLM, providing a general overview, detailing the relationships between specific activities, or even including constraints written in natural language. This flexibility allows the domain expert to tailor the input based on their unique insights and the specifics of the process at hand. It is important to note that the LLM is unaware of specific activity labels used in the recorded event data. The list of activities can be automatically derived from the event log, ensuring that all relevant actions are accurately captured in the generated constraints. Alternatively, the domain expert can provide the list of activities and add context to guide the LLM in relating the process description with the activity labels, resulting in constraints that involve the correct activity labels.

Feedback Integration. After generating the initial process model, it is presented to the domain expert for review. The domain expert is expected to examine the process model for accuracy, completeness, and practical alignment with real-life scenarios. In case of finding errors in the represented model, the domain expert can provide a written feedback and explain the behaviors that do not make sense in the real process. The LLM then adjusts and refines the declarative constraints based on this feedback.

Interactive Refinement. In typical scenarios, LLMs tend to provide answers that appear confident and definitive, often without indicating any uncertainty [7]. We facilitate a more detailed understanding of the provided textual descriptions by encouraging the LLM to express uncertainty and address it by asking questions. This stage involves a dynamic dialogue between the LLM and the domain expert. Should it encounter gaps in its knowledge or find ambiguities in the process descriptions, the LLM is encouraged to formulate and pose relevant questions. These questions are directed towards the domain experts, who then provide responses. The quality and precision of the responses provided by domain experts play a significant role in enhancing the quality of the generated constraints.

5.3 Rule Validation

An essential step in the framework is checking the extracted declarative constraints from the LLM's response. The LLM is instructed to encapsulate the constraints within specific tags in the response and to write them in a predefined language with no additional text or descriptions. Following extraction, the constraints undergo a validation process. This includes checking that the syntax of each constraint conforms to our predefined language, e.g., checking the type identifier and the number of activities specified within the constraint. Additionally, the labels of activities are verified against the activities recorded in the event log. If any errors are detected during validation, an error-handling loop is initiated. A new prompt specifies the problem and its location, prompting the LLM to adjust its output.

6   Case Study

A case study with the UWV employee insurance agency is conducted to demonstrate the usability of our approach in a real-life setting. UWV is responsible for managing unemployment and disability benefits in the Netherlands. For this case study, one of UWV's claim-handling processes is selected. Figure 4 depicts the normative model of this process, which was developed in collaboration with process experts who have a thorough understanding of the workflow. The event log used in this study contains 144,046 cases, 16 unique activities, and 1,309,719 events. Our GitHub repository provides the full prompting history and more readable process models[3].

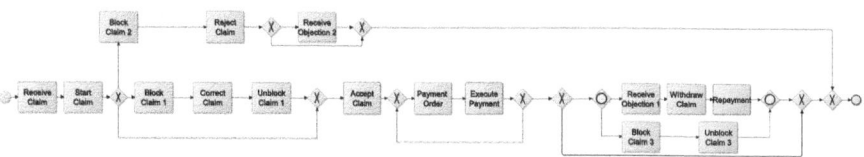

Fig. 4. Normative model of the UWV claim handling process, extracted manually in collaboration with domain experts [12].

6.1   Process Discovery Without Including Process Knowledge

Our initial attempt to discover a process model using the IMf algorithm with $f = 0.2$ resulted in the model shown in Fig. 5a. When compared to the normative model, significant differences are observed, e.g., *Receive Claim* and *Start Claim* are the first mandatory steps but the process model allows for skipping them or for many other activities occurring before them. Figure 5b illustrates the process model discovered using the IMr algorithm with $sup = 0.2$ and an empty set of input rules. Although this model shows more structural similarities to the normative model, it still contains some nonsensical differences. For instance, *Block Claim 1* should only be relevant if the claim is planned to be accepted, but this model permits it for rejected cases as well. Similarly, *Receive Objection 2* should only occur if the claim is rejected, yet the model allows it for accepted cases as well.

---

[3] https://github.com/aliNorouzifar/IMr-LLM/blob/master/files/prompts.pdf.

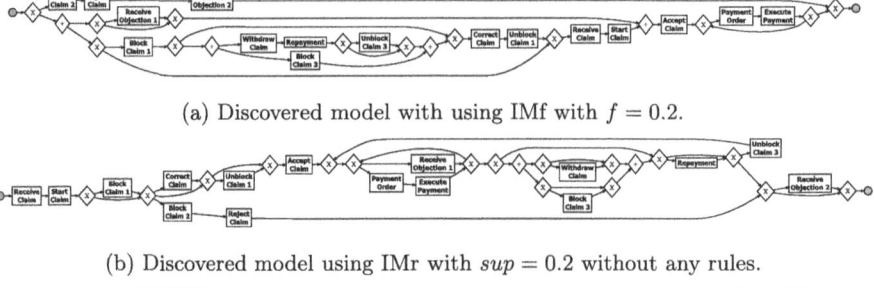

(a) Discovered model with using IMf with $f = 0.2$.

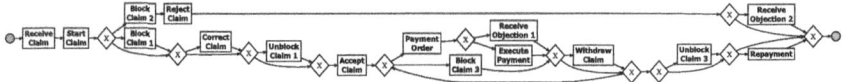

(b) Discovered model using IMr with $sup = 0.2$ without any rules.

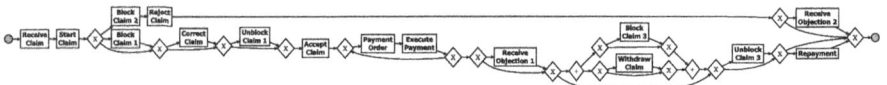

(c) Discovered model using IMr with $sup = 0.2$ and the rules provided by ChatGPT.

(d) Discovered model using IMr with $sup = 0.2$ and the rules provided by ChatGPT after incorporating the domain expert feedback.

Fig. 5. Discovered models from UWV event log using different strategies

6.2   Employing ChatGPT to Extract the Rules

We experimented with Gemini and various versions of ChatGPT to translate the process knowledge into declarative rules. ChatGPT-4o provided the best constraints and demonstrated a superior understanding of the task. By incorporating rules extracted by ChatGPT into the IMr framework, we obtained the process model shown in Fig. 5c. After iterating with feedback from domain experts, the model is further refined and improved, resulting in the process model depicted in Fig. 5d.

The process knowledge for this case study is incorporated as a set of activity labels and a process description provided by a domain expert. The blue boxes indicate the prompts given by domain experts, while the red boxes display the responses generated by ChatGPT.

Our process is designed to handle claim applications within an insurance agency, where applicants request various types of benefits, and our system processes these applications through a structured workflow. There are three types of blocking mechanisms: Block Claim 1, Block Claim 2, and Block Claim 3, each with corresponding resolutions: Unblock Claim 1, Unblock Claim 2, and Unblock Claim 3. The process begins once a claim is received. Block Claim 1 occurs when initial issues are detected, leading to necessary corrections before the claim is unblocked and accepted, resulting in the client receiving between one to three payments. Block Claim 2 is applied when the claim is deemed ineligible for receiving any payments, resulting in rejection and potentially an objection from the client. Block Claim 3 is triggered if there is missing information, such as incorrect bank account details, after a claim is accepted; the client is notified to provide the correct details, and the claim is unblocked upon receipt of this information. Additionally, clients may file an objection after receiving payments, leading to the withdrawal of the claim and repayment of the received money to the agency.

Following this process description, we encouraged ChatGPT to ask clarifying questions whenever it encountered uncertainties about the declarative rules. The box below provides an example of questions posed by ChatGPT.

Q1: Can "Block Claim 1", "Block Claim 2", and "Block Claim 3" occur in the same process instance, or are they mutually exclusive? **Q2:** Are the corresponding unblocking activities ("Unblock Claim 1", "Unblock Claim 2", and "Unblock Claim 3") strictly linked to their respective blocking activities, or can there be other sequences? **Q3:** Is there any restriction on the order of "Receive Objection 1" and "Receive Objection 2", or can they occur independently of each other? ...

The provided questions demonstrate the LLM's correct approach to understanding the declarative constraints between the activities in the process. For instance, the first question indicates that ChatGPT is attempting to discern whether there is a co-existence or non-co-existence relationship between block claim types. Question 3 is also pertinent, as the process description does not clearly distinguish between Receive Objection 1 and Receive Objection 2. The domain expert provided the following answers to these questions:

A1: Some cases may have both Block Claim 1 and Block Claim 3. But Block Claim 2 can never occur in the trace with Block Claim 1 or Block Claim 3. **A2:** Each blocking mechanism (Block Claim 1, Block Claim 2, Block Claim 3) has a corresponding unblocking mechanism (Unblock Claim 1, Unblock Claim 2, Unblock Claim 3), and each pair (Block, Unblock) occurs in a strict sequence. **A3:** Receive Objection 2 only occurs in the rejected cases and Receive Objection 1 only occurs in the accepted cases after they receive some payments. In general, Receive Objection 1 and Receive Objection 2 cannot occur together. ...

The set of declarative constraints extracted after answering the questions is:

not-co-existence(Block Claim 2, Block Claim 1), not-co-existence(Block Claim 2, Block Claim 3), co-existence(Block Claim 1, Unblock Claim 1), co-existence(Block Claim 2, Unblock Claim 2), co-existence(Block Claim 3, Unblock Claim 3), precedence(Block Claim 1, Unblock Claim 1), precedence(Block Claim 2, Unblock Claim 2), precedence(Block Claim 3, Unblock Claim 3), not-co-existence(Receive Objection 1, Receive Objection 2), precedence(Reject Claim, Receive Objection 2), precedence(Payment Order, Receive Objection 1), at-most(Correct Claim), precedence(Block Claim 1, Correct Claim), precedence(Correct Claim, Unblock Claim 1), response(Withdraw Claim, Repayment), responded-existence(Accept Claim, Payment Order), responded-existence(Payment Order, Execute Payment)

These rules are validated using a predefined program to ensure that the activity labels are consistent with the event log and that the declarative constraints are free of syntax errors. Then, the rules are used as input for the IMr framework in addition to the event log, and the process model represented in Fig. 5c is discovered. These rules are aligned with the process description and the follow-up questions and answers. For example, the answer to the first question (A1) led to the extraction of *not-co-existence(Block Claim 2, Block Claim 1)* and *not-co-existence(Block Claim 2, Block Claim 3)*, correctly illustrating the relationship between these activities. These rules help IMr avoid the incorrect positioning of *Block Claim 1* observed in Fig. 5b. Another improvement is achieved by considering *not-co-existence(Receive Objection 1, Receive Objection 2)*, which prevents *Receive Objection 1* and *Receive Objection 2* from occurring in the same trace. Figure 5b allows for *Block Claim 3* without the existence of *Unblock Claim 3*. The rule *co-existence(Block Claim 3, Unblock Claim 3)* guides IMr to avoid placing *Unblock Claim 3* as the re-do part of a loop. We presented this process model to domain experts for feedback. They identified some potential issues, which are then provided to ChatGPT to generate a better set of declarative templates.

The discovered process model is interesting but we observe some issues. For example, Receive Objection 1 should occur after all the payments are executed. This time of objection can occur at most one time per claim. Withdraw Claim can not be followed by another payment. Usually, after the payments are executed, the applicant has the option to send an objection, withdraw the claim, and repay the received benefits. Withdraw Claim only occurs at most once per claim.

After the consideration of the domain expert input, these constraints are added by ChatGPT to the previous set of constraints:

precedence(Execute Payment, Receive Objection 1), at-most(Receive Objection 1), not-succession(Withdraw Claim, Payment Order), not-succession(Withdraw Claim, Execute Payment), at-most(Withdraw Claim)

The discovered model using the modified set of constraints is illustrated in Fig. 5d. While this model differs from the normative model in Fig. 4, it better represents the actual process compared to the models in Fig. 5a and Fig. 5b, which were discovered without considering process knowledge. In comparison to Fig. 5c some improvements are achieved considering the provided feedback. The constraint *at-most(Receive Objection 1)* prevents *Receive Objection 1* from being included in a loop, and *precedence(Execute Payment, Receive Objec-*

tion 1) ensures it is positioned after *Execute Payment*. Additionally, *Withdraw Claim* is no longer in a loop due to the *at-most(Withdraw Claim)* constraint and is correctly positioned after *Payment Order* and *Withdraw Claim* because of the constraints *not-succession(Withdraw Claim, Payment Order)* and *not-succession(Withdraw Claim, Execute Payment)*.

7 Conclusion

The integration of process knowledge in the discovery of process models is often overlooked in the literature. In this paper, we leveraged advancements in large language models to demonstrate their capabilities in encoding textual domain knowledge into comprehensible rules for process discovery. Our proposed framework not only facilitates the integration of feedback from domain experts but also enables interactive improvement of process models. Through a comprehensive case study, we demonstrated the effectiveness of our framework in generating process models that better align with process knowledge. While the extracted set of declarative constraints from LLMs shows great promise, there is still room for improvement in precision and completeness. Future work focuses on expanding the range of declarative templates within the IMr framework and developing additional rule specification patterns. Additionally, providing more detailed examples in task definition steps helps LLMs capture a broader context, further enhancing the quality of the extracted constraints.

References

1. van der Aa, H., Ciccio, C.D., Leopold, H., Reijers, H.A.: Extracting declarative process models from natural language. In: Advanced Information Systems Engineering - 31st International Conference, CAiSE 2019. Lecture Notes in Computer Science, vol. 11483, pp. 365–382. Springer (2019)
2. van der Aa, H., Rebmann, A., Leopold, H.: Natural language-based detection of semantic execution anomalies in event logs. Inf. Syst. **102**, 101824 (2021)
3. Augusto, A., et al.: Automated discovery of process models from event logs: review and benchmark. IEEE Trans. Knowl. Data Eng. **31**(4), 686–705 (2019)
4. Dixit, P.M., Verbeek, H.M.W., Buijs, J.C.A.M., van der Aalst, W.M.P.: Interactive data-driven process model construction. In: Conceptual Modeling - 37th International Conference, ER 2018. Lecture Notes in Computer Science, vol. 11157, pp. 251–265. Springer (2018)
5. van Eck, M.L., Lu, X., Leemans, S.J.J., van der Aalst, W.M.P.: PM²: a process mining project methodology. In: Advanced Information Systems Engineering - 27th International Conference, CAiSE 2015. Lecture Notes in Computer Science, vol. 9097, pp. 297–313. Springer (2015)
6. Grohs, M., Abb, L., Elsayed, N., Rehse, J.: Large language models can accomplish business process management tasks. In: Business Process Management Workshops - BPM 2023 International Workshops, Revised Selected Papers. Lecture Notes in Business Information Processing, vol. 492, pp. 453–465. Springer (2023)
7. Huang, L., et al.: A survey on hallucination in large language models: Principles, taxonomy, challenges, and open questions. CoRR abs/2311.05232 (2023)

 8. Klievtsova, N., Benzin, J., Kampik, T., Mangler, J., Rinderle-Ma, S.: Conversational process modelling: state of the art, applications, and implications in practice. In: Business Process Management Forum - BPM 2023 Forum. Lecture Notes in Business Information Processing, vol. 490, pp. 319–336. Springer (2023)

 9. Kourani, H., Berti, A., Schuster, D., van der Aalst, W.M.P.: Process modeling with large language models. In: Enterprise, Business-Process and Information Systems Modeling - 25th International Conference, BPMDS 2024, and 29th International Conference, EMMSAD 2024. Lecture Notes in Business Information Processing, vol. 511, pp. 229–244. Springer (2024)

10. Leemans, S.J.J., Fahland, D., van der Aalst, W.M.P.: Discovering block-structured process models from event logs containing infrequent behaviour. In: Business Process Management Workshops - BPM 2013 International Workshops, Revised Papers. Lecture Notes in Business Information Processing, vol. 171, pp. 66–78. Springer, Cham (2013)

11. Maggi, F.M., Bose, R.P.J.C., van der Aalst, W.M.P.: Efficient discovery of understandable declarative process models from event logs. In: Advanced Information Systems Engineering - 24th International Conference, CAiSE 2012. Lecture Notes in Computer Science, vol. 7328, pp. 270–285. Springer (2012)

12. Norouzifar, A., Dees, M., van der Aalst, W.M.P.: Imposing rules in process discovery: an inductive mining approach. In: Research Challenges in Information Science - 18th International Conference, RCIS 2024, Part I. Lecture Notes in Business Information Processing, vol. 513, pp. 220–236. Springer (2024)

13. Polyvyanyy, A., van der Aalst, W.M.P., ter Hofstede, A.H.M., Wynn, M.T.: Impact-driven process model repair. ACM Trans. Softw. Eng. Methodol. **25**(4), 28:1–28:60 (2017)

14. Schuster, D., van Zelst, S.J., van der Aalst, W.M.P.: Utilizing domain knowledge in data-driven process discovery: a literature review. Comput. Ind. **137**, 103612 (2022)

15. Schuster, D., van Zelst, S.J., van der Aalst, W.M.P.: Cortado: a dedicated process mining tool for interactive process discovery. SoftwareX **22**, 101373 (2023)

16. Vidgof, M., Bachhofner, S., Mendling, J.: Large language models for business process management: Opportunities and challenges. In: Business Process Management Forum - BPM 2023 Forum. Lecture Notes in Business Information Processing, vol. 490, pp. 107–123. Springer (2023)

3rd International Workshop on Data-Driven Business Process Optimization (BPO 2024)

3rd International Workshop on Data-Driven Business Process Optimization (BPO 2024)

Business process management is a very promising paradigm for optimizing the way in which work is performed in an organization. Decision-making is key in managing processes, and supporting decisions by combining insights from data and corresponding process models was the focus of this workshop. Decisions in business processes include assigning resources to the tasks for which they are most suited, ordering the execution tasks to best meet customer deadlines, etc. While such questions are important in administrative processes, they are even more important in processes that have a physical component in domains such as transportation, production and healthcare. In these domains, assigning tasks to the wrong resource or performing them in the wrong order immediately leads to increased costs, dissatisfied customers and even health risks.

Traditionally, the research area of operations research has studied techniques for modeling and solving optimization problems in much detail. At the same time, the research area of business process management has studied techniques for aggregating the data that is needed for modeling, analyzing and in the end optimizing business processes. Combining techniques from both areas makes it possible to solve optimization problems from practice, using models that are based on real-world data, with fewer assumptions. In particular, it allows us to create clear and realistic data-driven models of the way in which customer orders pass through the organization and of the behavior and performance of resources. While this provides clear benefits in terms of more realistic models and analysis, it also brings challenges in terms of the computational complexity of the used analysis and optimization techniques.

The goal of the Data-Driven Business Process Optimization workshop was to bring together researchers from both the area of Business Process Management and the area of Operations Research as well as other related areas, with the overall goal of developing techniques for optimizing business processes in an organization based on models that are created from real-world data.

The workshop covered both presentations on techniques for optimizing business processes and applications of such techniques to real-world problems. It received four paper submissions, out of which two were accepted for presentation. It also received three abstract submissions, which were invited for presentations. In addition, Avigdor Gal delivered a keynote presentation on 'Trade offs in Responsible Business Processes' and a business process optimization competition was organized on which a report was presented.

We hope that the reader found the selection of papers useful to get an insight into how operations research and business process management can be combined to solve business process optimization problems.

Organization

Organizing Committee

Arik Senderovich   York University, Canada
Remco Dijkman   Eindhoven University of Technology,
The Netherlands

Business Process Optimization Competition Organization

Felix Schumann   Technical University of Munich, Germany
Matthias Ehrendorfer   Technical University of Munich, Germany
Michel Kunkler   Technical University of Munich, Germany
Remco Dijkman   Eindhoven University of Technology,
The Netherlands

Program Committee

Akhil Kumar   Penn State University, USA
Anna Kalenkova   University of Adelaide, Australia
Avigdor Gal   Technion, Israel
Chiara Di Francescomarino   University of Trento, Italy
Chiara Ghidini   Free University of Bozen-Bolzano, Italy
Cristina Cabanillas   Universidad de Seville, Spain
Han van der Aa   University of Mannheim, Germany
Henrik Leopold   Kühne Logistics University, HPI, Germany
Marcos Sepulveda   Pontificia Universidad Católica de Chile,
Chile

Martin Matzner   Friedrich Alexander Universität, Germany
Massimiliano de Leoni   University of Padua, Italy
Matthias Weidlich   Humboldt University, Germany
Orlenys López Pintado   University of Tartu, Estonia
Xixi Lu   Utrecht University, The Netherlands

The Business Process Optimization Competition

Felix Schumann[1], Matthias Ehrendorfer[1], Michel Kunkler[1], Kiran Busch[2],
Henrik Leopold[2], Leon Urny[1], Martin Schmauch[1], Olga Rodzik[1],
Stefanie Lanz[1], Efe Tıraş[1], Amgad Al-Zamkan[1], Jana El Kari[1],
and Remco Dijkman[3(✉)]

[1] Technical University of Munich, TUM School of Computation,
Information, and Technology, Garching, Germany
[2] Kühne Logistics University, Hamburg, Germany
[3] Eindhoven University of Technology, Eindhoven, The Netherlands
r.m.dijkman@tue.nl

Abstract. Optimizing organizational operations with respect to different decisions and to achieve a variety of objectives has been thoroughly studied. Examples include deciding which transport order to assign to which truck in order to minimize costs or deciding on how much stock to keep of a particular product in order to balance costs and delivery reliability. Business processes bring unique challenges to optimization problems, due to the complexity and uncertainty of the interdependencies between their tasks. The business process optimization competition presented such an optimization problem. Participants had to develop an algorithm that decided when to admit patients for hospital treatment, with various uncertain treatment tasks ahead. Eight different algorithms were submitted and compared to identify the most effective types of algorithm for business process optimization. Four algorithms outperformed the heuristic benchmark, with two of these being reinforcement learning algorithms. This suggests that reinforcement learning is a promising approach for optimizing business processes.

Keywords: Business Process · Optimization · Heuristics

1 Introduction

The optimization of organizational operations with respect to different decisions and to achieve a variety of objectives has been thoroughly studied. Examples of optimization problems include transportation resource allocation, which concerns deciding which transportation resource to assign to which transportation orders in order to minimize cost or CO2 emissions; job shop scheduling, which concerns deciding which production order to assign to which machine in which timeslot in order to minimize the amount of time required to finish all production orders; and spare parts planning, which concerns deciding how many spare parts

© The Author(s), under exclusive license to Springer Nature Switzerland AG 2025
K. Gdowska et al. (Eds.): BPM 2024 Workshops, LNBIP 534, pp. 61–72, 2025.
https://doi.org/10.1007/978-3-031-78666-2_5

to keep on hand to both minimize cost and maximize uptime of the machines that require the parts when they break down.

Business processes bring unique challenges to optimization problems due to the complexity and uncertainty of the interdependencies between their tasks [4]. Where business processes are specifically designed to consider multiple tasks in their relation to each other, operational optimization traditionally focuses on one or a few tasks. This narrow scope is understandable because solving an optimization problem for a single task is often already NP-hard. Extending this to multiple tasks increases computational complexity further, making it difficult to optimize processes within a feasible time frame. For example, in transport resource allocation, we typically allocate transportation orders to trucks separately from allocating trucks to drivers. This helps to manage complexity. However, considering these problems together can lead to better plans. For instance, if we separate these problems, we might assign a transport order to a truck for which a qualified driver is not available on assigned required day.

Job shop scheduling is probably most similar to business process optimization, in that job shop scheduling considers production processes that consist of multiple tasks (jobs) [5]. However, in job shop scheduling, the tasks that must be performed in a process are typically known in advance. In contrast, in business processes the tasks that must be performed for a customer are typically not known in advance. Therefore, business process optimization poses novel optimization problems, which in turn requires that we find new solution approaches.

The business process optimization competition presented such a novel optimization problem. In particular, participants had to solve a scheduling problem, in which they had to decide when to admit patients for hospital treatment, with various uncertain treatment tasks ahead. These treatment tasks used shared resources and if a patient was admitted at a time at which no resources were available for treatment, the patient would leave for another hospital, while the objective was to treat as many patients as possible. The goal of the study was to compare different algorithms to identify which types work best in the context of business process optimization, particularly under uncertainty of the tasks that needed to be executed in the business process. This paper reports the results of this comparison.

Against this background, the remainder of this paper is structured as follows. Section 2 describes the optimization problem that had to be solved in more detail. Section 3 presents the solution approaches that were proposed by the competition participants. Section 4 compares how the different solution approaches performed on the problem, Sect. 5 gives a short overview on related work and Sect. 6 presents the conclusions.

2   Problem Description

The objective of the BPO Competition 2024 was to build a patient scheduling mechanism for an artificial hospital. To do this, the competitors had to implement a 'plan' function that assigned patients to the moment in time at which they would be treated. During opening hours, two types of patients could arrive at the hospital:

1. Planned patients who have been given an appointment according to the implemented 'plan' function.
2. Emergency patients who show up randomly throughout the day.

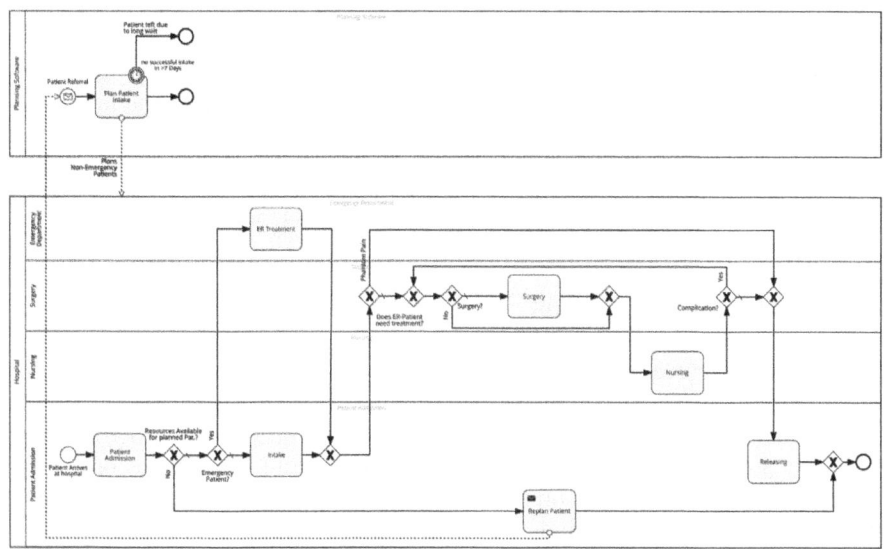

Fig. 1. A Hospital Treatment Process.

Patients will then proceed according to the process that is shown in Fig. 1. Upon arrival, ER patients will immediately proceed to the Emergency Department to receive emergency treatment. For some of the ER patients, surgery or treatment in a nursing ward may be necessary, while others can be released after having received the emergency treatment. The ER patients for which further treatment is necessary will receive treatment, e.g., surgery, with a higher priority than the planned patients. Planned patients are referred to the hospital with a known diagnosis. They must be assigned a time slot on which they will subsequently arrive to receive their treatment. If patients arrive at the hospital and it is apparent that timely treatment is infeasible, because there are insufficient resources such as hospital beds or operating rooms available, they will be sent home again. If treatment is feasible, the planned patients will proceed to intake. Depending on their diagnosis, some patients need surgery, while others need only nursing. After having finished their treatment, they will be released. For both patient groups, complications during their treatments can arise, which require additional surgery or nursing. The arrival rates of patients of different types is stochastic, as are the times that the tasks take for the different types of patients, and the probability that complications arise during treatment, such that the tasks that need to be performed for the patients are uncertain when they first enter the process.

The goal of the competition is to decide which patient to admit at which time. Patients must be planned for a timeslot at least 24 h before their hospitalization.

This must be done in such a way as to meet the hospital's objectives. The hospital's main objective is to process as many patients as possible. Planned patients who do not get their intake within 7 days after their referral will leave for another hospital and will not count towards this objective. In addition, the hospital aims to treat emergency patients within four hours and aims to minimize the number of planned patients who are sent home due to unavailability of resources. The number of patients who are treated within 7 days, the number of emergency room patients who are treated within four hours, and the number of patients who are sent home on intake are counted and the algorithms are scored according to the weighted average of these numbers.

More details about the number of resources and the probability distributions are provided in the problem description document [17].

3   Solution Approaches

The competition received eight submissions, which used different solution approaches. This section briefly summarizes the different approaches.

The submissions address the problem in either of two ways: Either they estimate a score for a solution candidate in advance and hence focus on finding a solution candidate with a good score, or they evaluate the score of a solution in retrospect. The first option finds application in the submissions that use meta-heuristics or constraint programming. All submissions that pursued this option chose different score functions, which will be described below. The other option to evaluate the score of a solution in retrospect was pursued in the submissions that use reinforcement-learning-based approaches.

3.1   Approach 1–3: Simulated Annealing

Simulated Annealing in general tries to (a) start from an initial solution, (b) find neighbours from this initial solution, and then evaluate the neighbor based on (c) the cost assigned to each solution and (d) a temperature which decreases over time. The decision if the new solution (i.e., the neighbor found) should be accepted as best solution is based on the costs of the current solution compared to the neighbor solution and the temperature. Three approaches (i.e., (1)–(3)) using simulated annealing to find a solution for the business process optimization competition have been submitted. They use different approaches for the abovementioned parts (i.e., (a)–(d)) of the algorithm.

(a) Initial solution
   (1) Random solution as initial solution (between 1 day and 7 days).
   (2) Solution of "naive planner" provided within the challenge as initial solution.
   (3) Scheduling patients at the earliest feasible times as initial solution.
(b) Finding neighbors
   (1) For each patient simulated annealing is applied individually - neighbor is found by increasing or decreasing the planned time based on a random value and the allowed limits.

(2) Decrease scheduled time of one random patient slightly (by a random amount).

(3) Increase or decrease scheduled time of one random patient slightly (by a random amount).

(c) Cost assigned to a solution

(1) Assigns random completion times (based on the distributions given in the problem description) to busy resources and then checks if these resources would still be busy at the planned time. The cost is then calculated based on for how many of the tasks resources would be available.

(2) Assigns a cost based on the difference between the time for which each individual patient is planned and the current (simulation) time. The further a patient is planned into the future the higher the cost gets. The cost is then summed up over the patients.

(3) Penalties based on resource overloads (squared difference between available resources and resources needed for the current solution) together with a penalty if patient's scheduled time exceeds the latest acceptable time (which is calculated by adding some time based on the diagnosis of the patient to the current (simulation) time) are used as cost.

(d) The new (neighbor) solution is accepted if the cost of the neighbor is smaller than the cost of the current solution. Additionally, there is a chance to accept the new solution even if the cost is worse than for the current solution: $probability = e^{\frac{current\_cost - neighbor\_cost}{current\_temp}}$ - the temperature is decreased in each iteration based on the following formula: $next\_temp = current\_temp * cooling\_rate$

(1) The following parameters are included in the documentation (the one in bold is the one contained in the uploaded code and therefore used for the results in Sect. 4): initial temperature: (1,**10**, 100, 1000), cooling rate: (0.8, 0.9, **0.95**, 0.99), maximum iterations: (**100**, 1000, 10000).

(2) The following parameters are extracted from the code: initial temperature: 100, cooling rate: 0.8, stopping temperature: 0.1.

(3) The following parameters are extracted from the documentation and code: initial temperature: 100, cooling rate = 0.9, maximum 500 iterations or minimal temperature 0.1.

3.2   Approach 4: Evolutionary Algorithm

This approach used a genetic algorithm to find a good intake time for a patient. In contrast to other approaches, this algorithm does not batch-process the arriving patients but encodes different possible planning slots in one genome. In the fitness function, an ER-Penalty is generated depending on the resource queue situation and the patient's diagnosis. The penalty for the sent-home score is calculated based on the working time during the planned time slot. The penalty for the processed score is based on the remaining days until the deadline. Through mutation and crossover, the best time slot is found within the hardcoded 10 generations of the algorithm.

3.3   Approach 5: Constraint Programming

The constraint programming approach submitted to the challenge used the Python library provided for Google OR-tools. The proposed approach waits until a certain batch size is reached and then schedules the intake of each patient in the batch. Therefore, the final batch size was set to 6. Yet, in the submission description, the competitors argue that no significant improvement was made when creating a bigger batch size. As constraints, the approach limits the intake capacity and considers on- and off-hours of the surgery resources. Nursing resources are estimated, and complications are adjusted. As an objective function, the overall system load is minimized.

3.4   Approach 6: Tabu Search

This submission uses the Tabu search meta-heuristic to schedule the patient intakes. An initial schedule is first obtained by drawing the planning times for every plannable patient in a time span from 72 h to 144 h in advance from a uniform distribution. The submission defines a neighbor to a solution candidate as a time shift of either minus or plus one hour for a single patients planning time. The score of a candidate schedule consists of a penalty for every patient not scheduled within 7 days and a penalty for every invalid planning, i.e., for patients scheduled in the past. The Tabu search algorithm is run for 100 iterations, after which the solution that yielded the lowest penalty score is chosen. The planning of the patients is then conducted according to the obtained solution.

3.5   Approach 7: Deep Reinforcement Learning

The submission uses Fitted Q-Iteration to train a Deep Q-Network for the patient scheduling. The approach uses a fully connected neural network that can schedule one patient at a time. For every new patient, the neural network takes the already planned patient intakes, the time of the day, the length of the surgery queue, and the type of the to-be-scheduled patient as input. The planned patient intakes are represented as an array where each array element represents a 30-minute interval from 25 to 168 h (7 days) in advance. The neural network then yields a time offset for which the patient should be planned. The neural network was pretrained for 50 000 iterations, where each iteration corresponds to 100 simulation days. The training loss consists of the problem three scores plus some additional scores that penalize meaningless planning decisions, e.g., when a patient is planned at night.

3.6   Approach 8: Reinforcement Learning Q-Table

This solution to the hospital patient planning problem is based on a Reinforcement Learning approach, specifically utilizing Q-Learning. The algorithm learns an optimal policy for scheduling patient intakes by selecting the best time slots based on the current state of the hospital. The hospital state is defined by key variables such as resource availability—intake staff, operating rooms, and nursing beds—and the number of planned activities over the next 168 h (7 days). Actions represent minute-level time slots within the next 7 days, and the reward function is designed to penalize delays in treating emergency patients, sending planned patients home due to resource constraints, and failing to process cases. This structure encourages the planner to align its scheduling decisions with the hospital's goals. The algorithm employs an epsilon-greedy strategy, balancing exploration of new scheduling options with exploitation of the best-known actions to continually improve performance. The planner first evaluates the current hospital's state. Based on this evaluation, it selects an action. Once a time slot is selected, the planner schedules patient intake. After each case is completed, the planner updates its Q-values using the Bellman equation, refining its decision-making process based on the outcomes. This continuous learning allows the RL-based planner to adapt dynamically to hospital conditions.

4   Comparison of Results

The performance of the different scheduling approaches was evaluated through three different scores: The "ER-Score" values the number of emergency patients that the hospital processed. The "Processed Score" values the number of processed planned patients. Lastly, the "Sent-Home Score" values the number of patients whose scheduled appointments could not be held due to resource shortage.

Table 1. Outcome of the BPO Competition 2024

Submission	Approach	No.	Score
Urny	RL Deep-Q-Network	7	0.180
Schmauch	Genetic Algorithm	4	0.186
Busch	RL Q-Table	8	0.201
Rodzik	Tabu Search	6	0.206
Reference	Plan 1 patient per hour		0.217
Lanz	Simulated Annealing	1	0.225
Reference	Plan 0 patients per hour		0.329
Tıraş	Constraint Programming	5	0.428
Reference	Plan 2 patients per hour		0.464
Al-Zamkan	Simulated Annealing	3	0.538
Reference	Plan 3 patients per hour		0.547
Reference	Plan any patient as early as possible		0.616
El-Kari	Simulated Annealing	2	0.633

For a formal definition of the scores, the reader is referred to [17]. For the final evaluation, each score was min-max normalized among the submissions and five reference planning approaches. Four of the five reference approaches schedule [0..3] patients per hour to an intake timeslot. The last reference approach schedules any arriving patient 24 h after its first arrival. After summing the three normalized scores for each submission, we received the scores shown in Table 1.

We can see that the reference approach of planning one patient every working hour performed already well and outperformed 4 submissions. Yet, the remaining 4 submissions outperformed all reference approaches. Two of the top three submissions were based on reinforcement learning. This indicates an advantage of reinforcement learning approaches in stochastic scheduling scenarios where the effect of a scheduling decision, e.g. scores, can hardly be estimated in advance. For the winning submission, it appears that introducing the penalty for scheduling patients at nighttime has led to even better scheduling policies. It shows that tailoring the score function to the specific problem is decisive to good performance.

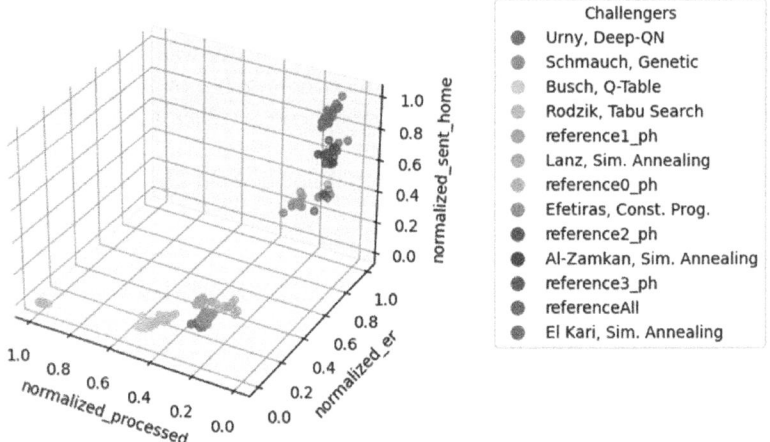

Fig. 2. All Submission and reference scores plotted with each score as one axis

Figure 2 shows the Pareto frontier between the Processed-Score and the ER-Score very well. As shown in Fig. 3, most of the highly ranked approaches achieved a Sent-Home-Score of close to 0. Therefore, it seems evident that this should be the primary objective of the approaches and that the main challenge was finding a good balance between patients and planned patients. Submissions with a high Sent-Home-Score usually scheduled too many patients into the respective timeslots and overexerted the processing capabilities of the hospital. This seemed to be a problem, especially for the approaches using simulated annealing, where one of the three submissions could at least balance this a bit better.

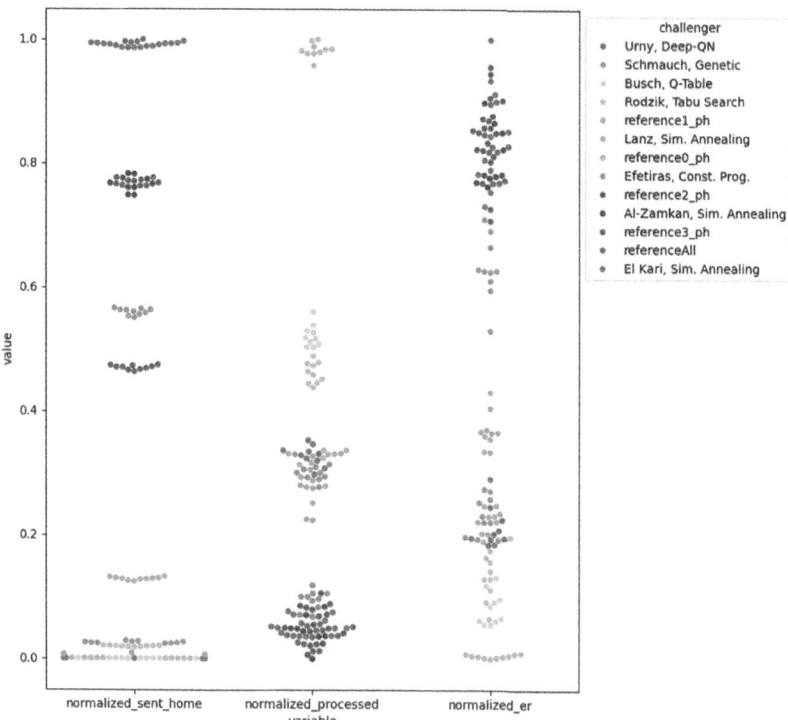

Fig. 3. Swarmplot of all Submissions and references for all three scores

5   Related Work

Business process optimization, particularly optimal allocation of resources in business processes, has gained significant attention recently. Rule-based methods (e.g., [1,6,11,18,20]) use process mining techniques to mine static resource allocation rules. Other techniques optimize resources on a tactical level, usually based on simulation (e.g., [12,16]). Dynamic methods, like those proposed by Park and Song [15], adapt to current process circumstances. Reinforcement Learning (RL) has also been applied, with methods like RLRAM [7] and DRL [13,19] improving scalability and efficiency. While the body of research on business process resource allocation is by now substantial, this paper studies scheduling in the context of business processes. On this little work has been done so far, although case-studies have been done [3].

Competitions have become important for benchmarking and advancing the state-of-the-art in multiple areas related to data science and optimization. These competitions provide a structured environment where researchers and practitioners can test their algorithms against challenging benchmarks and compare their performance with others in the field.

In the area of data science, Kaggle Competitions [10] are a prominent example. They host a variety of optimization challenges ranging from predictive modeling to more complex optimization problems. These competitions are often sponsored by companies who are looking to solve specific problems.

Competitions are also frequently a part of conferences, with some conferences even organizing multiple competitions. Prominent examples of conferences in the area of data science or optimization that organize competitions include the NeurIPS conference, which organizes competitions on a broad range of topics in the area of machine learning [14], and the IEEE Computation Intelligence Society, which organizes multiple conferences and competitions in the area of computational intelligence [8]. The Business Process Management Conference, of which this competition is a part also has a history of organizing competitions. In particular, the Business Process Intelligence Challenges [9]. Also, the Business Process Optimization Competition has been organized three times now since 2022 [2].

6   Conclusion

This paper presented the results of the Business Process Optimization Competition 2024. The competition aimed to compare different algorithms for optimizing business processes, specifically in the context of patient scheduling in a hospital. Participants were tasked with developing algorithms that decided when to admit patients for treatment, considering various uncertain treatment tasks and resource constraints, and optimizing the number of patients that received treatment.

Among the eight algorithms submitted, four outperformed the heuristic benchmark, demonstrating the potential for advanced optimization techniques in this domain. Notably, two of the most successful algorithms were based on reinforcement learning, suggesting that this approach holds significant promise for future research and application in business process optimization.

The problem was a multi-objective optimization problem, and striking the right balance between the different objectives turned out to be important for an algorithm to be successful. The most successful algorithms focused on minimizing the number of patients sent home while considering other objectives like treating emergency patients promptly and processing planned patients efficiently.

In conclusion, the Business Process Optimization Competition provided valuable insights into the challenges of business process optimization and the effectiveness of different algorithms. The results highlight the potential of RL and the importance of balancing multiple objectives. Future work on exploring different business process optimization problems, including optimal decision-making, resource allocation problems, and scheduling problems, is planned for future business process optimization competitions.

References

1. Arias, M., Rojas, E., Munoz-Gama, J., Sepúlveda, M.: A framework for recommending resource allocation based on process mining. In: Business Process Management Workshops, pp. 458–470 (2016)
2. BPO 2022: Data-driven BPO Competition. https://sites.google.com/view/bpo2022/competition. Accessed 28 Aug 2024
3. Di Cunzolo, M., et al.: Combining process mining and optimization: a scheduling application in healthcare. In: Business Process Management Workshops, pp. 197–209. Springer, Cham (2023)
4. Dijkman, R.: Resource optimization in business processes. In: van der Aa, H., Bork, D., Schmidt, R., Sturm, A. (eds.) Enterprise, Business-Process and Information Systems Modeling, pp. 3–9. Springer (2024)
5. Farahani, A., van Elzakker, M.A.H., Genga, L., Troubil, P., Dijkman, R.M.: Relational graph attention-based deep reinforcement learning: an application to flexible job shop scheduling with sequence-dependent setup times. In: Proceedings of International Conference on Learning and Intelligent Optimization (LION), pp. 347–362 (2023)
6. Havur, G., Cabanillas, C., Mendling, J., Polleres, A.: Resource allocation with dependencies in business process management systems. In: Business Process Management Forum, pp. 3–19. Springer, Cham (2016)
7. Huang, Z., van der Aalst, W.M.P., Lu, X., Duan, H.: Reinforcement learning based resource allocation in business process management. Data Knowl. Eng. **70**(1), 127–145 (2011)
8. IEEE Computational Intelligence Society: Competitions - IEEE Computational Intelligence Society. https://cis.ieee.org/activities/educational-activites/competitions. Accessed 28 Aug 2024
9. IEEE Task Force on Process Mining: BPI Challenges - IEEE Task Force on Process Mining. https://www.tf-pm.org/competitions-awards/bpi-challenge. Accessed 28 Aug 2024
10. Kaggle: Kaggle competitions. https://www.kaggle.com/competitions. Accessed 28 Aug 2024
11. Kuchař, S., Vondrák, I.: Automatic allocation of resources in software process simulations using their capability and productivity. J. Simul. **10**(3), 227–236 (2016). https://doi.org/10.1057/jos.2015.8
12. López-Pintado, O., Dumas, M., Berx, J.: Discovery, simulation, and optimization of business processes with differentiated resources. Inf. Syst. **120**, 102289 (2024)
13. Meneghello, F., et al.: Optimizing resource allocation policies in real-world business processes using hybrid process simulation and deep reinforcement learning. In: Proceedings of BPM, pp. 167–184. Springer (2024)
14. NeurIPS: NeurIPS 2024 competition track program. https://neurips.cc/Conferences/2024/CompetitionTrack. Accessed 28 Aug 2024
15. Park, G., Song, M.: Prediction-based resource allocation using LSTM and minimum cost and maximum flow algorithm. In: 2019 International Conference on Process Mining, pp. 121–128 (2019)
16. Peters, S.P.F., Dijkman, R.M., Grefen, P.W.P.J.: Resource optimization in business processes. In: 2021 IEEE 25th International Enterprise Distributed Object Computing Conference (EDOC), pp. 104–113 (2021)
17. Schumann, F., Ehrendorfer, M., Kunkler, M., Dijkman, R.: BPO 2024 - competition. https://sites.google.com/view/bpo2024/competition. Accessed 28 Aug 2024

18. Xu, J., Liu, C., Zhao, X.: Resource allocation vs. business process improvement: how they impact on each other. In: Business Process Management, vol. 5240, pp. 228–243 (2008)
19. Żbikowski, K., Ostapowicz, M., Gawrysiak, P.: Deep reinforcement learning for resource allocation in business processes. In: Process Mining Workshops, pp. 177–189 (2023)
20. Zhao, W., Yang, L., Liu, H., Wu, R.: The optimization of resource allocation based on process mining. In: Advanced Intelligent Computing Theories and Applications, pp. 341–353 (2015)

Ad-Hoc Subprocesses – The Missing Link Between Scheduling and Business Process Modelling

Asvin Goel[(✉)] [iD]

Kühne Logistics University, Hamburg, Germany
asvin.goel@klu.org

Abstract. This paper aims at bridging the gap between scheduling and business process modeling, two fields that have evolved independently from each other, even though both share common goals, i.e., the improvement of business operations. The paper shows how BPMN 2.0 can be used to model scheduling problems using a BPMN element that is often overlooked: ad-hoc subprocesses. It shows how ad-hoc subprocesses provide the basis for modeling scheduling requirements and demonstrates how well-known scheduling problems, such as the travelling salesperson problem and the job shop scheduling problem can be represented as BPMN models. Business problems containing scheduling problems can be directly modelled by domain experts and process owners, eliminating the need to consult experts in mathematical programming. Moreover, new avenues are opened for research in optimisation algorithms capable of operating directly on BPMN models.

Keywords: BPMN 2.0 · Modelling · Optimisation

1 Introduction

Scheduling can be described as the process of planning and organising activities to be conducted to achieve a given purpose subject to relevant restrictions. It involves determining when and in which sequence particular activities should take place, ensuring that required resources such as time, personnel, and equipment are allocated accordingly. The main purpose of scheduling is to improve the efficiency of business operations by using mathematical optimisation techniques to eliminate idle times and bottlenecks.

Business process modelling, on the other hand, is concerned with the representation of processes and workflows, and involves a description of decision points, information flows, and the sequence in which activities shall be conducted. The main purpose of such models is to provide a structured way to understand how different parts of an organisation work together to achieve specific goals or outcomes.

Although scheduling and business process modelling are both concerned with business operations, the respective scientific fields have been evolving in isolation.

© The Author(s), under exclusive license to Springer Nature Switzerland AG 2025
K. Gdowska et al. (Eds.): BPM 2024 Workshops, LNBIP 534, pp. 73–83, 2025.
https://doi.org/10.1007/978-3-031-78666-2_6

There is a good reason for this development: while business process models usually provide a specific sequence in which activities are conducted, scheduling focuses on determining the sequence in which activities shall be conducted. Thus, the respective goals appear to be contradictory.

This paper aims at building a bridge between scheduling and business process modelling with the goal of allowing the formulation of scheduling problems using business process models. This goal is achieved by leveraging a feature in the BPMN 2.0 specification that is often overlooked: ad-hoc subprocesses. The paper shows how BPMN 2.0 can be used to formulate well-known scheduling problems, like the travelling salesperson problem and the job shop scheduling problem.

2 Related Work

Scheduling problems are commonly used when resources with limited capacity have to conduct multiple tasks. In such cases, decisions have to be made when tasks are to be executed so that the capacity of the resource is not exceeded at any point in time. In many cases, a related decision is required, i.e., the decision which task is to be executed by which resource. The allocation of tasks to resources has found some attention in the literature related to business process management, e.g., [1,2,6,7], and a comprehensive survey is provided by [12]. Several suggestions have been made to add resource allocation information into process models. [2] and [1] present a resource assignment language and a graphical notation for resource assignments that can be used together with BPMN to provide information related to the assignment of (human) resources to tasks. Focusing on healthcare applications, [11] propose to introduce a dedicated BPMN element indicating that a task requires a resource with limited capacity. [8] propose so-called action-evolution Petri nets to model and solve dynamic task assignment problems using deep reinforcement learning. Deep reinforcement learning is also used by [9] to conduct resource allocations in business processes.

Above approaches have in common, that they assume that in order to be executed, a task needs to be assigned to one or more resources. A different perspective on resources is provided by [5], who does not model resources as entities or objects to which tasks are assigned. Instead, it is proposed to represent resources as processes or by processes managing the resource. Resource requirements can thus be represented by a collaboration between processes requiring the service of a resource and processes providing the service. Instead of including a task conducted by a resource into a process model, only a request for a service is included. The task conducted by the resource is then included in the process representing the resource, and when this task is completed, a message is sent back to the original process indicating that the resource has completed the task. Resource allocation decisions are therefore replaced by decisions regarding which request message shall be sent to which process instance. Compatibility constraints can easily be validated when making such decisions. To consider that most resources can only conduct one task at a time, [5] proposed so-called resource activities which require sequential execution of the tasks to be executed. As we will see, the original idea of [5] can also be applied without the introduction of resource activities. Instead, it is possible to use BPMN ad-hoc subprocesses to ensure that a resource only conducts one task at a time.

3   Ad-Hoc Subprocesses

The BPMN specification [10] defines ad-hoc subprocesses as follows:

> *An ad-hoc subprocess is a specialized type of subprocess that is a group of activities that have no required sequence relationships. A set of activities can be defined for the process, but the sequence and number of performances for the activities is determined by the performers of the activities.*

The specification restricts the BPMN elements that may be used within ad-hoc subprocesses:

- **Required elements:** ad-hoc subprocesses must contain at least one activity.
- **Allowed elements:** ad-hoc subprocesses may contain data objects, sequence flows, associations, data associations, groups, message flows (as a source or target), gateways, and intermediate events.
- **Prohibited elements:** ad-hoc subprocesses must not contain start events, end events, conversations (graphically), conversation links (graphically), and choreography activities.

The specification as well as other sources explaining ad-hoc subprocesses, e.g. [3], use examples similar to the one given in Fig. 1. The figure shows a subprocess with multi-instance marker ||| and adhoc-subprocess marker ∼. Several activities are included in this multi-instance ad-hoc subprocess, but not all of them are connected through sequence flows. According to the operational semantics, each activity without incoming sequence flows is enabled initially. A performer of an ad-hoc subprocess can select any of the enabled activities for execution. When any of these activities is completed, tokens are produced on each outgoing sequence flow and all tokens are forwarded as far as possible. Activities that receive a token become enabled and may be selected by the performer for execution.

Most examples known to the author describe use cases in which a certain degree of creativity or intuition is required and in which the sequence in which activities are conducted depend on some undefined behaviour of a human performer. The BPMN specification itself says that ad-hoc subprocesses do *"not contain a complete, structured BPMN diagram description"* and that they allow *"modeling of processes that are not necessarily executable"*. Camunda, a vendor of process automation software, provides the following reasoning on the use of ad-hoc subprocesses [3]:

> *Any party who executes this subprocess decides what to do and when to do it. You could say that the "barely structured" nature of what happens inside this subprocess reduces the whole idea of process modeling to an absurdity*

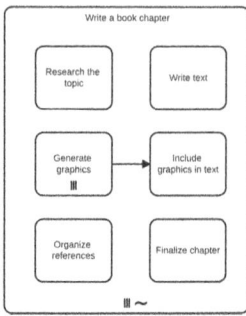

Fig. 1. An example ad-hoc subprocess

because what happens and when are the things we most want to control. On the other hand, this is the reality of many processes, and you can't model them without representing their free-form character. Frequent examples are when a process relies largely on implicit knowledge or creativity, or when different employees carry out a process differently. You can use the ad-hoc subprocess to flag what may be an undesirable actual state. Doing so could be a step on the path to a more standardized procedure.

Given the allegedly missing structure caused by the use of ad-hoc subprocesses, they are often not perceived to be an important modelling element for many use cases. For example, the repository provided by [4] contains over 600 BPMN models, but none of them contains ad-hoc subprocesses.

A feature of ad-hoc subprocesses that is easily overlooked is the possibility to specify that activities within ad-hoc subprocesses must not be conducted in parallel by adding the attribute `ordering="sequential"` to an ad-hoc subprocess. This attribute does not influence the visual appearance of the ad-hoc subprocess and the default value is `ordering="parallel"`. By specifying that activities in an ad-hoc subprocess must be conducted in sequential order, BPMN provides the means to require that a decision has to be made on the ordering of activities. Thus, ad-hoc subprocesses with sequential ordering can be used to model an essential feature required to model scheduling problems. BPMN does not specify how such decisions are made and states that *"performers determine when activities will start, what the next activity will be, and so on"*.

Performers are BPMN modelling elements without visual representation and define the resources responsible for conducting activities. They can be specified on process- or activity-level. A performer specified for a (sub-)process can be interpreted to be responsible for all activities within the (sub-)process. Moreover, a performer responsible for an ad-hoc subprocess is allowed to decide on how often an activity in the ad-hoc subprocess is conducted and when such an activity starts. BPMN does not restrict the behaviour of performers. Thus, it is possible to define a performer in a way that it starts each activity in an ad-hoc subprocess exactly once and freely decides on the start times of activities.

If such a performer is used for each ad-hoc subprocess with sequential ordering, we can use such ad-hoc subprocesses and performers to model scheduling
problems in BPMN. Instead of using ad-hoc subprocesses for parts of a model
that are lacking structure, we thus can use ad-hoc subprocess to define clearly
structured models in which there is a requirement that the sequence in which
activities are conducted needs to be carefully chosen to achieve operational efficiency. In general, such scheduling decisions should not be conducted at the time
of modelling a process, and require knowledge of data concerning each relevant
process instance.

In the following, two examples are given demonstrating how ad-hoc subprocesses with sequential ordering and respective performers can be used to model
well-known scheduling problems.

4   Travelling Salesperson Problem

The *travelling salesperson problem* can be described as the problem of finding
the shortest possible roundtrip visiting a given set of locations and returning to
the origin. Each location on the roundtrip must be visited exactly once and the
goal is to find the roundtrip that minimises the total distance travelled.

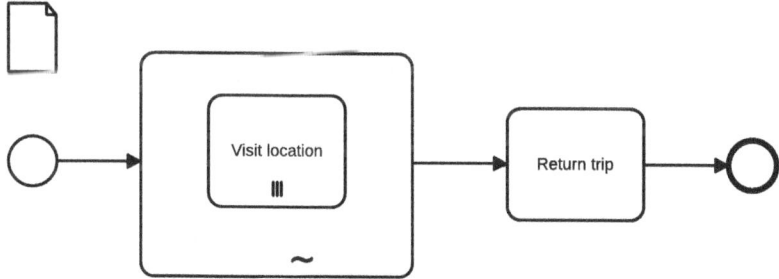

Fig. 2. The travelling salesperson problem as business process model

Figure 2 illustrates a business process model of the travelling salesperson
problem. The process has a performer representing the salesperson. It starts
with an ad-hoc subprocess containing a multi-instance task responsible for visiting each of the given locations. The order of visiting the locations is unspecified
and, therefore, the *Visit location* task is instantiated in parallel for each location. However, as the ordering of the ad-hoc subprocess is set to sequential, the
performer must not visit different locations in parallel. Thus, all locations are
visited one after another in a sequence that needs to be determined by the performer. After all locations have been visited, the process continues with a task
representing the return trip of the salesperson. A data object is used allowing
to store the origin, the current location, and the accumulated distance of the
salesperson as process variables.

To fully represent the travelling salesperson problem, the origin as well as the number of instantiations of the multi-instance activity and the respective locations must be provided as process variables. The distance to the next location can be determined using the process variables for the current location and the respective next location. The accumulated distance can be stored in another process variable to keep track of the objective value to be minimised.

An example of the travelling salesperson problem with a salesperson based in Hamburg who is required to visit Berlin, Cologne, and Munich is given in the following. Distances between the cities can be provided in a distance table as shown in Table 1.

Table 1. Distance table

From	To	Distance
Hamburg	Berlin	296
Hamburg	Cologne	432
Hamburg	Munich	778
Cologne	Berlin	573
Cologne	Hamburg	432
Cologne	Munich	575
Berlin	Cologne	573
Berlin	Hamburg	296
Berlin	Munich	585
Munich	Cologne	575
Munich	Hamburg	778
Munich	Berlin	585

When the process starts, the current location is set to Hamburg and the distance travelled is set to zero. Then, an instance of the *Visit location* task is created for Berlin, Cologne, and Munich. The performer, i.e. the salesperson, now needs to decide which of these tasks to perform first. For each option, it can lookup the distance from the current location in the distance table. When the next location is selected, the current location and the total distance are updated accordingly. After having visited all locations, the trip back to the origin, i.e. Hamburg, is conducted. If the performer always selects the location with the smallest distance, the approach resembles the nearest neighbor heuristic for the travelling salesperson problem.

5   Job Shop Scheduling Problem

The *job shop scheduling problem* is the problem of efficiently scheduling a set of jobs on a set of machines. Each job belongs to an order, requires a certain amount of processing time on a specific machine, and must be conducted without preemption. Jobs belonging to the same order must be processed in a particular sequence. Each machine can only conduct one job at a time. The goal is to find a schedule that minimises a specific criterion, such as total completion time or makespan, while adhering to precedence relationships between jobs of the same order and the limited availability of machines.

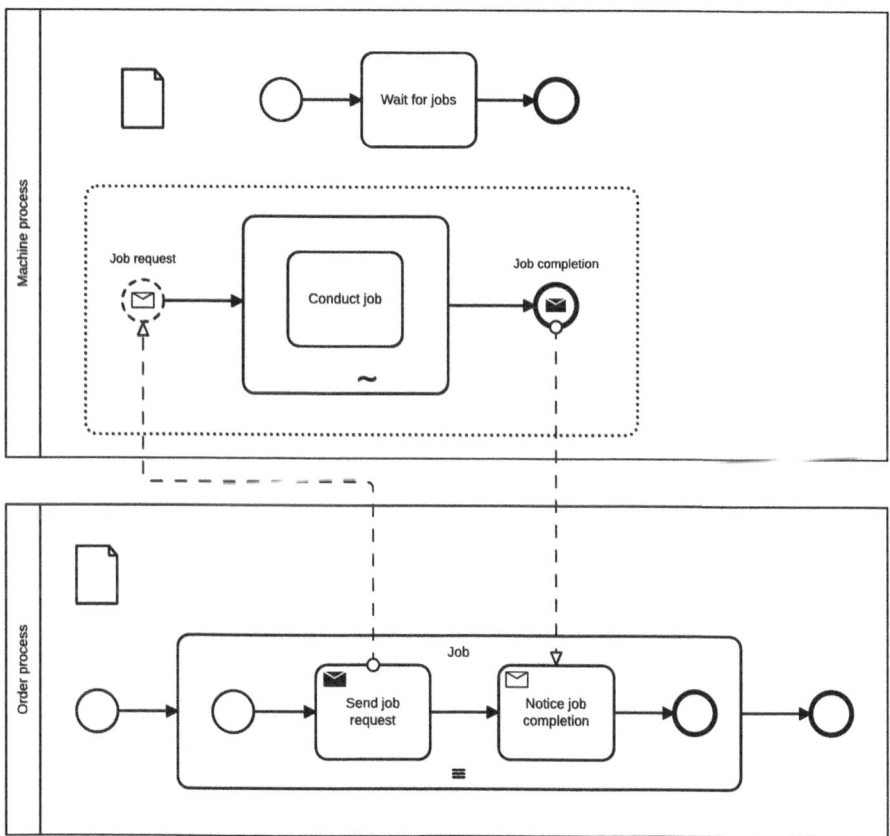

Fig. 3. The job shop scheduling problem as business process model

Figure 3 illustrates a business process model of the job shop scheduling problem. The model has two processes, one describing the *Order process*, another describing the *Machine process*.

The order process starts with a multi-instance activity describing each job belonging to the order. The sequential multi-instance marker is used to indicate that these jobs must be conducted in the given sequence. Each job consists of sending a message requesting to be processed by the respective machine and waits for the machine to notify when the job is completed.

The machine process has a performer representing the machine. It contains a single task having the sole purpose of ensuring that all job requests can be received. After all job requests are received, this task is expected to terminate. Job requests are handled through a non-interrupting event-subprocess which is initiated when a job request message is received. The event-subprocess contains an ad-hoc subprocess containing a single task responsible for conducting the requested job. The performer of the machine process is responsible for all ad-hoc subprocess of all active event-subprocesses. By embedding the *Conduct job* task within an ad-hoc subprocess with sequential ordering, we can inform the performer that these tasks must not be conducted in parallel even if they belong to different event-subprocesses. After completion of a job, the event-subprocess terminates with sending a completion notification to the job. When all event-subprocesses have completed and no further job request is expected, the machine process terminates.

To fully represent the job shop scheduling problem, the number of instantiations of the *Job* activity and the respective machine required and the processing time must be provided as process variables for each order process. For each machine process, the number of jobs expected must be provided to be able to identify when the *Wait for jobs* activity can be terminated. During process execution, the number of job requests received can be stored in a process variable. A global variable can be used to capture the objective criterion, e.g., the total completion time or makespan.

An example of the job shop scheduling problem with three machines and three orders is given in the following. For each order, the *Order process* is instantiated with the following data indicating the sequence of machines required and the respective processing times on these machines:

- **Order 1:** (Machine 1, 3),(Machine 2, 2), (Machine 3, 2)
- **Order 2:** (Machine 1, 2), (Machine 3, 1), (Machine 2, 4)
- **Order 3:** (Machine 2, 4), (Machine 3, 3)

For each machine, the *Machine process* is instantiated with the following data indicating the number of jobs to be conducted:

- **Machine 1:** 2
- **Machine 2:** 3
- **Machine 3:** 3

When an instance of the *Machine process* starts, the number of requests received is set to zero, the *Wait for jobs* task starts and must not be completed before the given number of jobs have been requested.

When an instance of the *Order process* starts, the *Job* activity is instantiated for each job. The first job of each order sends a request message including the required processing time to the respective machine and waits for the completion notification. Thereafter, the instantiation for the next job is executed and so on.

When a machine process receives a job request message, the respective event-subprocess starts and the required processing time of the job is stored in a local variable. The process variable representing the number of requests received is incremented. If the required number of requests have been received, the *Wait for jobs* task completes and the process terminates after all event-subprocesses have terminated.

During process execution, multiple event-subprocesses may be running for each machine. Additional event-subprocesses may be instantiated whenever a new job request is sent. Therefore, the performer of the machine process, not only has to make a decision which of the *Conduct job* tasks to conduct next, but also a decision whether to wait for new requests that could still arrive at the same or a later point in time.

Even though this example only has a few orders and machines, implementing a good decision making strategy for the job shop scheduling problem is non-trivial and underlines the necessity of developing efficient solution approaches. If such a decision making approach is available, we can use this approach to solve the job shop scheduling problem based on the BPMN model formulation provided.

6   Prototypical Implementation

To provide a proof of concept, a BPMN execution engine has been designed and implemented. The execution engine reads a BPMN model in which appropriate extension elements are provided specifying data attributes, restrictions on attribute values, and operators allowing to modify data attributes. Instance data, i.e. specific values e.g. for the number of locations and their coordinates in the travelling salesperson problem, or the number of orders, jobs, and machines in the job shop scheduling problem, are read independently from the process models allowing to use the same process model for a variety of instances.

The execution engine follows the token flow logic described in the BPMN specification and advances tokens as long as no decision is required. Whenever a decision is required which activity within an ad-hoc subprocess shall be conducted next, the execution engine requests such a decision from a dedicated component, called the controller. The controller is responsible for making all decisions required during process execution and different controllers may be implemented allowing different decision mechanisms to be applied.

Preliminary tests have demonstrated the equivalency of the business process modelling formulation of selected scheduling problems with the mathematical formulation. However, it must be noted that the quality of the solutions obtained with the prototypical implementation is not very high. At the time of writing, only basic controllers allowing to make decisions based on simple heuristic rules are implemented, e.g., a nearest neighbour heuristic for the travelling salesperson problem. It is planned to develop more sophisticated controllers based on constrained programming or deep reinforcement learning to improve the solution quality of the respective scheduling problems. However, the design of efficient controllers capable of dealing with a wide range of scheduling problems modelled as business processes is still an open research problem.

The implementation details of the execution engine and the controllers are out of scope of this paper and cannot be presented here due to page limitations. The source code including a comprehensive documentation and a collection of example models is available at https://bpmnos.telematique.eu.

7   Conclusion

Scheduling and business process modelling have traditionally been considered to be independent scientific fields and have evolved in isolation. While scheduling allows to improve the efficiency of business operations by eliminating idle times and bottlenecks, it usually requires experts in mathematical programming and, depending on the underlying business problem, optimisation algorithm engineers to develop dedicated approaches for solving the resulting scheduling problems. Business process modelling, on the other hand, allows domain experts and process owners without mathematical training to specify requirements of their business processes.

This paper shows how ad-hoc subprocesses with sequential ordering can be used to provide BPMN model formulations for scheduling problems and provides BPMN model examples for well-known scheduling problems. The modelling pattern allows domain experts and process owners to specify and formulate scheduling problems inherent to their use cases. Such process models can substantially facilitate the process of communicating the business problem to optimisation algorithm engineers, eliminating the need to formulate mathematical programs. Moreover, the possibility to formulate scheduling problems as BPMN models opens a new and exiting research direction of developing optimisation algorithms capable of working on a BPMN model to optimise a wide range of business problems. Simple algorithm using heuristic decision rules are already implemented, but more sophisticated decision-making approaches require further research.

References

1. Cabanillas, C., Knuplesch, D., Resinas, M., Reichert, M., Mendling, J., Ruiz-Cortés, A.: RALph: a graphical notation for resource assignments in business processes. In: International Conference on Advanced Information Systems Engineering, pp. 53–68. Springer (2015)
2. Cabanillas, C., Resinas, M., Ruiz-Cortés, A.: Ral: A high-level user-oriented resource assignment language for business processes. In: Daniel, F., Barkaoui, K., Dustdar, S. (eds.) Business Process Management Workshops, pp. 50–61. Springer (2012)
3. Camunda: BPMN 2.0 Symbol Reference - All BPMN 2.0 Symbols explained with examples (2024). https://camunda.com/de/bpmn/bpmn-2-0-symbol-reference
4. Corradini, F., Fornari, F., Polini, A., Re, B., Tiezzi, F., et al.: Reprostory: a repository platform for sharing business process models and logs. In: ITBPM@BPM, pp. 13–18 (2021)

5. Goel, A.: Towards a unifying framework for modeling, execution, simulation, and optimization of resource-aware business processes. In: Proceedings of the 2022 Winter Simulation Conference (2022). https://doi.org/10.1109/WSC57314.2022.10015245

6. Havur, G., Cabanillas, C., Mendling, J., Polleres, A.: Resource allocation with dependencies in business process management systems. In: International Conference on Business Process Management, pp. 3–19. Springer (2016)

7. Ihde, S., Pufahl, L., Völker, M., Goel, A., Weske, M.: A framework for modeling and executing task-specific resource allocations in business processes. Computing (2022). https://doi.org/10.1007/s00607-022-01093-2

8. Lo Bianco, R., Dijkman, R., Nuijten, W., van Jaarsveld, W.: Action-evolution Petri nets: A framework for modeling and solving dynamic task assignment problems. In: Di Francescomarino, C., Burattin, A., Janiesch, C., Sadiq, S. (eds.) Business Process Management, pp. 216–231. Springer, Cham (2023)

9. Middelhuis, J., Bianco, R.L., Scherzer, E., Bukhsh, Z.A., Adan, I.J.B.F., Dijkman, R.M.: Learning policies for resource allocation in business processes (2024)

10. Object Management Group: Business Process Model and Notation (BPMN) 2.0.2 (2013). http://www.omg.org/spec/BPMN/2.0.2/PDF

11. Onggo, B.S.S., Proudlove, N., D'Ambrogio, S., Calabrese, A., Bisogno, S., Levialdi Ghiron, N.: A BPMN extension to support discrete-event simulation for healthcare applications: an explicit representation of queues, attributes and data-driven decision points. J. Oper. Res. Soc. **69**(5), 788–802 (2018)

12. Pufahl, L., Ihde, S., Stiehle, F., Weske, M., Weber, I.: Automatic resource allocation in business processes: a systematic literature survey. arXiv preprint arXiv:2107.07264 (2021)

Applying Process Mining on Scientific Workflows: A Case Study on High Performance Computing Data

Zahra Sadeghibogar(✉) ⓘ, Alessandro Berti ⓘ, Marco Pegoraro ⓘ,
and Wil M. P. van der Aalst ⓘ

Process and Data Science, RWTH Aachen University, Aachen, Germany
{sadeghi,a.berti,pegoraro,wvdaalst}@pads.rwth-aachen.de

Abstract. Computer-based scientific experiments are becoming increasingly data-intensive, necessitating the use of High-Performance Computing (HPC) clusters to handle large scientific workflows. These workflows result in complex data and control flows within the system, making analysis challenging. This paper focuses on the extraction of case IDs from SLURM-based HPC cluster logs, a crucial step for applying mainstream process mining techniques. The core contribution is the development of methods to correlate jobs in the system, whether their interdependencies are explicitly specified or not. We present our log extraction and correlation techniques, supported by experiments that validate our approach, enabling comprehensive documentation of workflows and identification of performance bottlenecks.

Keywords: High Performance Computing · SLURM · Scientific workflow · Process mining

1 Introduction

A *workflow* is a description and automation of a process, in which data is processed by different logical data processing activities according to a set of rules. A *scientific workflow* is an ensemble of scientific experiments, described in terms of scientific activities with data dependencies between them [2]. Scientific workflows allow scientists to model and express the entirety of data processing steps and their dependencies. Figure 1 shows an example of a scientific workflow depicted as a flow chart, where each task is associated with a command.

Scientific workflows, characterized by their massive data processing needs, are often automated and necessitate parallel processing on platforms such as cloud

The authors gratefully acknowledge the German Federal Ministry of Education and Research (BMBF) and the Ministry of Education and Research of North-Rhine Westphalia for supporting this work/project as part of the NHR funding. Also, we thank the Alexander von Humboldt (AvH) Stiftung for supporting our research.

K. Gdowska et al. (Eds.): BPM 2024 Workshops, LNBIP 534, pp. 84–96, 2025.
https://doi.org/10.1007/978-3-031-78666-2_7

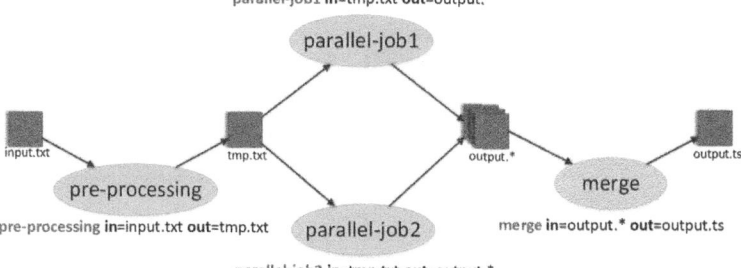

Fig. 1. Illustration of a Scientific Workflow: Starting with a single input file, the process involves pre-processing, parallel job execution, and merging outputs into a single final file.

or HPC clusters [2,9]. In this context, process mining emerges as an invaluable tool for workflow comprehension and detection of possible optimizations.

Over the past decades, there has been a growing interest in the field of process mining [1]. Process mining aims to extract information about processes from event logs, i.e., execution histories. This paper applies process mining to existing scientific workflows with the following goals:

- *Documentation of scientific workflows*: reporting which commands are executed and in which order. We pursue this goal by using process discovery techniques, one of the main branches of process mining [1]. Process discovery techniques assume that every record in the event log contains at least: (i) a reference to the executed activity, (ii) a reference to the identifier that associates an event with a particular execution of the process, and (iii) the timestamp at which the event occurred.
- *Detection of bottlenecks affecting the execution of scientific workflows*: We enrich the process model discovered in the previous step with the obtained performance results.

While the techniques proposed in this paper can be applied to any workflow system, we focus on the SLURM system to promote applicability. SLURM is a common choice for workflow management in HPC clusters, governing the RWTH HPC cluster[1], one of the most widely used platforms in the field.

The issue in examining the logs obtained from a given workflow system is the absence of a clearly defined case identifier that groups events associated with the same execution. To apply process mining to these logs, it is necessary to study the correlation between tasks that are running on the HPC cluster. Figure 2 shows an overall view of our approach. The RWTH HPC cluster is observed periodically, and an input log is generated. Based on how users execute their jobs on SLURM [8], we propose two different approaches to assign case IDs to events. Finally, we obtain an event log on which process mining techniques can be applied.

---

[1] https://help.itc.rwth-aachen.de/service/rhr4fjjutttf/.

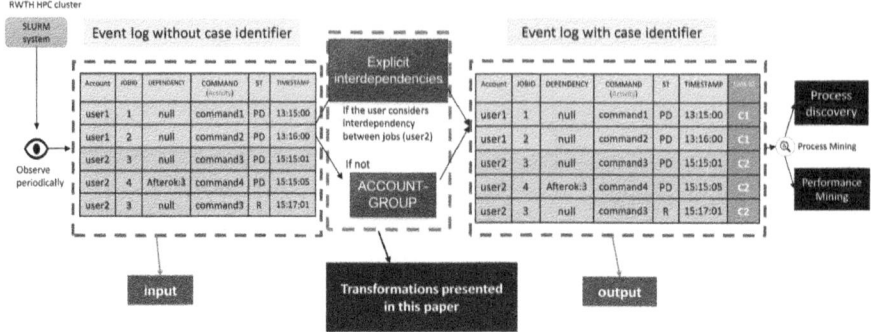

Fig. 2. An overall view of the proposed approach

The remainder of this paper is organized as follows. Section 2 reviews some related works. Section 3 shows some technical notions on how the SLURM system is implemented and which information is available to eventually form an event log. Section 4 explains our approach to apply process mining techniques to the scientific workflows running on SLURM-based HPC clusters. Section 5 introduces some analyses of the event log extracted from the SLURM system of RWTH Aachen University. Finally, Sect. 6 concludes the paper.

2   Related Work

Many studies have analyzed HPC behavior starting from data collected about the running jobs. In [4], an extension of miniHPC is proposed, to enable job-level monitoring to interpret anomalous behaviors such as load imbalance, CPU and I/O anomalies, or memory leaks. A framework for monitoring, analyzing, and predicting jobs running on PBS-based job scheduler HPCs is defined in [5]. The monitoring module captures data about the topology of in-use nodes while a job is running. This provides a deeper understanding of how the job is distributed across the HPCs node network. In [3], a software stack for center-wide and job-aware cluster monitoring to influence node performance is described.

Process mining techniques have been used to analyze scientific/business work-flow logs. In [9], a technique to mine scientific workflows based on provenance (i.e., the source of information involved in manipulating artifacts) is proposed. In [7], Scientific Workflow Mining as a Service (SWMaaS) is presented to support both intra-cloud and inter-cloud scientific workflow mining.

A limitation of [3–5] is that they examine the jobs regardless of their interdependencies. Moreover, in [7], it is assumed that the data source already contains all the necessary information to apply process mining, ignoring the situations in which no case notion is defined. This paper aims to introduce event correlation methods applicable to event data extracted from scientific workflows.

3   Preliminaries

We will focus on analyzing event data from the popular SLURM platform for HPC computing. Hence, in this section, we present some technical notions. To interact with SLURM, we have a set of possible commands. The most essential commands are listed here [8]:

- **srun**: runs a single job. We need to create a **srun** script, which can then be submitted on SLURM for real-time execution.
- **sbatch**: submits one or more **srun** commands for later execution on SLURM.
- **squeue**: reports the states of the running jobs. This command helps us to extract a log for process mining purposes.

3.1   Execution of a Single Job on SLURM

Any script runs on SLURM as a job. As mentioned above, the execution of a job on SLURM could be easily done with **srun** and sbatch containing one single **srun** command. Understanding the sequential stages a job undergoes for execution, and the data that can be extracted for each job running in the SLURM queue is valuable [8].

Typically, jobs pass through several states in the course of their execution. There are a total of 24 possible states for a job, however, three states are most common. In the *PENDING (PD)* state, the job awaits resource allocation; in the *RUNNING (R)* state, it is currently allocated, and in the *COMPLETING (CG)* state, the job is undergoing completion.

The SLURM scheduling queue contains all the information about running jobs. To view this information we use the **squeue** command. The most important features of the jobs that have been used in our study are listed in Table 1.

Table 1. Extracted features of running jobs on SLURM system [8].

Column title	Description
ACCOUNT	Account associated with the job
JOBID	An unique value as job identifier
DEPENDENCY	Specify the dependencies of the job on other jobs. This job will not begin execution until these dependent jobs are complete. In the case of a job that cannot run due to job dependencies never being satisfied, the full original job dependency specification will be reported. A value of NULL implies this job has no dependencies
COMMAND	The command to be executed
ST	Jobs typically pass through several states in the course of their execution. The typical states are *PENDING, RUNNING, SUSPENDED, COMPLETING*, and *COMPLETED*. ST is the compact state of the job
GROUP	Group name of the job. The project ID is reported as GROUP in SLURM

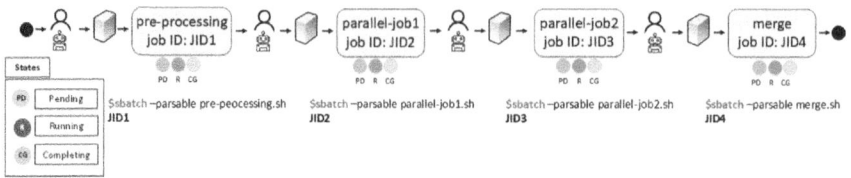

Fig. 3. Execution of a sequence of jobs without explicit interdependencies.

These features could be extracted with the `squeue -o "%a %i %E %o %t %g"` command on the SLURM system. This command shows the list of jobs in the SLURM scheduling queue along with their account, job ID, declared dependency, executed command, status, and project ID information [8].

3.2   Execution of a Sequence of Jobs on SLURM

To explain how to run a series of jobs (sequence of scripts) on the SLURM workflow system, we will go through an example. Consider a user who wants to run four scripts on SLURM, *pre-processing* as the first one, then *parallel-job1* and *parallel-job2*, which can be executed in parallel but must be executed after the *pre-processing* script, because they need its output. Finally, the *merge* script needs the output of the two parallel jobs for its execution. The user can run this sequence of jobs on SLURM in two ways: either manually (without explicit interdependencies) or automatically (with interdependencies).

Execution of a Sequence of Jobs without Explicit Interdependencies: In this case, the user runs the jobs manually—without declaring the inter-dependencies between jobs—and after submitting each job waits for its execution to be completed; then, executes the next job (Fig. 3). In this case, each job is executed as independent, and only the user knows that some of these jobs are logically dependent on each other.

Execution of a Sequence of Jobs with Explicit Interdependencies: In this scenario, the user uses the SLURM dependency management system and submits all jobs at once with correct inter-dependencies on the SLURM system, as shown in Fig. 4. Here, the user uses the `sbatch` command. This command is used to submit a job script for later execution using the `--dependency` option. The script typically contains one or more `srun` commands to launch parallel tasks. In this case, the user does not need to wait for the outputs of a single job, but can wait for the execution of all the tasks and retrieve the final results at completion (Fig. 4).

Table 2 shows the output of the `squeue` command where the user has declared explicit interdependencies between jobs. As one may see, the *DEPENDENCY* column in the *PENDING* state has a non-empty value. Conversely, when users manually initiate jobs, the *DEPENDENCY* column in the *PENDING* state remains empty, denoted as *(null) values*.

4   Approach

The input of most process mining algorithms is an event log, which contains at least a case, an activity, and a timestamp as attributes for each event. The majority of algorithms presume that the event data is fully accessible and has a clearly defined case notion.

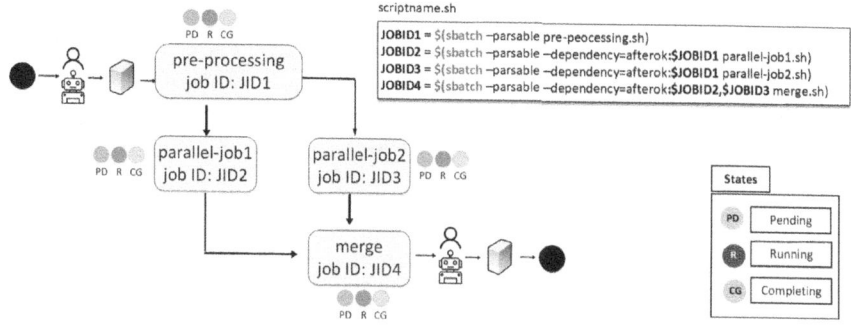

Fig. 4. Execution of a sequence of jobs with explicit interdependencies.

However, we cannot assume that we know the complete historical log, because of privacy issues and the required administrative privileges on the target workflow system. Instead, we aim to observe it for a limited amount of time as described in Sect. 4.1. However, this poses technical challenges. For instance, the observation interval may be too long, making it difficult to capture job information during rapid status changes.

Moreover, since the SLURM log does not contain any explicit case notion, in Sect. 4.2 we describe *event correlation* to assign a case to the different events and allow for process mining analyses.

4.1   Register SLURM Events

In order to extract an event log from the system, we perform the following operations periodically (we refer to this as *observing the system*):

1. Connect to the access node of the SLURM system
2. Observe the status (e.g., *PENDING, RUNNING, COMPLETING*) of the current jobs using the `squeue` SLURM command.
3. For each of the listed jobs (rows of the log file), one of the following situations occurs:
    - The *JOBID* is new: register an event related to the creation of the job.
    - The *JOBID* already exists, but the status has changed: register an event related to the status change.
    - The *JOBID* already exists, and the status has not changed: do nothing.

All the features mentioned in Table 1 are recorded for each job. Our log (as illustrated in the sample SLURM log in Table 2) includes a *TIME* column denoting the event observation time recorded by the script monitoring the SLURM system, and the *COMMAND* values are derived from the executed file path by extracting and using only the file name (the last part of the path).

Table 2. Sample SLURM Log

ACCOUNT	JOBID	DEPENDENCY	COMMAND	STATE	TIME	GROUP
userA	JID1	(null)	pre-processing.sh	PD	13:34:09	G1
userA	JID2	afterok:JID1 (unfulfilled)	parallel-job1.sh	PD	13:34:09	G1
userA	JID3	afterok:JID1 (unfulfilled)	parallel-job2.sh	PD	13:34:09	G1
userA	JID4	afterok:JID2 (unfulfilled), afterok:JID3 (unfulfilled)	merge.sh	PD	13:34:09	G1
userA	JID1	(null)	pre-processing.sh	R	13:35:00	G1
userA	JID1	(null)	pre-processing.sh	CG	13:49:10	G1
userA	JID2	(null)	parallel-job1.sh	R	13:52:09	G1
userB	JID5	(null)	Import_input.sh	PD	13:52:32	G2
userB	JID6	afterok:JID5 (unfulfilled)	Main_calculation.sh	PD	13:52:33	G2
userB	JID7	afterok:JID6 (unfulfilled)	Export_output.sh	PD	13:52:33	G2
userA	JID3	(null)	parallel-job2.sh	R	13:54:13	G1
userA	JID3	(null)	parallel-job2.sh	CG	14:12:10	G1
userB	JID5	(null)	Import_input.sh	R	14:12:12	G2
userA	JID2	(null)	parallel-job1.sh	CG	14:38:10	G1
userB	JID5	(null)	Import_input.sh	R	14:39:32	G2
userA	JID4	(null)	merge.sh	R	14:51:30	G1
userA	JID4	(null)	merge.sh	CG	14:53:30	G1
userB	JID6	(null)	Main_calculation.sh	R	14:54:12	G2
userB	JID6	(null)	Main_calculation.sh	CG	14:58:32	G2
userB	JID7	(null)	Export_output.sh	R	14:59:10	G2
userB	JID7	(null)	Export_output.sh	CG	15:10:10	G2

4.2   Event Correlation

Let us now obtain case IDs from SLURM. We extract a case ID with different techniques, depending on whether the jobs were executed with or without explicit interdependencies.

Case ID Extraction with Explicit Interdependencies: We utilize this technique when the user has specified the inter-dependencies among jobs. This declaration allows the inclusion of the *DEPENDENCY* column in the extracted log, indicating the jobs on which the current job depends. Note that the *DEPEN-DENCY* column for the job lists only the dependencies that have not been completed yet. Thus, the *DEPENDENCY* list would be naturally empty for a job that is in the *RUNNING* state.

To implement this method, a Directed Acyclic Graph (DAG) is generated for each chain of connected jobs in *PENDING* state by utilizing the *JOBID*, and *DEPENDENCY* columns. The vertices are job IDs and the edges show dependent job IDs, and then a unique case ID will be assigned to all of the connected job IDs as shown in Fig. 5b. In the table in Fig. 5, we observe that JID2 and JID3 are dependent on JID1, and JID4 depends on both JID2 and JID3. Based on these dependencies, we assign case ID JID4321, as indicated in the *CASE ID* column of the table. Different cases will be assigned to different discovered connected components. For instance, JID111098 is assigned to another execution of the same chain of commands as JID4321.

	ACC	JID	COMMAND	DEP	TIME	ST	CASE ID
1	userA	JID1	pre-processing.sh	(null)	13:35:00	PD	JID4321
2	userA	JID2	parallel-job1.sh	afterok:JID1 (unfulfilled)	13:52:09	PD	JID4321
3	userA	JID3	parallel-job2.sh	afterok:JID1 (unfulfilled)	13:52:11	PD	JID4321
4	userA	JID4	merge.sh	afterok:JID2 (unfulfilled), afterok:JID3 (unfulfilled)	13:54:13	PD	JID4321
5	userB	JID5	Import_input.sh	(null)	14:12:12	PD	JID765
6	userB	JID6	Main_calculation.sh	afterok:JID5 (unfulfilled)	14:51:30	PD	JID765
7	userB	JID7	Export_output.sh	afterok:JID6 (unfulfilled)	14:54:12	PD	JID765
8	userA	JID8	pre-processing.sh	(null)	14:56:15	PD	JID111098
9	userA	JID9	parallel-job2.sh	afterok:JID8 (unfulfilled)	14:59:10	PD	JID111098
10	userA	JID10	parallel-job1.sh	afterok:JID8 (unfulfilled)	15:10:10	PD	JID111098
11	userA	JID11	merge.sh	afterok:JID9 (unfulfilled), afterok:JID10 (unfulfilled)	16:05:17	PD	JID111098

(a)   (b)

Fig. 5. Case ID extraction with explicit interdependencies by studying *JOBID* and *DEPENDENCY*. ACC stands for ACCOUNT, DEP for DEPENDENCY, JID for JOBID, and ST for STATE. We use these shorter forms to make the table more compact.

Case ID Extraction without Explicit Interdependencies: In this case, we do not have explicitly defined job dependencies; therefore, we need to use the attributes at the event level to determine correlations and dependencies between the jobs. We use a combination of the following two attributes in order to define the case identifier:

- The *account* executing the job: it is reported as *ACCOUNT* in SLURM.
- The *group* of the given job: the project ID is intended to group the jobs belonging to the same project. The status should be empty or default if the user does not call the command with the project ID. The project ID is

reported as *GROUP* in SLURM. So, whenever we execute the same scripts several times, the project ID is reported as the same *GROUP* in SLURM.

We have a many-to-many relationship between accounts and groups. All the jobs executed by an account under a given group are therefore related to the same project. We can use *ACCOUNT-GROUP* as case ID; in this case, we are certain that the jobs executed by the same user under the same project are all collected. In this technique, we generate a unique case ID per each unique combination of *ACCOUNT* and *GROUP*, as shown in Table shown in Fig. 6a. The parallel relationship between *Parallel-job1* and *Parallel-job2* has been discovered based on their occurrence in rows 2, 4 and 8, 10, which show they can be executed in any order.

In this method, we may consider only the account instead of considering the combination of group and account, but the advantage of considering also the group is that the control flow of different projects of the same account is not combined. This technique also has limitations, considering loops where they shouldn't be, due to the inability to differentiate between consecutive executions of the same command related to different experiments. As a result, the precision is significantly reduced, because many different behaviors and command sequences are allowed by the resulting model.

	ACC	JID	COMMAND	ST	TIME	G	CASE ID
1	userA	JID1	pre-processing.sh	R	13:35:00	G1	userA-G1
2	userA	JID2	parallel-job1.sh	R	13:52:09	G1	userA-G1
3	userC	JID3	pre-processing.sh	R	13:52:11	G1	userC-G1
4	userA	JID4	parallel-job2.sh	R	13:54:13	G1	userA-G1
5	userB	JID5	Import_input.sh	R	14:12:12	G2	userB-G2
6	userA	JID6	merge.sh	R	14:51:30	G1	userA-G1
7	userB	JID7	Main_calculation.sh	R	14:54:12	G2	userB-G2
8	userC	JID8	parallel-job2.sh	R	14:56:15	G1	userC-G1
9	userB	JID9	Export_output.sh	R	14:59:10	G2	userB-G2
10	userC	JID10	parallel-job1.sh	R	15:10:10	G1	userC-G1
11	userC	JID11	merge.sh	R	16:05:17	G1	userC-G1

(a)

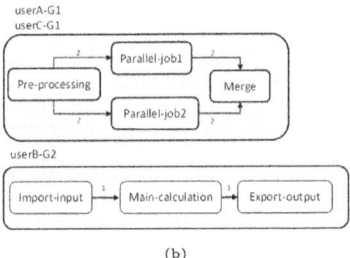

(b)

Fig. 6. Case ID extraction without explicit interdependencies by studying *ACCOUNT* and *GROUP*. ACC stands for ACCOUNT, JID for JOBID, ST for STATE and G for GROUP. We use these shorter forms to make the table easier to read and understand.

Table 3. Some event log statistics extracted from the SLURM system. The system was observed in a time interval from 2022-12-07 11:51:45 to 2022-12-09 10:49:07.

Number of events	81632
Number of unique submitted jobs	17997
Number of accounts	123
Percentage of accounts who submitted jobs with explicit interdependencies	0.06%
Percentage of jobs defined with explicit interdependencies	0.02%
Average number of allocated CPUs per job	13.71
Average amount of allocated RAM per job	5G

5   Experiments

In our previous work, we developed SLURMminer, a tool specifically designed for mining and analyzing process models from SLURM job logs [6]. The experiments presented in this paper utilize SLURMminer. SLURMminer was applied to the SLURM system at RWTH Aachen University multiple times, resulting in the extraction of an event log[2]. Table 3 presents key statistics derived from the event log.

Given the variety of research areas (including physics, chemistry, biology, and computer science) and executed purpose-specific scripts, generating a comprehensive process model containing the behavior of all the accounts proved impractical. Instead, we focused on individual account process models. These process models show the scientific workflows executed by a single user/research group. Moreover, the process model is annotated with performance information on the arcs, allowing for the detection of paths with high execution time (bottlenecks), and therefore fulfilling the second goal of finding root causes of performance problems.

To highlight different execution paradigms, we focus on two accounts:

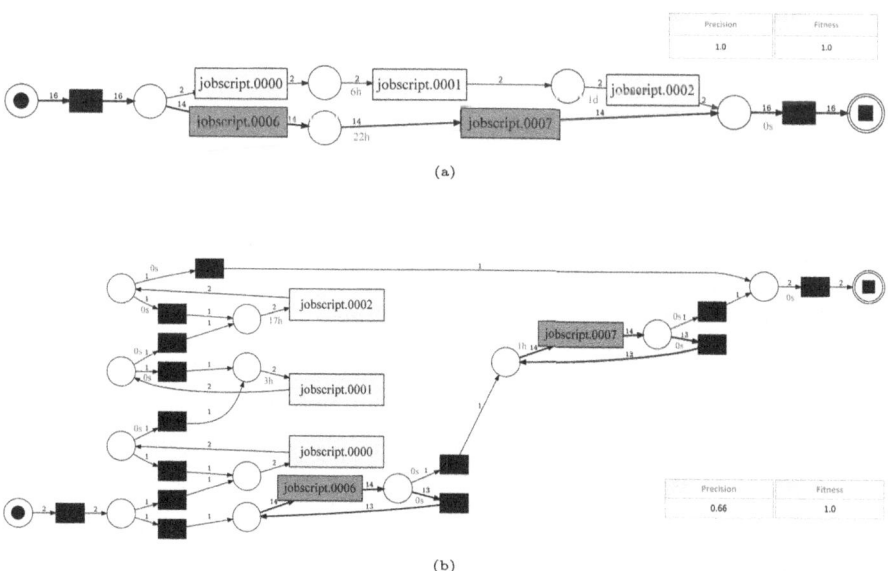

Fig. 7. The discovered process models for the account *jara0180* considering (a) explicit and (b) implicit interdependencies.

---

[2] For more details on the RWTH HPC cluster, visit: https://help.itc.rwth-aachen.de/ service/rhr4fjjutttf/. A sample event log can be downloaded at: https://www.ocpm. info/hpc_log.csv.

– *jara0180*: contains computations performed on a funded research project (Molecular dynamics simulations of P2X receptors).
– *thes1331*: contains scientific experiments performed for an MSc thesis.

The executions carried out by *jara0180* take advantage of explicit interdependencies (since HPC expertise is involved). Therefore, for *jara0180* we were able to develop a meaningful process model, as depicted in Fig. 7a. In this figure, we observe two distinct chains of commands. The first chain comprises commands from *jobscript_0000* to *jobscript_0002*, while the second chain includes the remaining commands related to two different projects and corresponding to 30 distinct cases in the event log. We could still obtain a model from the data (contained in Fig. 7b) without considering these interdependencies (considering the *ACCOUNT* and *GROUP* values lead to two distinct cases). However, this model is less precise because it relies solely on the temporal order of command execution, where every event belongs to the same case in the event log.

The executions performed by *thes1331* are defined without explicit interdependencies. The case extraction approach, relying on explicit interdependencies, leads to the assignment of a unique case ID to each execution (seven distinct cases). Consequently, the model depicted in Fig. 8a exhibits concurrency among all the executed commands, rendering it highly imprecise. For *thes1331*, it is more appropriate to focus on the models discovered without considering the explicit interdependencies (contained in Fig. 8b), which shows the temporal order of execution of the commands.

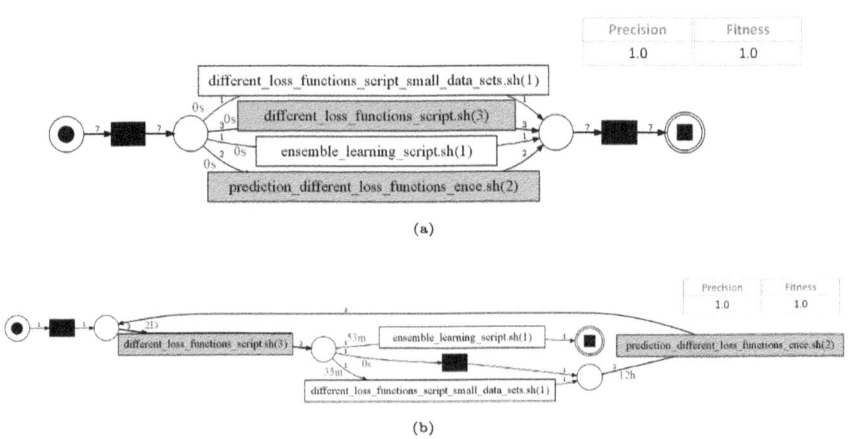

Fig. 8. The discovered models for the account *thes1331* considering explicit (a) and implicit (b) interdependencies.

The discovered models offer insights by visualizing control flow and execution frequency, aiding users in identifying bottlenecks for improvement. Table 3 shows that only a small fraction of HPC users submit jobs with explicit interdependencies, crucial for identifying connected jobs. Without these dependencies, resulting models may be imprecise, with events either belonging to different cases or all belonging to the same case.

6 Conclusion

In this paper, we propose an approach to extract and analyze process mining event logs of an HPC system (in particular, we focus on the SLURM system). While this is not the first application of process mining to HPC systems, existing techniques assume the case notion to be well-defined in the data source. This assumption is not satisfied by mainstream systems, and we propose two different case notions (using and not using explicit interdependencies). Moreover, we propose the *SLURMminer* as a tool to connect to the HPC system, extract an event log, and perform a process mining analysis. The analyses allow us to document the execution of scientific workflows for different accounts or research groups utilizing process models that are annotated with performance information (allowing us to detect bottlenecks). Therefore, we address our initial research question: How can process mining be applied to SLURM-based HPC clusters to document workflows and identify execution bottlenecks?

Our event logs are extracted from information that is publicly available in the SLURM system (including the command that is executed and the requested environment, i.e., the number of CPUs, RAM, and disk space required). However, we do not know the detailed content of the commands or have access to more advanced profiling options. This would require collaboration with the specific research groups operating in the HPC systems and availability to modify the execution of scientific workflows to accommodate more detailed process mining analyses.

Our process mining analyses rely on a single account or research group. Since the naming schema of the commands is quite arbitrary, we could not identify shared logical steps (e.g., pre-processing, training of ML model, testing of the model) between different accounts; therefore, we could not produce a generic process model. This is indeed a limitation that could not be tackled without properly naming the commands executed on SLURM and without having insights about the commands.

Overall, our approach succeeds in extracting an event log for process mining purposes from the SLURM HPC system, and we can respond to our basic analytical goals. However, given the arbitrary execution styles and naming conventions, we could not produce more general analyses, which remain as a goal for future work.

References

1. van der Aalst, W.M.P., Carmona, J. (eds.): Process Mining Handbook. LNBIP, vol. 448. Springer, Cham (2022)
2. Deelman, E., Gannon, D., Shields, M.S., Taylor, I.J.: Workflows and e-science: an overview of workflow system features and capabilities. Future Gener. Comput. Syst. **25**(5), 528–540 (2009)

3. Dietrich, R., Winkler, F., Knüpfer, A., Nagel, W.E.: PIKA: center-wide and job-aware cluster monitoring. In: IEEE International Conference on Cluster Computing, CLUSTER 2020, Kobe, Japan, 14–17 September 2020, pp. 424–432. IEEE (2020)
4. Kunz, P.: HPC Job-Monitoring with SLURM, Prometheus, and Grafana (2022)
5. Pal, A., Malakar, P.: MAP: a visual analytics system for job monitoring and analysis. In: IEEE International Conference on Cluster Computing, CLUSTER 2020, Kobe, Japan, 14–17 September 2020, pp. 442–448. IEEE (2020)
6. Sadeghibogar, Z., Berti, A., Pegoraro, M., van der Aalst, W.M.P.: Slurmminer: a tool for SLURM system analysis with process mining. In: Demonstration & Resources Forum at BPM 2023 co-located with 21st International Conference on Business Process Management (BPM 2023), 11–15 September 2023. CEUR Workshop Proceedings, vol. 3469, pp. 97–101. CEUR-WS.org (2023)
7. Song, W., Chen, F., Jacobsen, H., Xia, X., Ye, C., Ma, X.: Scientific workflow mining in clouds. IEEE Trans. Parallel Distrib. Syst. **28**(10), 2979–2992 (2017)
8. Yoo, A.B., Jette, M.A., Grondona, M.: SLURM: simple Linux utility for resource management. In: Feitelson, D., Rudolph, L., Schwiegelshohn, U. (eds.) JSSPP 2003. LNCS, vol. 2862, pp. 44–60. Springer, Heidelberg (2003). https://doi.org/10.1007/10968987_3
9. Zeng, R., He, X., van der Aalst, W.M.P.: A method to mine workflows from provenance for assisting scientific workflow composition. In: World Congress on Services, SERVICES 2011, Washington, DC, USA, 4–9 July 2011, pp. 169–175. IEEE Computer Society (2011)

8th International Workshop on Business Processes Meet Internet-of-Things (BP-Meet-IoT 2024)

8th International Workshop on Business Processes Meet Internet-of-Things (BP-Meet-IoT 2024)

The Business Process Management (BPM) discipline, as it is known today, emerged as the result of significant advances experienced since the mid-1990s in business methods, tools, standards and technology. Since then, this discipline has significantly evolved but mainly focused on the business domain with the objective of helping organizations to achieve their goals. However, the arrival of the Internet of Things (IoT) has put into play a huge amount of interconnected and embedded computing devices with sensing and actuating capabilities that are revolutionizing our way of living. The incorporation of this technology into the BPM field has the potential to make business processes (BPs) aware and adaptive to their execution environment and its changes. In addition, the proper combination of these two fields (IoT and BPM) can foster the development of innovative solutions not only in the business domain where BPM emerged, but also in many different application areas in which the IoT can be applied (e.g., smart cities, smart agriculture, smart factories, e-health).

Whereas the integration of IoT technology and BPM opens up plenty of opportunities, it also imposes a set of challenges that need to be addressed. In particular, research is needed to address questions such as:

- What is the impact of introducing IoT technology into the BPM lifecycle?
- How can the top-down and bottom-up paradigms on which BPM and the IoT respectively rely coexist and benefit each other when merged?
- How to bridge the gap between the low and the high level on which IoT and BPM operate respectively?
- How will BPM deal with the changing nature imposed by IoT technology?
- How will the real-time communication and collaboration required in IoT systems be supported by BPM?
- How to consider privacy aspects in data captured by IoT devices and analyzed with BPM?
- How can BPM be used to model behavior in IoT systems?

The objective of this workshop was therefore to attract novel research that tackles these challenges as well as to create a space for discussion and interactions between the research communities dealing with integration between the IoT and BPM fields.

The 8th edition of this workshop attracted seven international submissions, each of which was reviewed by three members of the program committee. The following four submissions (3 regular and 1 short paper) were finally accepted and selected for presentation:

"Check my Flow: Distributed Conformance Checking at the Source" authored by Julia Andersen, Patrick Rathje and Olaf Landsiedel (regular paper). In this paper, the authors discuss distributed conformance checking at the source in an IoT context. More specifically, they present an approach where local data stays local and in this way, privacy can be preserved.

"Sustainability in and through IoT-enhanced Business Processes" authored by Antoni Mestre, Manoli Albert, Ronny Seiger, Victoria Torres and Pedro Valderas (short paper). In this paper, the authors provide an overview of the opportunities and challenges that IoT devices introduce in BP when sustainability aspects need to be measured and improved.

"From IoT Event Logs to Human Routines via Community Detection Algorithms" authored by Massimo Callisto De Donato, Fabrizio Fornari and Abel Armas-Cervantes (regular paper). In this paper, the authors introduce a method to identify human routines from IoT event logs using community detection algorithms. The study aims to enhance the monitoring and analysis of human activities in smart environments by uncovering patterns in IoT data.

"LLM-based event abstraction and integration for IoT-sourced logs" authored by Mohsen Shirali, Mohammadreza Fani Sani, Zahra Ahmadi and Estefanía Serral (regular paper). In this paper, the authors propose an approach that uses LLM for generating an abstracted log from multiple sensor data, specifically tailored for a smart home environment.

"Machinery activity recognition in the industry based on heterogeneous data" authored by Marta Podobińska-Staniec, Marek Kęsek and Edyta Brzychczy (regular paper). In this paper, the authors present a systematic literature review on Machinery Activity Recognition (MAR) in industrial contexts, focusing on construction, mining, logistics and medicine. The authors highlight that MAR is prevalent in the construction industry but has significant applications in other domains.

Before these papers were presented, we had a very insightful keynote by Jan Mendling, Einstein Professor of Process Science with the Department of Computer Science at Humboldt-Universität zu Berlin. The keynote was titled: "Visualizing Business Process Distribution over Time and Space". This keynote made clear the importance of transforming data into understandable visualizations that aid in strategic decision-making, resource optimization and improving coordination between different parts of the organization.

The workshop took place on the 2nd of September 2024, and attracted about 40 participants who actively interacted during the workshop presentations and the discussions. The organizers of this event would like to specially thank the authors of the submitted papers as well as the keynote speaker and participants in the fruitful discussions. We would also like to thank the PC members for their valuable input and the conference organizers who facilitated the workshop. We hope that the reader finds the final selection of papers interesting and useful to get a better insight into the integration of IoT and BPM from both theoretical and practical points of view.

Organization

Organizing Committee

Agnes Koschmider	University of Bayreuth, Germany
Francesco Leotta	Sapienza Università di Roma, Italy
Massimo Mecella	Sapienza Università di Roma, Italy
Estefanía Serral	KU Leuven, Belgium
Victoria Torres	Universitat Politècnica de València, Spain

Program Committee

Adrian Mos	Naver LABS, Grenoble, France
Andrea Delgado	Universidad de la República, Uruguay
Andreas Oberweis	Karlsruhe Institute of Technology, Germany
Felix Mannhardt	Eindhoven University of Technology, the Netherlands
Jianwen Su	University of California at Santa Barbara, USA
Mathias Weske	University of Potsdam, Germany
Pnina Soffer	University of Haifa, Israel
Ronny Seiger	University of St. Gallen, Switzerland
Vicente Pelechano	Universitat Politècnica de València, Spain
Yannis Bertrand	KU Leuven, Belgium
Zakaria Maamar	Zayed University, UAE

Check My Flow: Distributed Conformance Checking at the Source

Julia Andersen, Patrick Rathje, and Olaf Landsiedel[⊠]

Kiel University, 24118 Kiel, Germany
{jro,pra,ol}@informatik.uni-kiel.de

Abstract. IoT networks in, for example, smart manufacturing, smart homes, and smart health demand process mining beyond traditional event logs. However, conventional process discovery and conformance checking algorithms expect data to be collected at a central location for mining. As a result, they struggle to handle the vast amounts of data generated by IoT networks, which often leads to privacy concerns. To address these challenges, this paper introduces Distributed Conformance Checking (DisCC). DisCC leverages a footprint-fitness method to perform distributed online conformance checks directly at the data source where an event is sensed, ensuring scalable conformance checking and enabling privacy by only sharing aggregated event data with a central entity. Our evaluation of DisCC demonstrates its effectiveness in accurately performing conformance checks at the event, trace, and log levels. We experimentally show the correctness of DisCC and how quickly its interim results converge to the correct fitness value. The system supports real-time alerts for non-conforming events and traces, detects concept drifts and temporary fitness losses through a sliding window implementation, and offers a scalable, privacy-enabling solution for process monitoring in IoT networks.

Keywords: Online Conformance Checking · Distributed Process Mining · Event Data Stream · Scalability

1 Introduction

Today, the field of process mining extends beyond the classic event log: Driven by the Internet of Things (IoT), new applications, for example on smart manufacturing, smart home, or smart health, collect large amounts of data for mining [1–4]. Existing process discovery and conformance checking algorithms assume logs to be readily available for mining at a central location [5,6]. However, the sheer volume and velocity of data generated by spatially distributed IoT devices, coupled with inherent privacy concerns associated with centralized data storage, make centralized data collection insufficient [7,8], necessitating distributed conformance checking directly at the data sources to address these issues and enable real-time analysis.

K. Gdowska et al. (Eds.): BPM 2024 Workshops, LNBIP 534, pp. 101–112, 2025.
https://doi.org/10.1007/978-3-031-78666-2_8

This paper introduces Distributed Conformance Checking (DisCC), a distributed variant of the footprint-fitness approach by Molka et al. [9], which is directly deployable on the source nodes. Unlike established alignment-based [10] or token-based recall [11] approaches, DisCC leverages the simplicity of footprint-based fitness. Alignment-based methods typically require a global view of the entire process model, making them impractical for distributed monitoring, especially for real-time analysis. Token-based recall, on the other hand, can mainly be performed locally, but also includes steps that require a global overview.

DisCC enables real-time conformance checking at event, trace, and log level. At the event level, nodes promptly alert a central entity upon locally detecting non-conforming events, enabling immediate corrective actions to maintain process integrity. Further, based on fitness aggregates DisCC provides an analysis for entire traces. At the log level, DisCC gives insights into overall process adherence by computing fitness metrics for entire event logs. Using sliding windows, DisCC enables the detection of concept drifts in process behavior over time and temporary fitness losses, ensuring adaptive and responsive process management.

DisCC is flexible in its choice of footprint-based fitness metrics at the log level and, for example, also supports the metric described by van der Aalst [6]. To address privacy concerns inherent in many IoT applications, such as central storage of full event logs, DisCC ensures that nodes transmit only highly aggregated event data to central entities. This approach protects sensitive information and maintains the efficiency and scalability necessary for large-scale deployments. Overall, DisCC leverages the local processing power of IoT devices for more privacy-aware and scalable conformance checking in IoT networks. In our evaluation, we show that the fitness of our distributed approach ultimately matches the value of the underlying centralized approach. For datasets with no large series of mismatching traces at the end of the event log, the fitness value quickly converges to the value of the underlying traditional approach, which means that we show the correctness of DisCC. This allows us to incorporate a sliding windows while still achieving meaningful fitness values and it enables us to see more precisely whether a reduction in fitness is a concept drift or just a temporary phenomenon. In summary, the contributions of this paper are as follows.

- We introduce DisCC, a distributed, online conformance checking algorithm for IoT networks based on footprint-fitness.
- We show how DisCC covers fitness evaluation on event, trace, and log level.
- We enable real-time alerts for non-conforming events and traces and the detection of concept drifts as well as temporary fitness losses on the log level.
- We experimentally show the correctness of our approach[1].
- We demonstrate how DisCC is not restricted to one kind of footprint-based fitness metric on the log level.
- We ensure that nodes only share aggregated event data with a central entity to enable privacy.

The work is structured as follows: Sect. 2 introduces the necessary background, while Sect. 3 presents related work on distributed conformance check-

---

[1] Available as open-source at: https://github.com/ds-kiel/DisCC.

ing. Section 4 presents DisCC's design. Section 5 holds our evaluation, and Sect. 6 concludes this paper.

2   Background

This section defines key concepts such as footprint matrices and introduces the footprint-based metric proposed by Molka et al. [9]. Activities, event logs, cases, and traces are defined using standard terminology [12,13].

A footprint matrix (FM) encodes all direct successions within a process model or event log. For event log L with activity set T_L, the corresponding footprint matrix FM is quadratic in $|T_L|$. Each activity in L is mapped to a specific row and column in the matrix. A unique ID in $\{0, \ldots, |T_L| - 1\}$ identifies each activity. This way, FM_{ij} describes whether activity i is directly followed by j.

$$FM_{ij} = 1 \quad \Leftrightarrow \quad j \text{ directly follows } i$$
$$FM_{ij} = 0 \quad \Leftrightarrow \quad j \text{ does not directly follow } i$$

In our approach, a partial FM denotes a single column of the entire FM. For $i, j \in \{0, \ldots, |T_L| - 1\}$, FM^i represents the FM consisting solely of column i, while FM^i_j denotes the entry of column i at row j. We proceed with the footprint-based fitness metric proposed by Molka et al. [9]. Let T_L be the activity set over log L and P represent the powerset over T_L. To capture all directly-follows relations within a log, we define the directly-follows function $d_L : T_L \to P$ which maps every activity $a \in T_L$ to the set of activities that are directly followed by a, i.e., which precede a in at least one trace in L; formally $d_L(a) = \{b \in T_L | \exists \sigma \in L \wedge \exists i \in \mathbb{N} \wedge \sigma_i = b \wedge \sigma_{i+1} = a\}$. Let D_L denote the set of activity tuples that includes all directly-follows relations of the log: $D_L : T_L \times T_L$ with $D_L = \{(a,b)|a, b \in T_L \wedge b \in d_L(a)\}$. For this approach, an additional activity \$ is introduced to the activity set T_L, denoting "no activity". It is added to the beginning and end of each trace in L, symbolizing the start and end of the trace. Hence, the fitness metric also accounts for correctly identifying start and end activities within a trace. Based on the directly-follows footprint, Molka et al. [9] define a recall fitness measure $fp_{recall} = \frac{|D_L \cap D_M|}{|D_L|}$, where D_M denotes the directly-follows relations present in the model's FM. This measure assesses the proportion of directly-follows relations correctly captured in the log L and the model M.

3   Related Work

Numerous works in the field of distributed conformance checking introduce decomposition techniques to manage large process models and event logs effectively. Munoz-Gama et al. [14] introduce the Single-Entry Single-Exit (SESE) decomposition technique, which divides large process models and event logs into smaller, independently analyzable subprocesses.

Van der Aalst and Verbeek [15] present an approach that utilizes the notion of passages to decompose process mining tasks, such as conformance checking. A passage is defined as a pair of non-empty sets of activities that share a direct successor-predecessor relationship. This decomposition enables the distribution of process mining problems over a network of computers, with each subproblem handled independently and contributing to an aggregated overall solution.

Leemans et al. [16] introduce the Projected Conformance Checking (PCC) framework for distributed conformance checking. This framework employs a divide-and-conquer strategy, decomposing the conformance checking problem into smaller subproblems, each focusing on a subset of activities. Recall, fitness, or precision measures are computed for these subsets, and the final conformance metrics are averages of these subset measures. This approach enables the efficient handling of large and complex event logs, but like the SESE and passages approaches, it remains centralized within its subprocesses and does not naturally support real-time conformance checking or ensure privacy.

In contrast to all three approaches, DisCC operates directly on distributed IoT sensor nodes, emphasizing the use of localized data sources for online conformance checking. While all approaches, including DisCC, utilize decomposition, DisCC decomposes the model into the smallest entities – individual activities – leveraging the localized processing power of distributed nodes. Unlike the other methods, DisCC's decomposition allows it to avoid centralization within subprocesses, thereby enhancing privacy and enabling real-time conformance checking.

4   Design

We first outline the assumptions and design setting for DisCC, followed by an overview of the basic concept, before formally presenting the DisCC algorithm.

4.1   Assumptions on Events

Following established conventions, we assume that events contain the attributes case identifier, activity name, and timestamp, with each case identifier being unique. These assumptions are both realistic and applicable in a variety of real-world contexts. For example, in a factory, each product may be assigned a unique identifier that serves as a case identifier, and events such as painting or forming a piece of metal are considered predefined activities.

4.2   DisCC Setting

We deploy DisCC in typical IoT scenarios, such as networks of sensor nodes integrated into machines in smart factories or smart healthcare systems. We assume that the sensor data has already been abstracted into events. The sensor nodes either communicate directly with each other or over the Internet. Further, we assume that a membership management system is in place that allows nodes to identify and connect to other nodes, which is a common feature in

IoT applications. This setup allows nodes to reliably determine the predecessor and successor of an event [17]. Furthermore, we assume that the model used for conformance checking is provided in the form of an FM, which DisCC then splits into partial footprint matrices and distributes them among the nodes; see Fig. 1. For simplicity, we assume that each node monitors a single type of activity. However, by using virtual nodes, this model can be easily adapted to allow each node to manage multiple activity types.

4.3 DisCC Algorithm: Overview

Our algorithm evaluates conformance at three distinct levels. (1) *Event Level:* A node alerts a central entity if an event fails to conform to its partial FM; see Fig. 2. (2) *Trace Level:* A node corresponding to an end event computes the fitness measure, specifically the fp_{recall}, for the entire trace; see Fig. 3, 4 and 5. (3) *Log Level:* Upon request, all nodes transmit their local data to a central entity, which then calculates the overall fitness of the entire log; see Fig. 6.

Note that while log-level compliance is assessed on demand, event- and trace-level compliance are triggered automatically when events are detected. This multi-level approach ensures immediate conformance assessment for multiple granularities. Note that it is not necessary to have a single central entity. It is possible to have multiple entities from which to request the log level fitness.

4.4 Conformance on Event and Trace Level

The node with activity identifier i locally stores two kinds of partial FMs: The first type is a column of the model's FM $mFM^i \in \{0,1\}^{|T_L| \times 1}$, where T_L comprises all activities in L and the "no activity" activity \$. Each node only stores one mFM column. The second type of FM is a column of the log's FM $lFM^i \in \{0,1\}^{|T_L| \times 1}$. In addition to the FM lFM^i, which comprises the whole event log, multiple log FMs l_kFM^i exist where k denotes the case identifier to which this specific partial FM applies. Note that merging all l_kFM^i for a fixed i results in lFM^i. However, it is more efficient to compute lFM^i at runtime. Each partial FM (mFM, lFM, and l_kFM) has a flag indicating whether the respective activity was part of an end event in the process model, log, or trace.

When node i detects an event e_i, it identifies the node j that detected its predecessor event e_j, e.g., by utilizing the assumed communication infrastructure [17]. Node j forwards the trace's current match and mismatch counts to i. Then, i checks if $mFM^i_j = 1$, to verify whether the directly-follows relation conforms to the model. If it conforms, i updates lFM^i and l_kFM^i accordingly, with k denoting the current case identifier. The match and mismatch counts do not need to be updated if the cell is already set to 1. However, regardless of whether the cell was already set, if there is a mismatch with the model's FM, the node informs a central entity about the mismatch to ensure it is immediately noticeable externally. Figure 2 shows an event-level alert based on the model displayed in Fig. 1. If the cell was not previously set to 1, either the match or mismatch count is incremented accordingly.

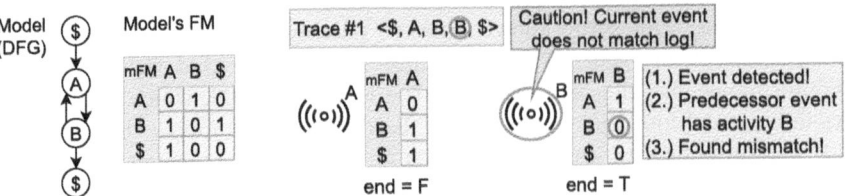

Fig. 1. Model of running example (Fig. 2, 3, 4, 5 and 6) in DFG form, along with its corresponding FM. The matrix is distributed column-wise across all nodes in the network. The model includes start and end activity $.

Fig. 2. Detecting a mismatch at event level. The node associated with activity B detects an event with a predecessor event also corresponding to activity B. It verifies whether the respective cell for activity B in the model's FM mFM_B^B is set to 1. In this example, it is not, prompting the node to alert a central entity about detecting a non-conforming event.

If no predecessor is found, the detected event e_i is identified as a start event. The procedure remains the same, except that the predecessor activity is set to $ and the match and mismatch counts are initialized to 0 for this trace. The node i must update the partial FMs lFM^i and $l_k FM^i$ accordingly. If the activity corresponding to node i is also a start activity in the process model, it increments the match count; otherwise, the mismatch count is incremented.

When a node detects the end event of a trace, it checks the model's end flag, potentially updates the trace and/or log end flag and increases the match or mismatch count. An event is considered an end event if a predefined timer expires without detecting a successor. If it was wrongfully set as an end node, it can rectify its end activity flag, revise its match and mismatch counts, and then forward the counters to the successor. Once the event is confirmed as an end event, the corresponding node computes the trace's fitness using the $fp_{recall} = \frac{|D_L \cap D_M|}{|D_L|}$ measure. Here, the nominator represents the number of directly-follows relations that are identical in the model's and the trace's FM, corresponding to the match count. The denominator describes the total number of distinct directly-follows relations in the trace, i.e., the sum of the match and the mismatch count. See Fig. 3, 4 and 5 for an example run with a model shown in Fig. 1.

This approach to calculate a trace's fitness enables real-time alerts and enhances privacy; at no point are full traces stored in a single location. Each node only possesses knowledge of its own and its predecessor or successor events. Nodes pass only the match and mismatch counts along the trace, aggregating the actual data. The node detecting the end event then uses these counts to calculate the trace's fitness. Depending on the use case, in the event of a mismatch or if a fitness threshold is exceeded, a designated entity may be notified.

DisCC incorporates a sliding window mechanism to manage the increasing number of partial FMs created per node. When the memory reaches a certain threshold and a new case begins, the data from the oldest case is deleted. The

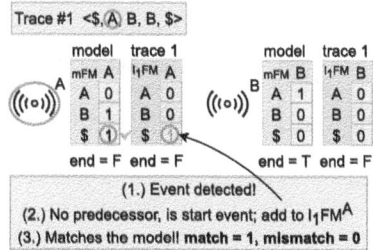

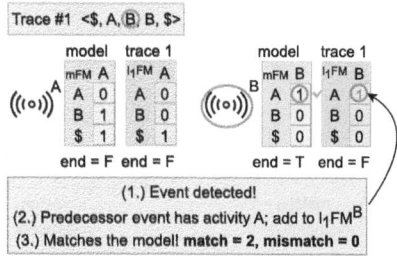

Fig. 3. Trace level example, part 1. Node of activity A detects a start event. It compares its log FM $mFM_\$^A$ with its trace FM $l_1FM_\$^A$. In this case, they conform, i.e., the activity is also a start activity in the model. Thus, the node sets the match count for this trace to 1.

Fig. 4. Trace level example, part 2. Node of activity B detects an event. The predecessor, the node of activity A, forwards the match and mismatch counts to the current node. The fact that B directly follows A conforms with the model. Thus, the node increments the match count.

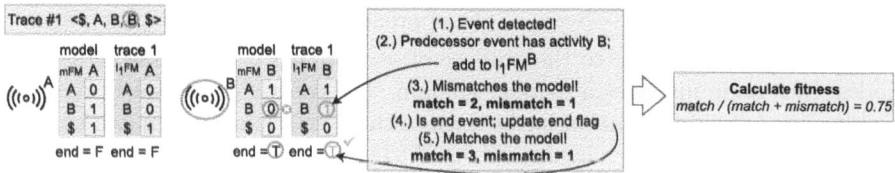

Fig. 5. Trace level example, part 3. The node of activity B detects an event. The predecessor event's activity is B. In the model B does not directly follow B. The node increments the mismatch count. As it is also the end event of the trace, the node verifies whether it also corresponds to an end activity in the model – it does not. The node increments the match count and computes the fitness value of the entire trace.

size of the sliding window is selected based on the sensor node's memory capacity. This method works at the trace level as it only counts mismatches when a trace's directly-follows relation does not exist in the model, and not the other way around. Otherwise, the dependence on a complete event log would be necessary, making trace-wise conformance impractical.

4.5   Conformance on Log Level

The fitness calculations discussed in the previous section occur upon event detection and operate on a trace basis, i.e., focusing on individual traces. Additionally, DisCC enables determining the log-level fitness, which considers all events up to the current point. To retrieve this value, querying the system is necessary: One of possibly multiple central entities requests data from all nodes, i.e., their respective match ($match_i$) and mismatch ($mismatch_i$) counts derived from comparing lFM^i with mFM^i, i denoting the activity identifier of the node; see Fig. 6.

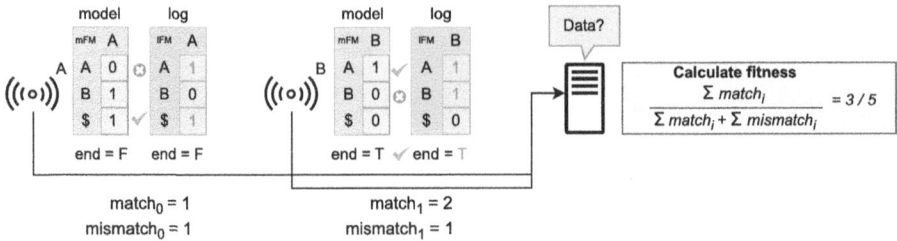

Fig. 6. Fitness computation on log level. On demand, a central entity queries all nodes for their match and mismatch counts of their log's FMs. For this, every cell that is set to 1/True is compared with the model's FM. The central entity sums up all match counts and all mismatch counts and calculates the fitness value of the current log.

They also compare the model's and log's end flags to obtain these counts. Subsequently, the central entity calculates the overall conformance of the entire log using the following formula, with n representing the number of activities in the current log: $\frac{\sum_{i=0}^{n-1} match_i}{\sum_{i=0}^{n-1} match_i + mismatch_i}$. The approach operates exclusively on match and mismatch counts due to the absence of overlapping partial FMs across the nodes; each node manages a distinct column. The central entity solely receives highly aggregated information in the form of these match and mismatch counts.

Note that the approach is not restricted to the fp_{recall} measure proposed by Molka et al. [9] at the log level. Another fitness measure that we can compute is the footprint-based fitness described by van der Aalst [6]. Here, matches and mismatches are defined differently: Let T_L represent all activities in the current log L. Then, $D_{all} := \{(a,b)|a,b \in T_L\}$ denotes the set of all possible tuple combination. Here, matches are defined as $(D_L \cap D_M) \cup ((D_{all}/D_L) \cap (D_{all}/D_M))$, including not only the directly-follows relations that are part of both log and model, but also the absences of direct-follows relations in both log and model. Thus, the fitness is defined as $\frac{D_L \cap D_M \ \cup \ D_{all}/D_L \ \cap \ D_{all}/D_M}{n^2}$. Note that this approach excludes the "no activity" element $; its corresponding row and column are omitted from the calculation.

In general, a distributed, online footprint-based conformance checking design offers several key benefits. First, it enables real-time alerts that allow immediate detection of non-compliant events, enhancing the responsiveness of process monitoring. Second, by pre-processing events locally at each sensor node, the system optimizes data handling efficiency, enhances scalability and makes it suitable for large-scale IoT deployments. Finally, the design prioritizes privacy by avoiding centralized storage of complete process traces, aligning with principles of decentralized data management and confidentiality in distributed systems.

5   Evaluation

We begin our evaluation of DisCC by introducing the datasets we use in our experiments. We then proceed with the experimental analysis.

5.1 Datasets

We use real-world event logs: the Hospital Log [18], Sepsis Cases [19], an IoT-Enriched Smart Factory Log[2] [3], BPI Challenge 2017 [20], and the Road Traffic Fine Management Process [21]. The first three logs are selected due to their inherent spatial distribution, aligning well with our distributed algorithm. The latter two logs are included to extend the breadth of our evaluation. To adapt these logs for conformance checking, we select the top k trace variants, in this case 85% of all variants (empirically chosen), and create a process model from these. As we compare footprint matrices, we construct the model as an FM.

5.2 Experimental Correctness

We first validate the correctness of DisCC by ensuring the computed fitness values match those obtained using the centralized approach by Molka et al. [9]. We validate our approach using both synthetic and real-world datasets.

5.3 Fitness over Time

We now compare the performance of all five datasets on trace-wise fitness, incremental fitness of partial event logs, and overall fitness using a sliding window.

Fitness per Trace. Figure 7 presents the fitness of individual traces across all datasets. For comparison, we normalize the case numbers. Each dataset exhibits distinct behavior. For instance, the Traffic Fine dataset has only one trace with less than 100% fitness, whereas the Hospital Log contains more noise, resulting in a higher number of non-perfect-fitness traces. In such cases, setting a threshold, such as 90%, could help the nodes to filter out noisy but not drastically mismatching traces. The Smart Factory event log shows two dense phases of non-perfect-fitness traces at around 75% and 85%, also impacting the overall and partial log fitness in subsequent analyses.

Incremental Log Fitness. Figure 8 illustrates the fitness of the event logs as the number of traces increases, showcasing DisCC's interim results which can be requested during runtime. These interim fitness values are compared to the centralized approach's final log fitness, demonstrating that while the final results align, trends and dips can be detected in real-time. All logs start with a perfectly fitting trace at 100% fitness. As faulty traces accumulate, the overall fitness declines. However, the addition of following traces with perfect fitness can still cause the curve to rise. Some datasets, such as BPIC 2017, Sepsis Cases and Traffic Fines, quickly converge to the full log's fitness. In contrast, the Hospital Log is characterized by significant noise and diversity, showing no clear convergence pattern. The Smart Factories dataset shows a pronounced dip in fitness at around 75% of the cases, indicating many different traces with non-perfect fitness.

---

[2] We only consider the `MainProcess.xes` file of the not cleaned version.

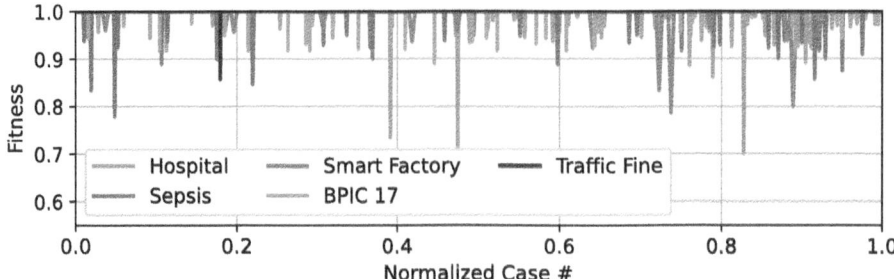

Fig. 7. Fitness for each trace across all datasets. To compare the datasets, we normalize the case numbers. For example, we can see that the Hospital Log is noisy with numerous fitness values below 1.0, and that the Traffic Fine dataset only has a single outlier.

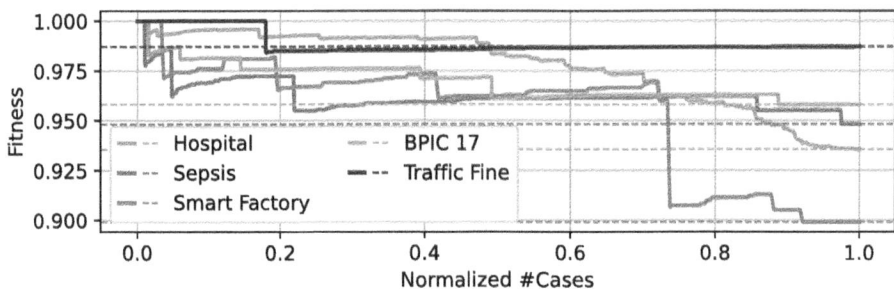

Fig. 8. DisCC's interim fitness results on log level. The figure depicts the evolution of log-level fitness results as the number of cases in the log increases (bold). For each dataset, a dotted line indicates the centralized approach, i.e., the fitness of the full log.

Fitness with Sliding Window. Figure 9 demonstrates the behavior of partial log fitness using a sliding window approach. This method involves analyzing a fixed number of cases at a time and progressively shifting the window. We set the window size to 1% of the total number of cases in the respective log and set the step to 0.5%. For the Traffic Fine dataset, it again shows where the single non-fitting trace is. With a sliding window, however, it does not have a long-term impact on the fitness. A persistently low fitness level over time would suggest a concept drift in the process.

5.4   Discussion

While our approach enhances privacy by eliminating the need for central storage of events, it has certain privacy limitations. At the event level, when a mismatch is detected, a designated entity is notified. Depending on the use case, sensitive information may need to be transmitted to this entity, potentially reducing privacy. At the trace level, if the case identifier is linked to an individual, the fitness values could inadvertently reveal personal information, such as an

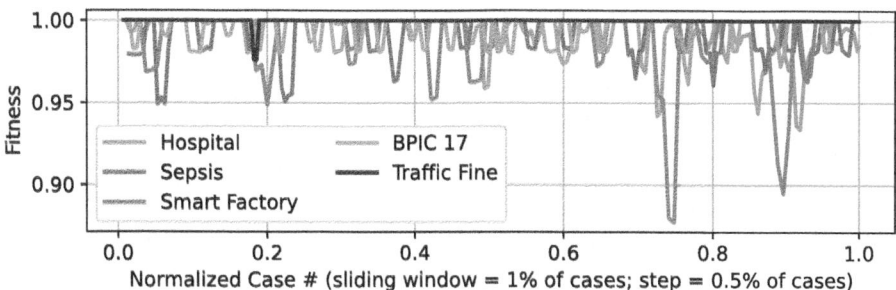

Fig. 9. The figure shows the log's fitness using a sliding window approach with a window size of 1% and a step size of 0.5% of all cases for each dataset. This method focuses on current data and remains unaffected by data from the distant past.

employee's work performance. Similarly, at the log level, if activities are associated with specific individuals or small groups, the designated entity could infer personal information from the match and mismatch counts. These limitations are context-dependent and vary with different use cases.

6   Conclusion

This paper presents Distributed Conformance Checking (DisCC), an approach that addresses the unique challenges of conformance checking in IoT networks. Leveraging a footprint-fitness methodology, DisCC performs online, distributed conformance checks directly at the data sources, ensuring privacy by sharing only aggregated event data with central entities. Our evaluation of DisCC on various real-world datasets demonstrates its ability to accurately perform conformance checks at multiple levels, including event, trace, and log level. DisCC effectively generates real-time alerts for non-conforming events and traces, detects concept drifts and identifies temporary fitness losses, packaged in a scalable and privacy-enabling framework. Further, we demonstrate that DisCC is not restricted to one kind of footprint-based fitness metric on the log level but that the metrics are, in fact, interchangeable. In our evaluation, we verify that the fitness of DisCC matches the fitness of the underlying centralized approach. For datasets without large series of mismatching traces, the fitness value quickly converges to the fitness value of the centralized variant, making it possible to incorporate a sliding window and still achieve meaningful fitness values. This also allows for a more accurate determination of whether a decrease in fitness indicates a concept drift or is just a temporary phenomenon. DisCC utilizes the computational resources of sensor nodes that are neglected in traditional conformance checking algorithms and removes the need for centralized data storage for privacy-enabling conformance checking.

Acknowledgments. This work received funding by the Deutsche Forschungsgemeinschaft (DFG, German Research Foundation), grant 496119880.

References

1. Bertrand, Y., De Weerdt, J., Serral, E.: A bridging model for process mining and IoT. In: Munoz-Gama, J., Lu, X. (eds.) Process Mining Workshops. Springer, Cham (2022)
2. Fernandez-Llatas, C., et al.: Process mining methodology for health process tracking using real-time indoor location systems. Sensors **15**(12) (2015)
3. Malburg, L., Grüger, J., Bergmann, R.: Dataset: An IoT-Enriched Event Log for Process Mining in Smart Factories (2022). https://doi.org/10.6084/m9.figshare.20130794
4. Janssen, D., et al.: Process model discovery from sensor event data. In: International Conference on Process Mining. Springer (2020)
5. Van Der Aalst, W.M.P.: Process Mining: Discovery, Conformance and Enhancement of Business Processes. Springer, Heidelberg (2011)
6. Van Der Aalst, W.: Process Mining: Data Science in Action. Springer (2016)
7. Janiesch, C., et al.: The Internet of Things meets business process management: a manifesto. IEEE Syst. Man Cybern. Mag. **6**(4), 34–44 (2020)
8. Devi, M., et al.: Data science for Internet of Things (IoT). In: Second International Conference on Computer Networks and Communication Technologies: ICCNCT 2019. Springer (2020)
9. Molka, T., et al.: Conformance checking for BPMN-based process models. In: Proceedings of the 29th Annual ACM Symposium on Applied Computing (2014)
10. Adriansyah, A., van Dongen, B.F., van der Aalst, W.M.: Conformance checking using cost-based fitness analysis. In: IEEE 15th International Enterprise Distributed Object Computing Conference, pp. 55–64. IEEE (2011)
11. Rozinat, A., Van der Aalst, W.M.: Conformance checking of processes based on monitoring real behavior. Inf. Syst. **33**(1), 64–95 (2008)
12. van der Aalst, W.M.P.: Process mining: overview and opportunities. ACM Trans. Manag. Inf. Syst. **3**(2), 1–17 (2012)
13. van der Aalst, W.M.P., et al.: Process mining manifesto. In: Business Process Management Workshops: BPM 2011 International Workshops, Clermont-Ferrand, France, Revised Selected Papers, Part I 9, pp. 169–194. Springer (2012)
14. Munoz Gama, J., Carmona, J., van der Aalst, W.M.: Single-entry single-exit decomposed conformance checking. Inf. Syst. **46**, 102–122 (2014)
15. van der Aalst, W.M., Verbeek, H.: Process discovery and conformance checking using passages. Fund. Inform. **131**(1), 103–138 (2014)
16. Leemans, S.J., Fahland, D., Van der Aalst, W.M.: Scalable process discovery and conformance checking. Softw. Syst. Model. **17**, 599–631 (2018)
17. Andersen, J., Rathje, P., Landsiedel, O.: EdgeAlpha: bringing process discovery to the data sources (2024). arXiv:2405.03426 [cs.DB]
18. van Dongen, B.: Real-life event logs - Hospital log (2011). https://doi.org/10.4121/uuid:d9769f3d-0ab0-4fb8-803b-0d1120ffcf54
19. Mannhardt, F.: Sepsis Cases - Event Log (2016). https://doi.org/10.4121/uuid:915d2bfb-7e84-49ad-a286-dc35f063a460
20. van Dongen, B.: BPI Challenge 2017 (2017). https://doi.org/10.4121/uuid:5f3067df-f10b-45da-b98b-86ae4c7a310b
21. de Leoni, M., Mannhardt, F.: Road Traffic Fine Management Process (2015). https://doi.org/10.4121/uuid:270fd440-1057-4fb9-89a9-b699b47990f5

From IoT Event Logs to Human Routines via Community Detection Algorithms

Massimo Callisto De Donato[1]([✉]) (ID), Fabrizio Fornari[1]([✉]) (ID),
and Abel Armas-Cervantes[2] (ID)

[1] University of Camerino, 62032 Camerino, MC, Italy
{massimo.callisto,fabrizio.fornari}@unicam.it
[2] University of Melbourne, Melbourne, VIC 3010, Australia
abel.armas@unimelb.edu.au

Abstract. In our everyday lives, we are exposed to IoT devices that sense the environment and produce raw events. Especially, in smart home scenarios such raw events can be used to detect and monitor human activities and behaviors. This could be of a great help in context that involve fragile people such as elderly or patients, to monitor their condition and support their daily life activities. With our contribution we aim at defining an approach that can be applied on smart-home IoT event logs and support the discovery and monitoring of human routines.

In this work we present our approach that relies on the application of community detection algorithms for the discovery of the routines and the adoption of process mining techniques for their inspection. Especially, we report on a first implementation and validation of the approach with respect to a well known IoT event log.

Keywords: Human Routines · IoT · Event Log · Approach · Community Detection · Process Mining

1 Introduction

Our world is increasingly linked through a large number of connected devices forming what is called the Internet of Things (IoT) [2]. Such devices are typically equipped with sensors and actuators that enable the perception of environmental conditions by collecting and exchanging data via various communication networks. Such data is then analyzed and manipulated to derive possible actions that devices can perform.

While the research in IoT has been growing rapidly, there are still many open challenges related to: the handling of heterogeneous IoT devices, protocols, and data; the different level of granularity in which IoT data is stored [12,20,24]; the integration of IoT in organization business processes [18]; the lack of standardization and interoperability that limits the reusability of the provided IoT solution [10,11,14]; technological barriers, such as the computational limitations of embedded systems and connectivity to back-end systems; and data privacy.

© The Author(s), under exclusive license to Springer Nature Switzerland AG 2025
K. Gdowska et al. (Eds.): BPM 2024 Workshops, LNBIP 534, pp. 113–124, 2025.
https://doi.org/10.1007/978-3-031-78666-2_9

Additional challenges might arise from the specificity of the considered scenario. Considering smart home scenarios, Fig. 1, one of the challenges is to adopt IoT sensors and analyze their produced data for supporting humans e.g., in the case of elderly people or patients, they might need monitoring and support for conducting daily activities [9]. In the case that IoT event logs are available, we can use them as a starting point for the discovery of human routines - patterns of behavior or activities that individuals engage in regularly. However, human routines are personal and might change over a life-span, which makes them non-trivial to discover and monitor.

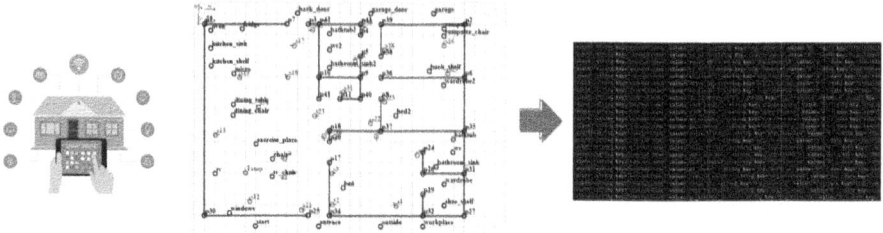

Fig. 1. IoT event log derived from a smart home scenario.

With our contribution we aim at defining an approach that can be applied to smart-home IoT event logs and supporting the discovery and monitoring of such human routines. In this work we present a first version of the approach that relies on the application of community detection algorithms for the discovery of routines and on the adoption of process mining techniques [1], coming from the Business Process Management area [13], for their inspection. Community detection is the task of decomposing a complex network topology into meaningful node clusters, is one of the oldest and most discussed problems in complex network analysis [16]. The intuition behind the definition of our approach is that an IoT event log can be represented as a Directly-Follows Graph, and Community Detection techniques could be applied to such a graph to discover clusters. Such clusters correspond to correlated events that, in the case of an IoT event log of a smart home scenario, correspond to human routines.

The rest of the paper is organized as follows. In Sect. 2 we illustrate and describe the proposed approach. In Sect. 3 we describe in details the implementation and application of the approach on a well-known IoT event log. In Sect. 4 we discuss the results obtained, compare them to a related work and highlight limitations and possible improvements. In Sect. 5 we report conclusions and future works.

2   Discovering PM Event Logs from IoT Logs

In this section we present the conceptualization of our approach, illustrated in Fig. 2, describe each step and the techniques that it relies on. First, we describe the part related to community detection, and then the part related to the actual routine discovery.

Community Detection. Our approach starts from an IoT event log from which we discover a Directly-Follows Graph (DFG), *Step 1* in Fig. 2. The DFG treats events in the IoT log as nodes, and the connections among nodes represent sequential relationships; two nodes are connected if one event is recorded directly after the other. A frequency value can be associated with such relations to count how many times these relations occur.

Starting from the DFG, we aim at identifying groups of events that are related to each other. This task can be related to the Community Detection problem, which involves decomposing a complex network topology into meaningful node clusters. Formally, in the Community Detection problem, given a network G, a community C is defined as a set of distinct nodes: $C = v1, v2, ..., vn$. The community detection problem aims to identify the set C of all communities in G. In our approach, each node of the DFG is a node in G, and a community in the DFG represents a group of events that frequently occur in sequence or are related through similar operational contexts. Therefore, after the application of a Community Detection algorithm, *Step 2* in Fig. 2, a set of communities is detected.

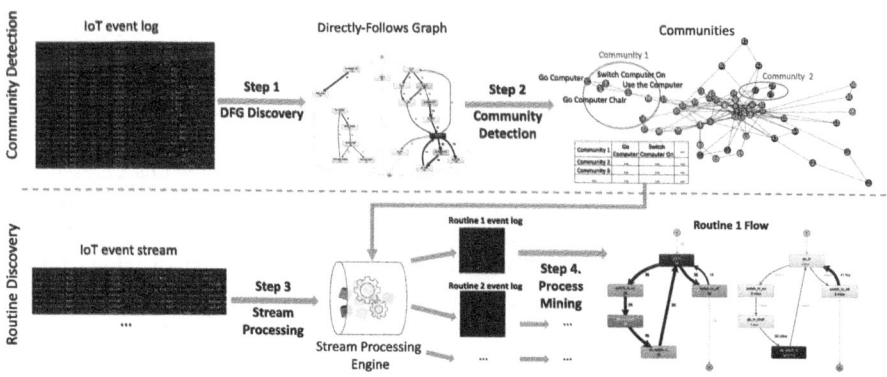

Fig. 2. Illustration of the proposed approach to discover routines.

Routine Discovery. Once events have been grouped into communities, we can use them as a reference to scan an incoming IoT event stream that could be recorded from the same smart-home scenario. This stream of events could be handled by a stream processing engine with the objective of filtering and grouping together events that appertain to the same community, constructing an event log that corresponds to a routine, *Step 3* in Fig. 2.

Based on the obtained event logs, standard process mining techniques can be applied to support the user in analyzing the discovered routines [15], *Step 4* in Fig. 2. For instance, performance metrics such as average processing times, waiting times, and the frequency of different paths can be used to detect the person's behavior during the observed period, including which activities are performed and for how long. Conformance checking techniques can also be applied to establish whether a given routine matches the expected one, highlighting anomalies and deviations from the norm.

3   Implementation of the Approach on a IoT Event Log

In this section we describe the implementation and application of the approach illustrated in Sect. 2. We apply our approach to a well-known IoT event log made available for the challenge proposed during the 4th International Workshop on the BP-Meet-IoT[1] aimed at exploring the interplay between IoT and process mining. The code that we developed for implementing our approach is accessible from the PROS Lab website[2].

3.1   IoT Event Log

The IoT event log that we considered refers to the event data collected from a smart-home environment. The dataset represents a realistic set of simulated logs of a smart home, describing habits, activities, and actions performed by humans inside a physical environment, together with the sensor measurements detected. The scenario named *sim_ 22d1p* describes the daily activities of a single user for 21 consecutive days (three consecutive weeks). Three XES[3] files are distributed: a *Sensor Log* (*SensorLog.xes*), which refers to the lowest level log reporting measurements from sensors; an *Event Log* (*EventLogXESNoSegment.xes*) that lists IoT events describing atomic actions performed by a human (e.g., *go_ computer*, *switch_ computer_ on*) and a *Segmented Event Log* (*EventLogXES.xes*) that reports the same events but grouped based on the routine that they relate to, e.g., the *use_ computer* routine groups events such as: *go_ computer*, *switch_ computer_ on*, *go_ computer_ chair*, *use_ the_ computer*, *switch_ computer_ off*. The IoT event log reports 17 routines repeated a total of 348 times over 21 days. We inspected the *Segmented Event Log* and reported in Table 1 the list of routines along with their corresponding events.

Some of these routines describe activities that last for an extended period of time during the day. For example, *Go work* usually occurs during the week, starting around 8:00 AM and continuing until returning home after 6:30 PM. More frequent routines include *Drinking*, which can occur several times in the

---

[1] https://bp-meet-iot.webs.upv.es/2020.

[2] https://pros.unicam.it/RoutineDetector.

[3] https://xes-standard.org/.

evening, both before and after meals. We used the extracted information as the ground truth to evaluate the communities identified with the Community Detection algorithms.

Table 1. Routine identified within the Segmented Event Log.

Routine	Events
Cook and Eat	go_fridge, get_ingredients_from_fridge, go_kitchen_shelf, get_ingredients_from_shelf, go_oven, use_oven, go_dining_table, eat_warm_meal, pack_food, put_meal_to_fridge, go_kitchen_sink, put_plate_to_sink
Prepared sleep	go_wardrobe, get_clothes, go_bathtub, have_bath, go_bathroom_sink, brush_teeth, go_bed, sleep_in_bed, change_clothes, go_to start
Eat cold	go_bathroom_sink, wash_hands, go_fridge, get_food_from_fridge, go_kitchen_shelf, get_bread, go_dining_table, eat_cold_meal, go_kitchen_sink, put_plate_to_sink
Eat warm	go_bathroom_sink, wash_hands, go_fridge, get_cold_warm_food, go_micro, use_micro, go_dining_table, eat_warm_meal, go_kitchen_sink, put_plate_to_sink
Shop	go_wardrobe, change_clothes, go_shoe_shelf, dress_up_outdoor, go_outside, get_food, dress_down_outdoor, go_kitchen_shelf, pack_goods
Go work	go_wardrobe, change_clothes, go_shoe_shelf, dress_up_outdoor, go_workplace, work, dress_down_outdoor, finish_walk
Do walk	go_wardrobe, change_clothes, go_shoe_shelf, dress_up_outdoor, go_outside, walk_outside, dress_down_outdoor, finish_walk
Drink	go_kitchen_shelf, get_glass, go_kitchen_sink, get_water, drink_water
Use computer	go_computer, switch_computer_on, go_computer_chair, use_the_computer, switch_computer_off
Watch tv	go_tv, switch_tv_on, go_tv_chair, do_watch_tv, switch_tv_off
WC	go_wc, wc_do, wc_flush, go_bathroom_sink, wash_hands
Airing home	go_windows, raise_blinds, open_windows
Airing home end	go_windows, close_windows, lower_blinds
Someone at entrance	go_entrace, interact_with_man
Exercise	go_exercise_place, do_exercise
Do the dishes	go_kitchen_sink, wash_dishes
Rest	go_chair, rest_in_chair

3.2 Community Detection from the IoT Event Log

We applied the proposed approach starting from the previously described IoT *Event Log*. Before proceeding with the first step of the approach, we conducted some pre-processing which regarded the removal of so called "noise" events. We did so by filtering out events starting with the name *noise_*. After that, to discover a DFG from historical IoT events we used PM4PY[4], an open-source library written in Python that facilitates the handling and manipulation of event logs [4]. PM4PY provides various features and algorithms for working within the realm of process mining, including import, manipulation, and visualization

---

[4] https://pm4py-source.readthedocs.io/en/stable/index.html.

functionalities that facilitate the analysis of event logs. In our approach, we use PM4PY to import the event logs, and construct the corresponding DFG object by applying a standard algorithm from the library.

To discover the communities from the DFG object we used CDlib[5], a comprehensive Python library designed to support network clustering. CDlib offers different algorithms for extracting, comparing, and evaluating communities from complex networks [21]. To apply community detection algorithms with CDlib we have to translate the DFG into a NetworkX graph G. The set of nodes in G corresponds to the set of events in the DFG object and an edge between two nodes in G is added if the corresponding events in the DFG object are subsequent events. The weights of the edges in G correspond to the number of occurrences of this sequential relationship.

For selecting the most appropriate Community Detection algorithms we conducted an evaluation based on the F1 score obtained by comparing the communities identified by each available algorithm against the ground truth, reported in Table 1, generated from the *Segmented Event Log*. The *infomap* algorithm resulted in the highest F1 score (0.711) with 15 communities recognized. Infomap is a crisp algorithm that leverages principles from information theory. It models the movement of random walks within a network to represent the flow of information in the actual system. By analyzing this probability flow, the algorithm partitions the network into modules, effectively compressing the representation of the information flow [22]. After applying the infomap algorithm from CDlib we obtained the list of detected communities reported in Table 2. We report in Sect. 4 an extensive discussion on the obtained results.

3.3   Human Routine Discovery from the IoT Event Log

To filter a stream of IoT event logs based on the detected communities, we implemented a streaming analytics job using the open-source distributed processing engine Apache Flink[6]. The job is configured to load the .txt file containing the communities. Then, the job streams the *Event Log* which is filtered against the communities and generates the filtered event logs. In this initial implementation, we used the *Event Log* to replicate the event stream.

To identify instances of each routine from the filtered event logs, we consider a time window of 24 h starting at 03:00 a.m. We observed that some routines could start before midnight and end after midnight. For our case, we found it more appropriate to choose such a time window to avoid recording events that pertain to the same routine but appear in different traces of the event log. A similar splitting strategy has also been adopted in [19].

For inspecting each routine in a more user-friendly way, we applied a standard process mining tool such as Disco[7] to mine the process maps. In Fig. 3, we report an example of a discovered process map that an interested user could inspect to derive insights on a specific routine. Such a process map refers to community

---

[5] https://cdlib.readthedocs.io/en/latest/index.html.

[6] https://flink.apache.org/.

[7] https://fluxicon.com/disco/.

Table 2. List of the communities identified using the infomap algorithm.

ID	Community Events
1	go_wardrobe, change_clothes, go_shoe_shelf, dress_up_outdoor, go_workplace, work, dress_down_outdoor, finish_walk, go_outside, get_food, walk_outside
2	go_kitchen_shelf, get_glass, go_kitchen_sink, get_water, drink_water, put_plate_to_sink, pack_goods, wash_dishes
3	go_bathroom_sink, switch_computer_off, go_wc, wc_do, wc_flush
4	wash_hands, go_fridge, get_food_from_fridge, get_ingredients_from_fridge, put_meal_to_fridge
5	go_tv, switch_tv_on, go_tv_chair, do_watch_tv, switch_tv_off
6	go_windows, raise_blinds, open_windows, close_windows, lower_blinds
7	brush_teeth, go_bed, sleep_in_bed, go_to start
8	go_computer, switch_computer_on, go_computer_chair, use_the_computer
9	get_bread, go_dining_table, eat_cold_meal, eat_warm_meal
10	get_ingredients_from_shelf, go_oven, use_oven, pack_food
11	get_clothes, go_bathtub, have_bath
12	get_cold_warm_food, go_micro, use_micro
13	go_entrace, interact_with_man
14	go_exercise_place, do_exercise
15	go_chair, rest_in_chair

5 which groups events related to the activity of watching TV. While the left part of Fig. 3 describes how many times the events forming the routine occurred and in which order, by looking at the right part, the user can observe that the subject watches TV for an average of 112.6 min a day. Useful information can be derived based on the characteristics of the person. For example, if the person is sedentary with obesity issues, ensuring that the amount of time spent watching TV is kept below a certain threshold could improve his quality of life.

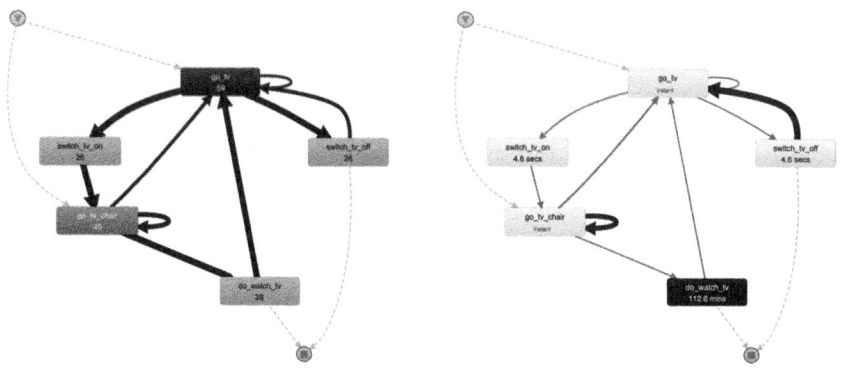

Fig. 3. Process map of the discovered routine *watch tv* showing the absolute frequency of the event (on the left) and the mean duration of the event (on the right).

4   Discussion

In this section, we discuss the results obtained by applying our approach to the IoT scenario.

Comparison with Ground Truth. First, we inspected the communities that were discovered and compared them with those extracted from the segmented file. Then, we compared the process maps obtained from the filtered event logs with those of the ground truth. To compare the ground truth and the detected communities the reader can refer respectively to Table 1 and Table 2. It can be noticed that some communities match perfectly. For example, community 15 groups all events that describe the person's resting which relate to the routine *Rest*. Community 14 groups events about physical exercise activities that relate to the routine *Exercise*. Community 13 groups events involving interaction with another person, which relate to the routine *Someone at Entrance*, while community 5 groups events of watching TV, which relate to the routine *Watch TV*. In some cases, the discovered communities miss events. For example, community 8 groups events related to computer usage that relate to the routine *Use computer*. However, the event *switch_computer_off* has not been included. Such event has been included in community 3, which mainly groups events related to the routine *WC*. The anomalies we observed in communities 8 and 3 demonstrate that the algorithm does not always group events perfectly. This can lead to errors in recognizing routines during the analysis of event log streams. For better visualizing this issue, we display the process map obtained by filtering for community 8 and community 3, respectively, in Fig. 4 and Fig. 5.

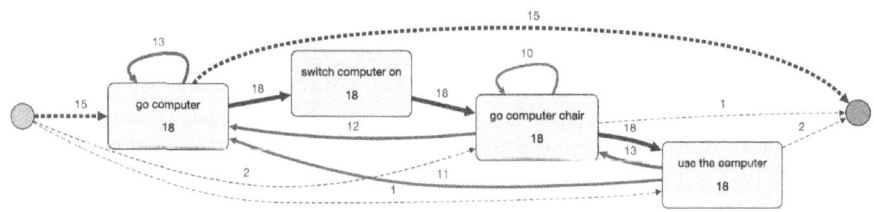

Fig. 4. Process map of the discovered routine *Use Computer*.

To solve such an issue we might consider implementing a hybrid approach involving an expert user or the subject that performs those routines. The user could participate in steps 2 and 3 of our approach, with the task of inspecting and manually adjusting the identified groups of events that form the communities as soon as they are detected. However, this might be possible only with IoT event logs that report events labeled in a human readable and comprehensible manner such as in the case we considered.

In this first implementation of the approach, we adopted the infomap algorithm which does not associate a node with more than one community simultaneously. However, in an IoT event log, this situation can occur; therefore we plan

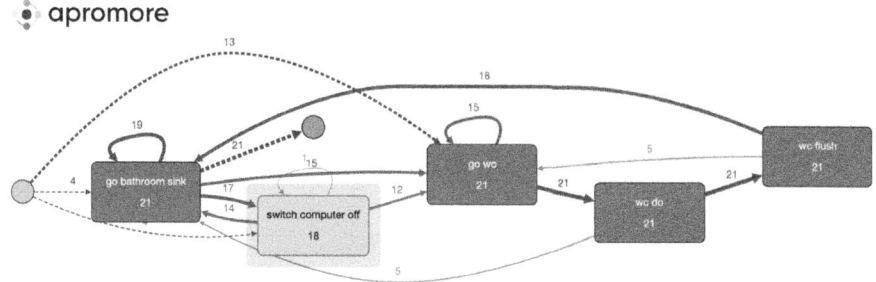

Fig. 5. Process map of the discovered routine *WC*.

to further investigate the choice of the community detection algorithms and the
metrics used for comparing them.

Comparison with a Related Work. In de Leoni et al. [19] an approach is presented that operates on the same IoT event logs, converts them into vectors via
frequency-based encoding and duration-based encoding, and then applies and
evaluates KMeans and DBScan as clustering algorithms to support the learning
of daily human habits. Despite we share the objective of grouping IoT events,
we do that differently by applying Community Detection algorithms starting
from a Directly-Followed Graph as described in Sect. 2 and Sect. 3. In addition,
in [19] no actual comparison with a ground truth is provided and the number
of detected routines is 10 out of 17 which is less then our 15 detected routines.
Especially, comparing the model of routine *go_ bed* that is reported in their
contribution, Fig. 6 here, with the one we obtained, Fig. 7, we can spot major
differences. In fact, their approach includes in the *go_ bed* routine several events
that are not related to that routine such as: *eat_ cold_ meal, go_ kitchen_ sink,
go_ entrance, interact_ with_ man, put_ plate_ to_ sink*. With the same application of the Inductive miner algorithm, our approach instead allowed discovering
a more precise routine with events that have a clear connection with the *go bed*
routine.

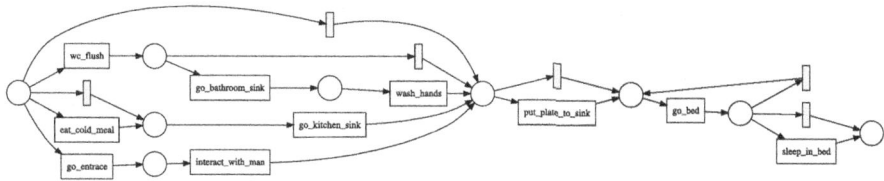

Fig. 6. Petri Net representing the go_bed routine from [19].

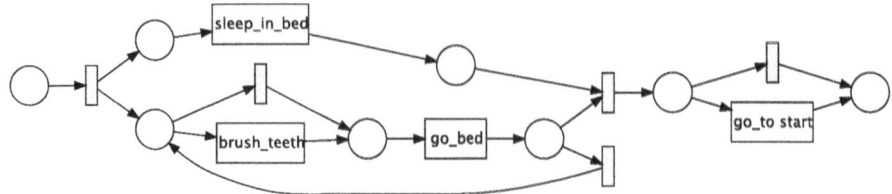

Fig. 7. Petri Net obtained from the community 7 associated to *go_ bed*.

5   Conclusion and Future Work

In this paper, we introduced an approach to detect human routines from a IoT event logs. A Directly-Follows Graph is mined from the IoT event log and given as input to a Community Detection algorithm to detect event logs related to each other. The intuition in this approach is that the discovered communities helps to automatically identify patterns in the streams of IoT event logs corresponding to human routines. To do so, we employed a stream processing engine to reconstruct the event log of each community and create separate event logs. Those event logs can be analyzed via standard Process Mining techniques and tools to obtain process maps describing human routines that can be, in turn, employed to monitor the condition of the human, to make decisions and plan interventions. We applied our approach on a well-known IoT event log and discussed the obtained results. We also provided a preliminary comparison with one, of the already available approaches [5], that applies clustering algorithms. The obtained results as discussed in Sect. 4 are promising and despite some reported limitations, the approach can provide support to those that want to inspect a stream of IoT event log for monitoring human routines.

As future work, to test our approach in different contexts we plan to apply it on other event logs, even creating realistic datasets for various scenarios, from smart homes and schools [6]. Especially, we plan to set up a real test environment within the project called VITALITY[8], a research project funded by the National Recovery and Resilience Plan (Piano Nazionale di Ripresa e Resilienza, PNRR), which focuses on the innovation and safety of living environments in the digital transition era. Our approach can support the project's activities related to the identification of inhabitants' activities in contexts, such as homes, and the related habits that can contribute to fostering their well-being in various ways. In this case a further effort will be required for the abstraction of low level sensor data acquired from the real environments to event logs. To handle the abstraction process various approaches exist that could be investigated [24].

While waiting for a real scenario to be available, we plan also to simulate realistic environments with contextual data. For this case we could investigate the application of simulation techniques such as in [23]. In particular we also plan to evaluate the role that our approach could have in a Digital Twin context such

---

[8] https://vitality-spoke6.unicam.it/en/.

as the one reported in [7,8] in which we propose a solution for the representation and simulation of IoT devices and the scenarios they are involved in.

In our implementation we mined process maps to graphically show some of the resulting routines and for comparison with another approach we applied the inductive miner to discover a Petri Net. In the future, we could apply other process mining algorithms and evaluate the resulting models of the routine together with an evaluation of precision recall to assess the accuracy of process models automatically discovered [3]. In addition, in this first version of the approach, we did not tackle directly the problem of deriving a label to assign to an identified community. However, we could try investigating different techniques such as the one adopted in [19] in which the identified clusters are named with the dominant activity name.

To conclude, recent research reports on the potential of process mining for assessing and analysing sustainability in business processes [17]. Monitoring human routines through process mining and combining them with energy consumption data could also raise awareness regarding the costs and environmental impact of human routines. This could translate into suggestions to improve such routines to reduce costs and the ecological footprint.

Acknowledgment. This work has been partially supported by the European Union – NextGenerationEU - National Recovery and Resilience Plan (NRRP), Mission 4 Education and Research - Component 2 From research to business - Investment 1.5, ECS_00000041- VITALITY - CUP J13C22000430001.

References

1. van der Aalst, W.M.P., Adriansyah, A., et al.: Process mining manifesto. In: International Conference on Business Process Management, pp. 169–194. Springer (2011)
2. Ashton, K., et al.: That 'Internet of Things' thing. RFID J. **22**(7), 97–114 (2009)
3. Augusto, A., Armas-Cervantes, A., Conforti, R., Dumas, M., Rosa, M.L.: Measuring fitness and precision of automatically discovered process models: a principled and scalable approach. IEEE Trans. Knowl. Data Eng. **34**(4), 1870–1888 (2022)
4. Berti, A., van Zelst, S.J., Schuster, D.: Pm4py: a process mining library for python. Softw. Impacts **17**, 100556 (2023)
5. Bertrand, Y., den Abbeele, B.V., Veneruso, S., Leotta, F., Mecella, M., Serral, E.: A survey on the application of process discovery techniques to smart spaces data. Eng. Appl. Artif. Intell. **126**, 106748 (2023)
6. Callisto De Donato, M., Corradini, F., Fornari, F., Re, B.: Safe: an ICT platform for supporting monitoring, localization and rescue operations in case of earthquake. Internet of Things **27**, 101273 (2024)
7. Callisto De Donato, M., Corradini, F., Fornari, F., Re, B., Romagnoli, M.: Design and development of a digital twin prototype for the SAFE project. In: EDOC 2023 Workshops. LNBIP, vol. 498, pp. 107–122. Springer (2024)
8. Callisto De Donato, M., Corradini, F., Fornari, F., Re, B., Romagnoli, M.: Enabling 3D simulation in thingsboard: a first step towards a digital twin platform. In: EDOC 2023 Workshops. LNBIP, vol. 498, pp. 325–330. Springer (2024)

9. Corradini, F., Angelis, F.D., Re, B., Anceschi, E., Callisto De Donato, M., Iddas, P.: Private assisted house for smart living. J. Ambient Intell. Smart Environ. **9**(6), 725–741 (2017)

10. Corradini, F., Fedeli, A., Fornari, F., Polini, A., Re, B.: FloWare: a model-driven approach fostering reuse and customisation in IoT applications modelling and development. Softw. Syst. Model. **22**(1), 131–158 (2023)

11. Corradini, F., Fedeli, A., Fornari, F., Polini, A., Re, B., Ruschioni, L.: X-IoT: a model-driven approach to support IoT application portability across IoT platforms. Computing **105**(9), 1981–2005 (2023)

12. Culmone, R., Falcioni, M., Giuliodori, P., Merelli, E., et al.: AAL domain ontology for event-based human activity recognition. In: MESA 2014, pp. 1–6. IEEE (2014)

13. Dumas, M., Rosa, M.L., Mendling, J., Reijers, H.A.: Fundamentals of Business Process Management, 2nd edn. Springer (2018)

14. Fedeli, A., Fornari, F., Polini, A., Re, B., Torres, V., Valderas, P.: FloBP: a model-driven approach for developing and executing IoT-enhanced business processes. Softw. Syst. Model. 1–30 (2024)

15. Federico, G.D., Fernández-Llatas, C., Ahmadi, Z., Shirali, M., Burattin, A.: Identifying variation in personal daily routine through process mining: a case study. In: ICPM 2023 Workshops. LNBIP, vol. 503, pp. 223–234. Springer (2023)

16. Fortunato, S.: Community detection in graphs. Phys. Rep. **486**(3–5), 75–174 (2010)

17. Graves, N., Koren, I., van der Aalst, W.M.P.: Rethink your processes! a review of process mining for sustainability. In: International Conference on ICT for Sustainability, ICT4S 2023, Rennes, France, 5–9 June 2023, pp. 164–175. IEEE (2023)

18. Janiesch, C., et al.: The Internet of Things meets business process management: a manifesto. IEEE Syst. Man Cybern. Mag. **6**(4), 34–44 (2020)

19. de Leoni, M., Pellattiero, L.: The benefits of sensor-measurement aggregation in discovering IoT process models: a smart-house case study. In: International Conference on Business Process Management, pp. 403–415. Springer (2021)

20. Mangler, J., et al.: From Internet of Things data to business processes: challenges and a framework. arXiv preprint arXiv:2405.08528 (2024)

21. Rossetti, G., Milli, L., Cazabet, R.: CDLIB: a Python library to extract, compare and evaluate communities from complex networks. Appl. Netw. Sci. **4**(1), 52:1–52:26 (2019)

22. Rosvall, M., Bergstrom, C.T.: Maps of random walks on complex networks reveal community structure. Proc. Natl. Acad. Sci. **105**(4), 1118–1123 (2008)

23. Veneruso, S., Bertrand, Y., Leotta, F., Serral, E., Mecella, M.: A model-based simulator for smart homes: enabling reproducibility and standardization. J. Ambient Intell. Smart Environ. **15**(2), 143–163 (2023)

24. van Zelst, S.J., Mannhardt, F., de Leoni, M., Koschmider, A.: Event abstraction in process mining: literature review and taxonomy. Granular Comput. **6**, 719–736 (2021)

Machinery Activity Recognition in the Industry Based on Heterogeneous Data

Marta Podobińska-Staniec[1]([⊠]) [iD], Marek Kęsek[1] [iD], and Edyta Brzychczy[2] [iD]

[1] Faculty of Civil Engineering and Resource Management, AGH University of Krakow, Kraków, Poland
{mstaniec,kesek}@agh.edu.pl

[2] Faculty of Mechanical Engineering and Robotics, AGH University of Krakow, Kraków, Poland
brzych3@agh.edu.pl

Abstract. Nowadays, machinery is getting much more complex and equipped with various sensors providing data about actual process execution. One of the directions of machinery data usage is activity recognition, which can be used in process modeling and analysis, enabling process improvements. We aimed to summarize machinery activity recognition (MAR) implementations in the industry. We formulated four research questions concerning MAR applications in the industrial domains, data sources, analytic approaches, and techniques. We carried out a Systematic Literature Review based on renowned publication databases. We started with 812 papers from Scopus and ISI Web of Science, and after detailed analysis, we finally ended with 29 papers used for data extraction. We discovered that the MAR is a relatively common data-oriented task in the construction industry and can also be noticed in other domains like mining, logistics, and medicine, proving that this kind of analytics has wide applications. Due to the nature of MAR, many papers present a supervised approach with various classifiers, among them, one can find neural networks as the most popular and effective techniques.

Keywords: Activity Recognition · Machinery · Sensor data · Process analysis

1 Introduction

Machinery and equipment play an essential role in industrial processes, especially in heavy industries. Their high performance and efficient use are the main goals of the maintenance process and have a substantial impact on the processes in which these machines are involved.

In Industry 4.0 conditions, machinery becomes much more complex, equipped with various sensors that provide data about actual process execution performed with its usage [1]. The data from sensors can be used in two ways. Firstly, data can be analyzed using data-oriented techniques to analyze statistics of parameters and dependencies of parameters and predict machinery state (degradation) or fault diagnostics. In this group, one can also consider various operational reports, including business statistics (e.g., Power BI dashboards). Secondly, data can be analyzed with process-oriented methods

© The Author(s), under exclusive license to Springer Nature Switzerland AG 2025
K. Gdowska et al. (Eds.): BPM 2024 Workshops, LNBIP 534, pp. 125–137, 2025.
https://doi.org/10.1007/978-3-031-78666-2_10

focusing on analyzing process execution as machinery activities workflow [2]. Data-oriented methods present relatively static information about machinery usage (in the form of reports), contrary to process-oriented methods, providing dynamic analyses of machinery behavior. However, both approaches and conducting static or dynamic analyses of machinery behavior require translating low-level data from machinery monitoring systems into higher-level activities describing machinery behavior. For this purpose, machinery activity recognition (MAR) is needed.

Activity recognition is quite widespread in human behavior analysis, known as human activity recognition topic (HAR). HAR is widely used in various domains due to sensor-based analytics enabling cutting-edge technologies to detect, recognize, and monitor human activities [3]. Main HAR applications include healthcare, ambient assisted living, gesture recognition, behavior analysis, sports, and event analysis, as well as Human-Computer Interaction (HCI) [3, 4].

There are common challenges in MAR and HAR related to data collection and data processing methods. The general classification of data sources in HAR comprises vision-based and sensor-based sources [3]. Similarly, in MAR, computer vision and sensor-based sources (including kinematic-based and audio-based) are used [5]. The main challenges in data collection are related to the volume of data (e.g., in the case of vision-based data) as well as the cost of data sources (e.g., sensors vs. vision-based capturing devices) [3]. The most important challenges in data processing include heterogeneous sensor data representation, meaningful feature extraction, imbalanced and overlapping data classes, and accuracy and robustness in activity recognition [6].

Monitoring of processes and machinery activities analysis is crucial for process management and its improvement, and various analytic approaches can be applied to use machinery data in this scope efficiently. Thus, our research investigates MAR applications in industrial settings with a systematic literature review.

We defined the following research questions (RQs):

RQ1: Which industries present applications of MAR?
RQ2: What kind of data sources are dominant in MAR?
RQ3: What kind of techniques are commonly used in MAR?
RQ4: What analytic approach is dominant in the use of MAR results?

Answers to RQs enable us to summarize the state of the art in using MAR described in the literature and provide guidelines for MAR conducting in the industry based on the described practices.

Our paper is structured as follows: Sect. 2 introduces the systematic literature review (SLR) and SLR protocol used in our research. Section 3 presents a summary of the conducted SLR with answers to RQs. Section 4 provides a summary of practices for MAR in industry domains. The final section covers the conclusions and highlights future works.

2 Materials and Methods

2.1 Systematic Literature Review (SLR)

We prepared our SLR based on the guidelines presented in [7]. We developed a review protocol comprising (1) a literature search including a selection of databases for the searching process, a definition of keywords, a sampling strategy, and results refinement, (2) screening for exclusion with defined criteria, (3) quality assessment with defined criterion and (4) data extraction.

Details of our protocol are as follows:

1. Literature search:

   - Databases for searching—as the primary source, we selected Scopus, the largest database of peer-reviewed literature, and ISI Web of Science (WoS) for cross-checking search results.
   - Keywords - we defined the following words and phrases for the searching process: ("activity recognition" OR "state recognition") AND ("machinery" OR "equipment").
   - Sampling strategy - we assumed (a) searching of keywords in title, abstract, or authors keywords, (b) we considered journal papers, review papers, conference papers, and book chapters.
   - Results refinement - we refined the results by (a) removing duplicates, (b) removing papers written in languages other than English, (c) in case of papers published in a journal and at the conference, we preferred the journals, (d) since in Scopus indexed keywords are also taken into account in search, we checked the results once again with the defined phrase.

2. Screening for exclusion - we defined the following topics and reasons to exclude the papers based on abstract screening: (a) HAR, (b) fault diagnosis, (c) machinery health monitoring (HM) or structural health monitoring (SHM), (d) abstract not related strictly to MAR and (f) full paper is not accessible.
3. Quality assessment - we defined the main quality criterion: a paper contains an example of industrial machinery activity recognition.
4. Data extraction - we defined the following categories to describe the selected papers: (a) Industry (construction, mining, medicine,...), (b) Machinery type, (c) Data sources (sensor-based (audio, kinematic, other), vision-based (video, image), (d) Type of sensor or collection device (e.g., accelerometer, camera, microphone), (e) Data types (numerical, categorical, mixed), (f) Used methods (supervised, semi-supervised, unsupervised), (g) Used techniques (type of classifiers, type of clustering etc.), (h) Analytic approach for MAR results usage (data-oriented, process-oriented).

All screening procedures in SLR were conducted independently by two researchers. In case of doubts, the third person, who was not involved in the paper assessment, made the final decision.

2.2 SLR Statistics in Our Study

SLR statistics obtained after the implementation of the protocol review are presented in Fig. 1.

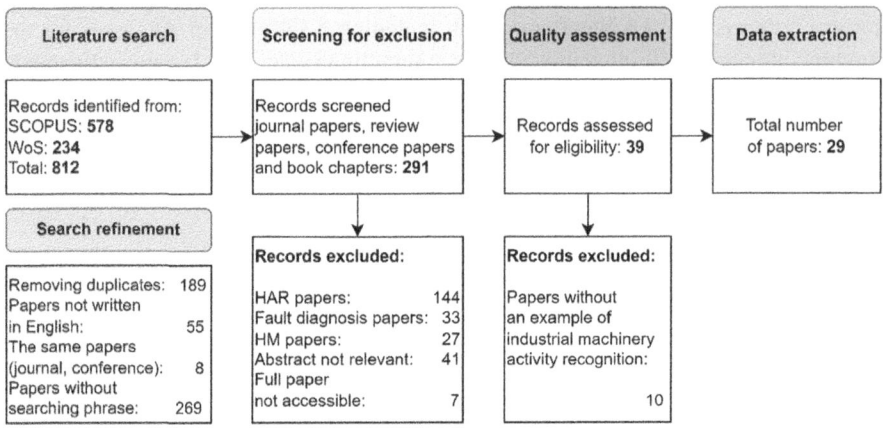

The results of the papers' analysis, considering formulated RQs, are presented in Sect. 3.

3 Results and Discussion

This section summarizes the paper dataset and findings concerning formulated RQs.

3.1 General Characteristics of the Dataset

Our final dataset contains 29 papers. Their distribution in time and types of publication are presented in Fig. 1.

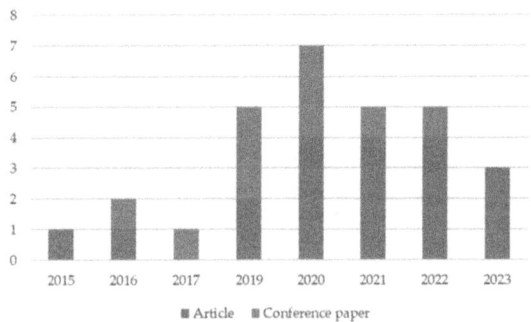

Fig. 1. Paper distribution in time with a type of publication

In the dataset, we have 19 journal articles and 10 conference papers. The most popular journals include Automation in Construction (5 articles) and Algorithms (2).

Conference papers have been published mainly in Computing in Civil Engineering (3 papers) and the International Symposium on Automation and Robotics in Construction (3). The initial analysis of the publishers suggests that our dataset has a predominance of MAR examples in the construction industry. One can see growing interest in the MAR topic since 2019, with the top in 2020 (7 papers). In the following years, the number of papers is quite similar (4–5).

The presented analysis confirms initial insights about the dominant content of MAR examples in the construction industry. We also observed the usage of machine learning models for activity recognition. One of the main goals of MAR presented in the papers is to analyze performance and productivity. Table 1 summarizes data extraction from selected papers.

Table 1. The main characteristics of papers in our dataset

Ref.no	Industry	Machinery types	Data sources[a]	Data type[b]	Used methods[c]	Approach[d]
[8]	construction	loader	SB	N	S	D
[9]	industrial forestry	excavator mastication	SB	M	S	D
[10]	construction	excavator	VB	M	S	D
[11]	construction	excavator	VB	M	S	D
[12]	logistics	forklift trucks	SB	M	S	D
[13]	logistics	forklift trucks	SB	M	SE	D&P
[14]	construction	loader, excavator, bulldozer	SB	N	S	D
[15]	construction	compactor	SB	N	S	D
[16]	construction	loader, excavator, compactor	SB	N	S	D
[17]	construction	rotary drilling rigs	SB	N	S	D
[18]	construction	forklift trucks	SB	M	SE	D
[19]	mining	loader	SB	M	U	P
[20]	construction	lifting machine	SB	N	S	D

(continued)

Table 1. (*continued*)

Ref.no	Industry	Machinery types	Data sources[a]	Data type[b]	Used methods[c]	Approach[d]
[21]	construction	lifting machine	SB	N	S	D
[22]	construction	excavator, compactor, dozer, breaker	SB	N	S	D
[23]	construction	excavator	VB	M	S	D
[24]	construction	excavator	SB & VB	M	S	D
[25]	construction	excavators, roller	SB	N	S	D
[26]	construction	excavator	SB	M	S	D
[27]	construction	excavator	SB & VB	M	S	D
[28]	construction	excavator	SB & VB	M	S	D
[29]	construction	loader, excavator	SB	M	S	D
[30]	construction	excavator	SB	N	S &U	D
[31]	medicine	surgery robot	VB	M	S	D
[32]	construction	loader, truck	VB	M	S	D
[33]	construction	loader, excavator, bulldozer, dozer, lifting machine	VB	M	S	D
[34]	construction	roller	SB	M	S	D
[35]	medicine	surgery robot	SB & VB	N	S	D
[36]	mining	shearer	SB	N	S	D

[a] SB – sensor-based, VB – vision-based, [b] N – numerical, C-categorical, M – mixed, [c] S – supervised, U – unsupervised, SE – semi-supervised, [d] D – data-oriented, P – process-oriented

3.2 Industrial Application of MAR (RQ1)

Aiming to answer the first RQ, we analyzed the kinds of industries present in the paper dataset. The most prevalent industry in our dataset is construction and civil engineering (22 papers). Other works describe MAR in mining (2), medicine (2), logistics (2), and industrial forestry (1). Given the diversity of applications and specificity of these industries, one can state that MAR can be applied in various industries and processes, not only in heavy types.

The types of machinery that are MAR subjects are strongly related to the industry branch (Fig. 2).

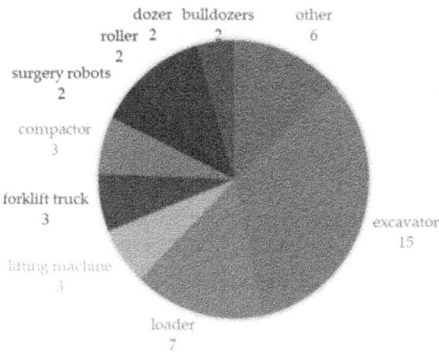

Fig. 2. Type of machinery used in MAR

The most often analyzed machines were excavators and loaders, mainly in the construction industry. However, they are also used in the mining and forestry industries. The forklift trucks were analyzed mainly in logistics. Obviously, surgery robots were subject to MAR in medicine applications. Other machines not presented in the figure include rotary drilling rigs or breakers (also used in the construction industry) and shearers in the mining industry.

3.3 Data Sources in MAR (RQ2)

Regarding the second RQ, we analyze the data sources and type of sensors presented in the paper dataset.

MAR in analyzed papers relied primarily on sensor-based sources (19 papers). Vision-based sources are present in 6 papers; however, we also noticed mixed usage of vision-based and sensor-based sources in 4 papers. In Fig. 3, we summarized the most often used sensors in MAR.

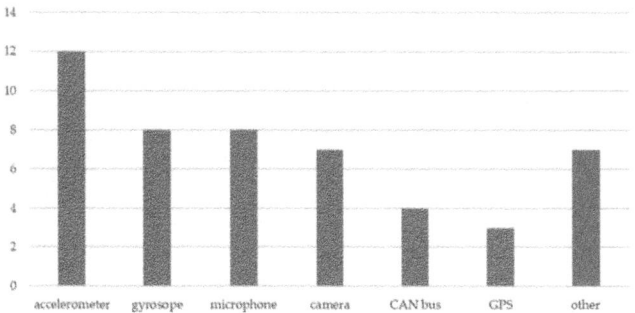

Fig. 3. Sensor types used in the analyzed papers

Considering sensor types in MAR, we observe the dominant usage of accelerometer and gyroscope sensors, microphones, and cameras. The presented analysis confirms the primary usage of sensor-based data sources (kinematic and audio). The other machinery sensors used come from the CAN bus, and the most frequent analyses cover speed, pressure, torque, and weight readings [12, 13, 17–19]. Other sensors used in MAR comprise, among others, angular velocity sensors [33], 3D images [35], GNSS for rotational and translation motion tracking [25], current and temperature [36], and lidar [9].

We observed that cameras play a double role in MAR. The first and the most common use of cameras relies on recording the image to later label the moments of activity occurrence that should be automatically recognized. Parallel to the image, the values of selected machine parameters are recorded. The combination of labels with values is the input material for machine learning. Examples of such solutions are presented, among others, in [9, 12, 13, 26, 28–30, 32, 33].

The second application of the cameras relies on using the image they recorded to detect machines on them and recognize the activities they carried out. This is definitely a more difficult task, which may solve the MAR problem. Such use of cameras can be found in the papers [10, 11, 23, 24, 31].

3.4   Methods and Techniques Used in MAR (RQ3)

Considering the third RQ, we analyze the data types, methods, and techniques applied in the paper dataset.

The primary data in the analyzed papers are of mixed type (numerical and categorical). Only numerical data were presented in 12 papers. Because of the nature of MAR and the classification type of data exploratory task, the main approach to analysis is supervised with various classifiers. One paper [19] presents clustering as an unsupervised example of activity recognition followed by expert labeling of discovered states. The relevant summary is presented in Fig. 4.

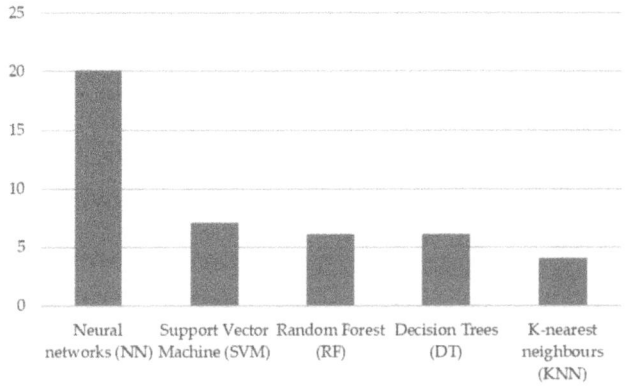

Fig. 4.   Methods and techniques used in MAR

A summary of the techniques used in MAR confirms the massive interest in neural networks (NN) observed nowadays in data mining practice. Simpler techniques, like SVM [8, 14–16, 27, 28, 33], DT [8, 19, 22, 27, 28, 30], RF [9, 12, 13, 22, 25, 27], or KNN [8, 22, 27, 28], were used less frequently.

Looking at the application of neural networks, one can observe various types of them like Convolutional Neural Network (CNN) [17, 23, 24, 26, 32, 34], Multi-Layer Perceptron (MLP) [18, 22], Recurrent Neural Network (RNN) [17, 26, 29], Deep Neural Network (DNN) [26], Probabilistic Neural Network (PNN) [36], Long Short-Term Memory (LSTM) [18, 29, 31, 34, 35], Region-based Convolutional Neural Networks (R-CNN) [11], 3D Residual Neural Network (3D ResNet) [11].

Looking at other methods used in MAR, we noticed RandomSubSpace, KStar [22], logistic regression [8], linear classifiers [27], fractional calculus-based feature augmentation [25], and fuzzy logic [36]. However, the usage of the latter two is mainly related to improving applied methods (NN or RF).

3.5 Analytic Approaches Used in MAR (RQ4)

Considering the analytic approaches in the papers (Table 1), the most dominant approach in MAR is data-oriented.

Activity recognition is mainly used for machinery utilization and performance analysis. An exceptional example of the pure process-oriented approach is presented in [19]. The authors applied unsupervised activity recognition (clustering with DBSCAN) and used the activities to model the process of the wheel loader operation in the mining process. The second paper considering process context is [13]. Authors after MAR checked an order of activities; however, they did not use discovered activities to create process models.

4 Summary of Practices in MAR

The most natural and straightforward approach to recognizing machine activity is regularly measuring selected machines' operating parameters. The current prevalence of various types of measurement transducers and data transmission capabilities allows for seamless data transmission to database servers. This enables two actions: firstly, the recorded data can be used to analyze the completed part of the process, and secondly, it can be used to monitor machine activity in real-time.

Four sources of acquiring data can be identified in the analyzed papers: (1) Measurements of physical quantities (temperature, pressure, etc.), (2) Measurements of angular velocity, acceleration, and orientation between sensors installed (IMU - Inertial Measurement Unit), (3) Sound recording, and (4) Image recording.

The nature of the machine's work determines the choice of data acquisition method. The first method is applied for machines built in a compact form (e.g., drills, trucks, etc.). The second method is used for machines with moving parts (e.g., excavators, cranes, etc.). The third method, which typically involves analyzing the spectrum of sounds produced by the machines, is appropriate when the machines operate in confined spaces. The fourth method is suitable for machines that move in the field or perform visibly recognizable movements, as image recognition algorithms can be applied. It's essential to note that the first two methods are independent of machine operating conditions. On the other hand, recognizing activities based on a recorded video in foggy, snowy, or rainy conditions is challenging. In such situations, methods based on sound recordings have an advantage. However, vision-based methods would outperform audio-based methods in scenarios where multiple machines of the same type operate simultaneously (e.g., several identical excavators). Additionally, data acquisition using microphones or cameras does not require additional infrastructure for sensor installation and data transmission, making them a natural choice for further development.

Various classification methods are used to distinguish machine activities. The predominant part of the discussed papers applies different variations of neural networks.

The preprocessing methods used before the actual classification process can significantly affect the classification quality. Examples of such preprocessing methods include fuzzing [36], Short-Time Fourier Transform [14–16], and Fractional calculus-based feature augmentation [25]. Similar to the selection of data sources, the choice of classifiers should also depend on their suitability for a specific case.

We believe that the presented spectrum of MAR applications can help future researchers more easily select appropriate solutions to their use cases and needs.

5   Conclusions, Limitations, and Future Works

In the paper, we presented a Systematic Literature Review of MAR in industry. Initially, we formulated four RQs related to the industrial domains, data sources, analytic approaches, and techniques. To answer these questions, we have reviewed 812 papers at two levels of detail (abstract/full paper). As a result, we selected 29 papers that we found relevant for further exploration.

We found out that MAR is quite a common data-oriented task in the construction industry, and a significant number of papers are devoted to MAR and workers' operation recognition. However, MAR can also be noticed in other domains like mining, logistics, and medicine, proving that this kind of analytics has wide applications. Due to the nature of MAR, most papers present a supervised approach with various classifiers, among them, one can find neural networks as the most popular and effective technique.

Regarding the limitations of our study, used searched phrase enabled us to find many papers on activity recognition and state monitoring, also on topics out of our interest, such as HAR or HM, confirming that we formulated quite an exhaustive search. However, some of the interesting papers, i.e., from the mining domain, have not been found in a systematic way, e.g., [2, 37, 38], mainly due to title content and used keywords (e.g., artificial intelligence, hyperparameters optimization, operational regimes, ore transportation, wheeled transport, signal segmentation, process mining, process improvement). These results lead us to further work on more domain-oriented SLR on MAR using searching phrases containing domain-specific technical vocabulary (e.g., cycle identification).

References

1. Mei, S., Yuan, M., Cui, J., Dong, S., Zhao, J.: Machinery condition monitoring in the era of industry 4.0: a relative degree of contribution feature selection and deep residual network combined approach. Comput. Ind. Eng. **168**, 108129 (2022). https://doi.org/10.1016/j.cie.2022.108129

2. Brzychczy, E., Gackowiec, P., Liebetrau, M.: Data analytic approaches for mining process improvement—machinery utilization use case. Resources **9**(2), 17 (2020). https://doi.org/10.3390/resources9020017

3. Arshad, M.H., Bilal, M., Gani, A.: Human activity recognition: review, taxonomy and open challenges. Sensors **22**(17), 6463 (2022). https://doi.org/10.3390/s22176463

4. Dang, L.M., Min, K., Wang, H., Piran, J., Lee, C.H., Moon, H.: Sensor-based and vision-based human activity recognition: a comprehensive survey. Pattern Recogn. **108**, 107561 (2020)

5. Sherafat, B., et al.: Automated methods for activity recognition of construction workers and equipment: state-of-the-art review. J. Constr. Eng. Manag. **146**(6), 03120002 (2020). https://doi.org/10.1061/(ASCE)CO.1943-7862.0001843

6. Ni, Q., Hernando, A.G., De la Cruz, I.: The elderly's independent living in smart homes: a characterization of activities and sensing infrastructure survey to facilitate services development. Sensors **15**(5), 11312–11362 (2015). https://doi.org/10.3390/s150511312

7. Xiao, Y., Watson, M.: Guidance on conducting a systematic literature review. J. Plan. Educ. Res. **39**(1), 93–112 (2019)

8. Akhavian, R., Behzadan, A.H.: Construction equipment activity recognition for simulation input modeling using mobile sensors and machine learning classifiers. Adv. Eng. Inform. **29**(4), 867–877 (2015). https://doi.org/10.1016/j.aei.2015.03.001

9. Becker, R.M., Keefe, R.F.: A novel smartphone-based activity recognition modeling method for tracked equipment in forest operations. PLoS ONE **17**(4), e0266568 (2022). https://doi.org/10.1371/journal.pone.0266568

10. Chen, C., Xiao, Bo., Zhang, Y., Zhu, Z.: Automatic vision-based calculation of excavator earthmoving productivity using zero-shot learning activity recognition. Autom. Constr. **146**, 104702 (2023). https://doi.org/10.1016/j.autcon.2022.104702

11. Chen, C., Zhu, Z., Hammad, A.: Automated excavators activity recognition and productivity analysis from construction site surveillance videos. Autom. Constr. **110**, 103045 (2020). https://doi.org/10.1016/j.autcon.2019.103045

12. Chen, K., Pashami, S., Nowaczyk, S., Johansson, E., Sternelöv, G., Rögnvaldsson, T.: Forklift truck activity recognition from CAN data. In: Gama, J., et al. (eds.) ITEM/IoT Streams 2020. CCIS, vol. 1325, pp. 119–126. Springer, Cham (2020). https://doi.org/10.1007/978-3-030-66770-2_9

13. Chen, K., Rögnvaldsson, T., Nowaczyk, S., Pashami, S., Johansson, E., Sternelöv, G.: Semi-supervised learning for forklift activity recognition from controller area network (CAN) signals. Sensors **22**, 4170 (2022). https://doi.org/10.3390/s22114170(2022)

14. Cheng, C.-F., Rashidi, A., Davenport, M.A., Anderson, D.: Audio signal processing for activity recognition of construction heavy equipment. In: ISARC 2016 - 33rd International Symposium on Automation and Robotics in Construction, pp. 642–650 (2016)

15. Cheng, C.-F., Rashidi, A., Davenport, M.A., Anderson, D.V.: Evaluation of software and hardware settings for audio-based analysis of construction operations. Int. J. Civil Eng. **17**(9), 1469–1480 (2019). https://doi.org/10.1007/s40999-019-00409-2

16. Cheng, C.F., Rashidi, A., Davenport, M.A., Anderson, D.V., Sabillon, C.A.: Acoustical modeling of construction jobsites: hardware and software requirements. Comput. Civil Eng. Smart Saf. Sustain. Resilience **2017**, 352–359 (2017)

17. Fischer, A., Bedrikow, A.B., Kessler, S., Fottner, J.: Equipment data-based activity recognition of construction machinery. In: IEEE International Conference on Engineering, Technology and Innovation, ICE/ITMC 2021 – Proceedings (2021). https://doi.org/10.1109/ICE/ITMC52061.2021.9570272

18. Fischer, A., Bedrikow, A.B., Tommelein, I.D., Nübel, K., Fottner, J.: From activity recognition to simulation: the impact of granularity on production models in heavy civil engineering. Algorithms **16**(4), 212 (2023). https://doi.org/10.3390/a16040212

19. Gackowiec, P., Brzychczy, E., Kęsek, M.: Enhancement of machinery activity recognition in a mining environment with GPS data. Energies **14**(12), 3422 (2021). https://doi.org/10.3390/en14123422

20. Harichandran, A., Raphael, B., Mukherjee, A.: A hierarchical machine learning framework for the identification of automated construction. J. Inf. Technol. Constr. **26**, 591–623 (2021). https://doi.org/10.36680/j.itcon.2021.031

21. Harichandran, A., Raphael, B., Mukherjee, A.: A Robust framework for identifying automated construction operations. In: Proceedings of the 37th International Symposium on Automation and Robotics in Construction, ISARC 2020, pp. 473–480 (2020)

22. Jeong, G., Ahn, C.R., Park, M.: Constructing an audio dataset of construction equipment from online sources for audio-based recognition. In: Proceedings - Winter Simulation Conference, pp. 2354–2364 (2022). https://doi.org/10.1109/WSC57314.2022.10015388

23. Kim, I.-S., Latif, K., Kim, J., Sharafat, A., Lee, D.-E., Seo, J.: Vision-based activity classification of excavators by bidirectional LSTM. Appl. Sci. **13**(1), 272 (2022). https://doi.org/10.3390/app13010272

24. Kim, J., Chi, S., Ahn, C.R.: Hybrid kinematic–visual sensing approach for activity recognition of construction equipment. J. Build. Eng. **44**, 102709 (2021). https://doi.org/10.1016/j.jobe.2021.102709

25. Langroodi, A.K., Vahdatikhaki, F., Doree, A.: Activity recognition of construction equipment using fractional random forest. Autom. Constr. **122**, 103465 (2021). https://doi.org/10.1016/j.autcon.2020.103465

26. Mahamedi, E., Rogage, K., Doukari, O., Kassem, M.: Automating excavator productivity measurement using deep learning. Proc. Inst. Civil Eng. Smart Infrastruct. Constr. **174**(4), 121–133 (2021). https://doi.org/10.1680/jsmic.21.00031

27. Rashid, K.M., Louis, J.: Automated activity identification for construction equipment using motion data from articulated members. Front. Built Environ. **5**, 144 (2020). https://doi.org/10.3389/fbuil.2019.00144

28. Rashid, K.M., Louis, J.: Construction equipment activity recognition from IMUs mounted on articulated implements and supervised classification. In: Computing in Civil Engineering 2019: Smart Cities, Sustainability, and Resilience - Selected Papers from the ASCE 2019, pp. 130–138 (2019). https://doi.org/10.1061/9780784482445.017

29. Rashid, K.M., Louis, J.: Times-series data augmentation and deep learning for construction equipment activity recognition. Adv. Eng. Inform. **42**, 100944 (2019). https://doi.org/10.1016/j.aei.2019.100944

30. Rashid, K.M., Louis, J.: Window-warping: a time series data augmentation of IMU data for construction equipment activity identification. In: Proceedings of the 36th International Symposium on Automation and Robotics in Construction, ISARC 2019, pp. 651–657 (2019)

31. Sharghi, A., Haugerud, H., Daniel, Oh., Mohareri, O.: Automatic operating room surgical activity recognition for robot-assisted surgery. In: Martel, A.L., et al. (eds.) Medical Image Computing and Computer Assisted Intervention – MICCAI 2020: 23rd International Conference, Lima, Peru, October 4–8, 2020, Proceedings, Part III, pp. 385–395. Springer International Publishing, Cham (2020). https://doi.org/10.1007/978-3-030-59716-0_37

32. Sherafat, B., Rashidi, A., Asgari, S.: Sound-based multiple-equipment activity recognition using convolutional neural networks. Autom. Constr. **135**, 104104 (2022). https://doi.org/10.1016/j.autcon.2021.104104

33. Sherafat, B., Rashidi, A., Lee, Y.-C., Ahn, C.R.: Automated activity recognition of construction equipment using a data fusion approach. In: Computing in Civil Engineering 2019: Data, Sensing, and Analytics - Selected Papers from the ASCE 2019, pp. 1–8 (2019)

34. Slaton, T., Hernandez, C., Akhavian, R.: Construction activity recognition with convolutional recurrent networks. Autom. Constr. **113**, 103138 (2020). https://doi.org/10.1016/j.autcon.2020.103138

35. Sun, Yu., Wang, Li., Jiang, Z., Li, B., Ying, Hu., Tian, W.: State recognition of decompressive laminectomy with multiple information in robot-assisted surgery. Artif. Intell. Med. **102**, 101763 (2020). https://doi.org/10.1016/j.artmed.2019.101763
36. Jing, Xu., Wang, Z., Tan, C., Liu, X.: A state recognition approach for complex equipment based on a fuzzy probabilistic neural network. Algorithms **9**(2), 34 (2016). https://doi.org/10.3390/a9020034
37. Skoczylas, A., Rot, A., Stefaniak, P., Śliwiński, P.: Haulage cycles identification for wheeled transport in underground mine using neural networks. Sensors **23**(3), 1331 (2023). https://doi.org/10.3390/s23031331
38. Koperska, W., Skoczylas, A., Stefaniak, P.: A simple method of the haulage cycles detection for LHD machine. In: Hernes, M., Wojtkiewicz, K., Szczerbicki, E. (eds.) ICCCI 2020. CCIS, vol. 1287, pp. 326–337. Springer, Cham (2020). https://doi.org/10.1007/978-3-030-63119-2_27

LLM-Based Event Abstraction
and Integration for IoT-Sourced Logs

Mohsen Shirali[1,2(✉)] [ID], Mohammadreza Fani Sani[3] [ID], Zahra Ahmadi[2] [ID],
and Estefanía Serral[2] [ID]

[1] Research Centre for Information Systems Engineering (LIRIS), KU Leuven,
Warmoesberg 26, 1000 Brussels, Belgium
[2] Department of Computer Science and Engineering, Shahid Beheshti University,
Tehran, Iran
m_shirali@sbu.ac.ir, {zahra.ahmadi,estefania.serralasensio}@kuleuven.be
[3] Microsoft Development Center Copenhagen, Copenhagen, Denmark
mfanisani@microsoft.com

Abstract. The continuous flow of data collected by Internet of Things
(IoT) devices, has revolutionised our ability to understand and interact
with the world across various applications. However, this data must be
prepared and transformed into event data before analysis can begin. In
this paper, we shed light on the potential of leveraging Large Language
Models (LLMs) in event abstraction and integration. Our approach aims
to create event records from raw sensor readings and merge the logs from
multiple IoT sources into a single event log suitable for further Process
Mining applications. We demonstrate the capabilities of LLMs in event
abstraction considering a case study for IoT application in elderly care
and longitudinal health monitoring. The results, showing on average an
accuracy of 90% in detecting high-level activities. These results highlight
LLMs' promising potential in addressing event abstraction and integration challenges, effectively bridging the existing gap.

Keywords: Internet of Things · Large Language Models · Event
Abstraction · Log Integration · Multi-modality

1 Introduction

The IoT has the potential to create a "cyber-physical" world where "things"
interact with the physical world [11]. Smart devices and low-cost sensors have
been integrated into applications ranging from smart homes to industrial IoT
systems. However, raw sensor data is often unstructured and non-informative,
requiring transformation into events to make information discoverable [20].
Without proper pre-processing, even advanced data mining techniques cannot
provide meaningful analysis, leading to a gap in abstraction between the data
and the insights they can offer [4,17].

Process Mining (PM) is a data analysis discipline that requires structured
event logs, where each event represents a distinct activity at a particular time. In

K. Gdowska et al. (Eds.): BPM 2024 Workshops, LNBIP 534, pp. 138–149, 2025.
https://doi.org/10.1007/978-3-031-78666-2_11

many cases, the available data is too granular for immediate use, as seen in smart home systems where events are logged at the sensor trigger level, complicating the application of PM techniques [4]. Additionally, the data is often scattered across various databases and information systems, making data integration a necessary step to create a comprehensive view for analysis.

The process of preparing event logs is typically manual and can be error-prone, especially in complex systems involving multiple data sources [8]. This manual effort requires significant domain knowledge and introduces challenges such as format and time inconsistencies and data quality issues. Furthermore, data can vary in granularity and context, complicating the analysis process [23]. Therefore, pre-processing is essential to ensure meaningful event logs before any actual analysis can be applied [4].

Log abstraction and integration are two key pre-processing tasks [26]. Log abstraction simplifies data by grouping low-level records into higher-level event-based representations [4,16]. Log integration combines data from multiple sources to create a unified view for analysis. These processes are vital for extracting meaningful insights from IoT data. Existing techniques for event abstraction are either unsupervised or supervised, each with its own limitations [26]. Unsupervised methods rely on control-flow similarities in data, while supervised methods require extensive domain knowledge and manual input [16].

To address these challenges, we propose the use of LLMs for automating data pre-processing and event log generation. LLMs, trained on extensive data, are capable of processing text and performing natural language tasks, making them suitable for abstracting sensor data into meaningful event logs. Our approach utilises LLMs to analyse IoT sensor logs, generate meaningful event records, and merge logs from multiple sources, facilitating PM applications.

In this work, we explore how an LLM can abstract low-level sensor readings into event logs with minimal user input. By providing a prompt with event label samples, we guide the LLM to generate abstracted event logs. Our contributions include automating the detection and labelling of sensor data, developing a method for event abstraction, and integrating data from multiple sources into a unified event log. Leveraging LLMs reduces the manual effort and expertise required, and improves the consistency and accuracy of event log generation.

The remainder of the article is structured as follows: Sect. 4 reviews existing event abstraction and integration methods, along with the multi-source IoT dataset used for evaluation. Section 3 outlines the proposed LLM-based approach for event log abstraction and integration. Section 4 evaluates the results and performance of the approach, while Sect. 5 discusses the advantages of using LLMs with IoT systems. Finally, Sect. 6 concludes the article's findings.

2   Background Knowledge

2.1   Existing Works for Event Abstraction and Integration

The journey from raw data to event logs suitable for PM has been widely researched. For instance, van Zelst *et al.* [26] and Fahland *et al.* [6] used log abstraction to group events into higher-level activities for simplified visualisation and analysis. A hierarchical framework for event abstraction based on activity instances was proposed in [12], while Senderovich *et al.* [20] presented a knowledge-driven approach to transform raw sensor data into standard event logs. Mangler *et al.* [14] proposed a semi-automatic framework to convert low-level IoT sensor data into higher-level process events, while Di Federico and Burattin [3] developed CvAMoS to identify recurring activity sequences and their context. Additionally, [19] introduced activity signatures to train supervised models for detecting higher-level process activities in larger datasets.

Building a ground truth using domain knowledge is crucial in early IoT data analysis stages, facilitating automated activity detection [19]. Human expertise helps map sensor data to activity events, as seen in activity recognition methods [23], where sensor readings, such as opening a refrigerator, are mapped to activities like "preparing meal." Predefined rules have also been used to translate sensor readings into events [13,22], while studies like [2,7] incorporate sensor types for recognising daily living activities (ADLs).

To minimise the need for domain knowledge, [16] introduced a method to automatically suggest event groups for abstraction, and [15] proposed unsupervised methods for transforming user interactions into event streams. Additionally, unsupervised learning techniques, such as [10], discretize sensor data into activities for smart home scenarios. Recently, LLMs have been applied for event abstraction, as seen in [8], where tasks were grouped into high-level activities and labelled based on their similarity.

For log integration, [5] combined logs from different departments for end-to-end process analysis, while [18] used a time-based heuristics miner to discover high-level and low-level process models in parallel from multi-source logs.

2.2 Multi-source IoT Dataset

Our study aimed to assess the effectiveness and accuracy of using LLMs on IoT-sourced datasets to create event logs suitable for Process Mining. We applied our proposed LLM to sensor logs from an IoT dataset collected in an Ambient Assisted Living scenario. This dataset contains 146 days of data, capturing the daily activities of a 60-year-old woman living independently in an apartment [7]. The data was collected using ambient sensors installed in the house, a wristband, and a smartphone.

The ambient sensors included PIR sensors placed on furniture and appliances, a power usage sensor indicating TV usage, contact sensors for detecting door openings and closings, and a gas detection sensor for identifying cooking activity. These sensors were positioned in six different areas of the house, as shown in Fig. 1 and recorded event data with timestamps, sensor names, and sensor values (e.g., on/off states). The participant also wore a Xiaomi Mi Band-3 wristband to collect sleep-related information such as sleep time, duration, and quality. Additionally, smartphone usage data, including timestamps and the names of used applications, were collected using an application.

The ambient sensors, such as the TV sensor and kitchen appliance sensors, are triggered when the participant performs specific activities. As a result, their measurements are at a low level of abstraction representing the presence of the subject near the appliances, the open or closed state of doors, and the use of the stove. By knowing the timestamp of their triggered states, we can discover the corresponding activities to express the exact start and end times and the duration of the conducted activity. These activities are mainly known as instrumented Activity Daily Livings (iADL), the activities performed using specific instruments and are investigated in several behaviour modelling and analysis studies and projects, like [2,21].

With the entire sequence of captured records for a specific duration, we can infer events related to the person's presence in different areas of the house, the performed activities, and the sequence of their movements. For this purpose, an event abstraction step is needed to elevate the abstraction level of sensor raw readings into recognisable activity events in an event log. The events related to smartphone usage should also be converted and inserted into the log, while the names of applications can be replaced with their type or with a general label such as 'Using Smartphone' to preserve the participants' privacy. Furthermore, the sleep-related information collected by the wristband will complete the preferred event log.

In the current case study, the event log includes location events, such as *Bedroom, Bathroom, Kitchen,* and the activity events like *sleeping, meal preparation, praying,* watching tv. A snapshot of sensor logs and the required event log, including the information of all modalities (the ground truth version), is depicted in Fig. 2. We fed these sensor logs obtained from devices' reports in our multi-modal IoT dataset into the proposed LLM model, asking it to create a PM-friendly event log. The resulting event log was then compared with a ground truth event log that was generated through a pre-processing step by applying multiple event detection rules determined by domain experts and then

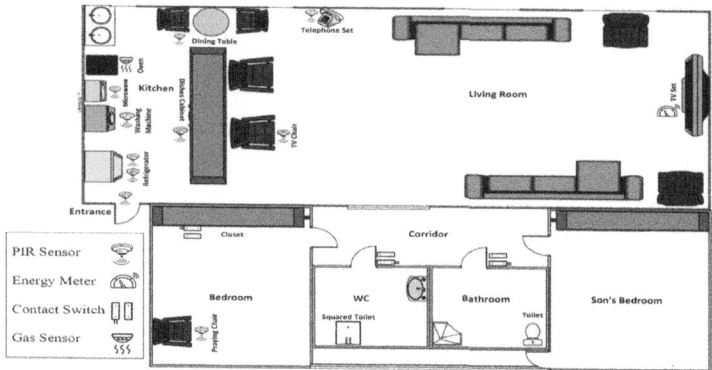

Fig. 1. The house floor plan, location and types of ambient sensors

inspecting manually to correct any falsely detected events, as described and used in [21].

Date	TV	Telephone	TV_Chair	Dining_Chair	Bathroom	WC	Closet	Entrance	PrayingChair	Refrigrator	Fridge	Washing Machine	Microwave	Stove	Kitchen Cupboard	Smartphone	Wristband Sleep	SensorStates	Ambient Label	Smartphone Label	Smartphone AppType	Wristband Label	Wristband Type
1/9/2020 14:14:52	1	0	0	0	0	0	0	0	0	0	0	0	0	0	0	0	0	10000000000000000					
1/9/2020 14:15:32	1	0	0	1	0	0	0	0	0	0	0	0	0	0	0	0	0	10010000000000000	meal begin				
1/9/2020 14:34:01	1	0	0	0	0	0	0	0	0	0	0	0	0	0	0	0	0	10000000000000000	meal end				
1/9/2020 14:37:35	1	0	0	0	1	0	0	0	0	0	0	0	0	0	0	0	0	10001000000000000	toilet begin				
1/9/2020 14:39:02	1	0	0	0	0	0	0	0	0	0	0	0	0	0	0	0	0	10000000000000000	toilet end				
1/9/2020 14:39:43	1	0	0	0	0	0	0	0	0	0	0	0	0	0	1	0	0	10000000000000100					
1/9/2020 14:39:46	1	0	0	0	0	0	0	0	0	0	0	0	0	0	0	0	0	10000000000000000					
01/09/20 14:42:23	1	0	0	0	0	0	0	0	0	0	0	0	0	0	0	1	0	10000000000000010		Start Using Smartphone	Call		
01/09/20 14:43:04	1	0	0	0	0	0	0	0	0	0	0	0	0	0	0	0	0	10000000000000000		End Using Smartphone			
01/09/20 14:43:36	1	0	0	0	0	0	0	0	0	0	0	0	0	0	0	1	0	10000000000000010		Start Using Smartphone	Messaging		
01/09/20 14:43:37	1	0	0	0	0	0	0	0	0	0	0	0	0	0	0	0	0	10000000000000000		End Using Smartphone			
01/09/20 14:43:44	1	0	0	0	0	0	0	0	0	0	0	0	0	0	0	1	0	10000000000000010		Start Using Smartphone	Call		
01/09/20 14:43:44	1	0	0	0	0	0	0	0	0	0	0	0	0	0	0	0	0	10000000000000000		End Using Smartphone			
1/9/2020 14:51:42	1	0	0	0	0	0	0	0	0	0	0	0	0	0	1	0	0	10000000000000100					

(a) We have 17 sensors and three categories of labels.

Source	Location/Events	Date	StartTime	EndTime	Duration
Location	Kitchen	1/9/2020	1/9/2020 14:11:00	1/9/2020 14:36:59	0:25:59
Ambient	eating and watching TV	1/9/2020	1/9/2020 14:15:32	1/9/2020 14:34:01	0:18:29
Location	LivingRoom	1/9/2020	1/9/2020 14:37:00	1/9/2020 14:37:01	0:00:01
Location	Bathroom	1/9/2020	1/9/2020 14:37:04	1/9/2020 14:38:59	0:01:55
Location	LivingRoom	1/9/2020	1/9/2020 14:39:02	1/9/2020 14:39:03	0:00:01
Location	Kitchen	1/9/2020	1/9/2020 14:39:04	1/9/2020 15:05:59	0:26:55
Smartphone	Call	1/9/2020	1/9/2020 14:42:23	1/9/2020 14:43:04	0:00:41
Smartphone	Messaging	1/9/2020	1/9/2020 14:43:36	1/9/2020 14:43:37	0:00:01
Smartphone	Call	1/9/2020	1/9/2020 14:43:44	1/9/2020 14:43:44	0:00:00

(b) Activities and presence in locations can be concurrent and be happened on different days.

Fig. 2. A view of (a) sensor logs and (b) resulted event log (the ground truth version) for the experimented IoT dataset.

3 Event Log Abstraction and Integration by LLMs

As previously stated, PM techniques rely on having an event log as an input, and without it, gaining insights is not possible. While many scientific works assume the availability of the event log, the procedure of pre-processing raw data, detecting events, abstracting them into activities, and generating an event log can be time-consuming in real scenarios. Abstracting events from sensor data, particularly when dealing with streams of input data and potential concurrent events and activities recorded by a combination of sensors, can be quite challenging [14]. In this section, we explain how we can use Generative AI and LLMs to (1) abstract the events from sensor data to activities, and (2) generate an event log from different resources.

3.1 Abstraction of Events Into Activities

We propose to use LLMs to detect activities from sensor data. In this regard, we give some basic explanations about the sensors, possible activity labels, and a few examples to LLM to provide us with a label for the detected change in the status of sensors. In other words, we call an LLM whenever we have a change in sensor values. The changes in the binary values of the sensors have the highest significance regarding process-level events, as it is stated in the [14]. Thus, each sensor log entry should denote a change in the status of sensors. Therefore, we can consider this task as a classification task in that LLM will be the classifier and activity labels are the classes. The number of classes is the number of possible desired activities plus one as we may encounter sensor changes that do not correspond to any specific activity. However, there are instances where sensor changes occur without any corresponding activities due to noises, irrelevant sensors, etc. In such cases, a blank class is defined to address these situations.

For this task, we utilised GPT-4 as the model and applied few-shot learning [25] and chain of thought [24] techniques. In the prompt template, we incorporated placeholders for the label of activities, the sensor explanation, and the output format[1].

In general, the prompt can be modified if we have more information based on the rules that we have in business. For instance, here the output of all sensors in our dataset are binary values (i.e., 0 and 1), and it is important to note that we take this advantage to save tokens and reduce the cost by just passing the aggregated sensor information in a binary format to LLM (the number of consumed tokens in a generated a prompt is one of the limitations in using LLMs). Hence, the LLM will need to recognise which value corresponds to which sensor based on the order of binary digits in the sensor state. Some examples of SensorStates are given in Fig. 2a. We provide the SensorStates for the current and previous state in our prompt to simplify the task for the LLM. The LLM will first identify what has changed in these two values and then provide the activity name.

3.2 Integration and Event Log Generation

We can have several information sources for one event log. Note that for each activity we have two events; the first one corresponds to the start of the activity and the second one indicates that the activity has ended. In real world, we may have more events for each activity considering the complete life cycle of activities [1]. We propose to use LLM to gather information from different resources and combine them into one event log. However, we have some limitations:

– We cannot give the whole information to LLM as we have a limitation on the number of tokens,
– We do not have a caseID[2] as all the information related in our dataset is related to one process instance.

---

[1] The sources and examples for our implemented LLM with more detailed information are publicly available at https://github.com/mfanisani/LLM4IoT.
[2] caseID is one of the main columns in any event log that represents the process instance.

To overcome the above limitations, we considered date as caseID. This will allow us to analyse the process of activities that have been done each day. Therefore, our data will include the date as the case ID, the activity, start time, and end time. However, it is possible that an activity start on one day and end on the next day, e.g. sleeping from 2022-11-23 22:00 to 2022-11-24 07:00. In this case, we will provide LLM with the information for activities that span two days and ask it to trim the event at midnight, and adjust the event log for each day accordingly. To handle this, we introduce another column with binary values indicating whether the activity is completed on the same date or not. If the activity crosses over to the next day, we will not include the end time for that next day. While we assumed activities span at most two consecutive days to mitigate the constraint on the number of tokens per LLM call, it's important to recognise that real-world scenarios may not adhere to this assumption and may have longer execution times.

Our prompt template includes possible activities, two-shot examples, the information for the date of interest and the following day, and the format of the desired output. In this way, we can increase the performance of the event log generation task, as we will call LLM once per day. Note that the output does not contain the header of event log columns. We use a basic script to collect the output information generated by LLM for different dates and compile it into a single event log.

The proposed framework for event log abstraction and integration can be used on different sensors or data logs with different prompts and on various LLMs. The samples for the appropriate labels are given to the model and it returns the expected output for the stream of data inputs immediately. Therefore, the framework is also able to deal with streams of data sources and the output of this layer is sufficient for the PM tools without any further involvement of the users and interventions.

4   Evaluation

It is important to assess the effectiveness of LLMs in event abstraction by analysing the results and evaluating the model's accuracy in generating event records and assigning the appropriate labels. This section presents the results of proposed LLM-based event abstraction on the real-world IoT-sourced dataset for our case study. We have conducted the evaluation in two steps. First, the abstracted events for each data source are examined and compared with the intended labels to measure the accuracy of the LLM model in identifying and abstracting the raw sensor records for each data source. This comparison will highlight the LLM model's potential in dealing with different IoT sources. In the next step, the outcome of merging three logs from ambient, smartphone and wristband into the final integrated PM-friendly event log is cross-checked with the ground truth event log.

4.1   Event Abstraction and Integration Results

The accuracy results of the LLM in the abstraction of events foreach of the three modalities in our dataset are provided in Table 1. For ambient sensors, we just consider activities with an occurrence of at least 50 times. The 21 labels in this source correspond to 10 activities with a start and end plus a label that corresponds to no activity. For smartphones and wristbands, we have respectively 13 and 3 labels indicating the type of used application or day/night sleep. For each activity, we have start, continue and end labels. Similar to ambient sensors, we have a label for no activity.

The accuracy for ambient sensor data is much lower than the other datasets as the number of related sensors and the number of labels are higher. For all three datasets, the lowest accuracy is for *None* label that corresponds to no activity. In several cases, we predicted an activity (e.g., *Watching TV* start/end) instead of *None*. We randomly checked some of the wrongly assigned labels and in almost all of them, the reasoning part produced by LLM was wrong.

4.2   Evaluation of LLM-Generated Event Log

The evaluation of the LLM-based event log in terms of the accuracy of the recognised events is crucial to assess the success of LLMs in the abstraction and integration task. This evaluation involves comparing two event logs that contain various events with their respective start and end times. A snapshot of the event log generated using LLM is presented in Table 2. Therefore, the matching of events labels, events orders, and the duration and time precedence of events are the important factors to be considered for evaluation. We have utilised the number of correctly generated events, the ratio of perfectly aligned date and Edit Distance Alignment (EDA) [9] metrics to measure the success rate of our proposed LLM-based approach.

In our experiment, we provided the GPT4-0613 model with sensor states and their original labels, tasking it to generate an event log. Notably, we used the ground truth labels, not those predicted by LLM. The ground truth dataset contains 17,165 events, while the LLM-generated event log has 15,420 events. Among these generated events, 12,741 were correctly detected, with 11,365 having matching start and end activities. In addition, LLM is able to generate all the unique activities that are defined for it.

In our analysis, we observed that self-loops—consecutive events with the same label—play a significant role. After removing these self-loops, we found that the generated event log contains 11,248 events, while the ground truth

Table 1. The accuracy of our activity detection method for different sensors

	Ambient sensors	Smartphone	wristband
Number of labels	21	13	3
Accuracy	0.81	0.97	0.92

Table 2. The snapshot of the LLM-generated event log from three sources.

Date	Type	Activity	Start Time	End Time	Next date
1/8/2020	Ambient	sleeping	23:04:33	08:28:59	True
1/8/2020	Ambient	toilet	23:05:39	23:07:21	False
1/8/2020	Smartphone	messaging	23:17:28	23:18:31	False
1/8/2020	Smartphone	messaging	23:18:32	23:20:28	False
1/8/2020	Smartphone	call	23:20:32	23:20:45	False
...	...	...	...	...	...

Table 3. Behavioral comparison of event logs with/without considering self loops.

Edit Distance Alignment		Perfectly Aligned Dates(%)	
All events	Without Self-loops	All events	Without Self-loops
0.83	0.94	0	33

dataset has 11,431 events. Interestingly, LLM sometimes aggregates self-loops, particularly when the user interacts with her smartphone, where the interactivity times between self-loops are negligible.

To show how the generated event log is aligned with the ground truth event log, we applied EDA between similar dates in both event logs using the technique proposed in [9]. This metric measures how similar the sequences of events are, where a value closer to *one* indicates higher similarity. The results are presented in Table 3 and indicate that the event log generated by LLM captures similar behaviour compared to the ground truth event log but it aggregated the self-loops.

5   Discussion

The results of this case study indicate that utilising LLMs can significantly alleviate the challenges associated with event log abstraction and integration for IoT-sourced logs. LLMs can bridge the gap between data collection and the demand for event logs at appropriate abstraction levels for data analysis techniques, including Process Mining. Although, traditional methods for event abstraction and integration are often manual, error-prone, and time-consuming, but, LLMs can automate and speed up these tasks while reducing human errors. They can swiftly pre-process large volumes of raw sensor data, identify and abstract events with high accuracy, and uniformly label them, thus freeing up human resources for more strategic tasks. However, it's important to acknowledge that the output of our proposed solution may not always be highly reliable. Therefore, we recommend incorporating human oversight in applications that require a high level of reliability.

Key advantages of using LLMs in event log generation and integration for IoT systems include:

- **Automation and efficiency.** LLMs can automate the pre-processing of raw data and make the process more efficient by reducing manual effort and errors.
- **Reduced need for domain knowledge.** Expertise is crucial in supervising the abstraction process and interpreting IoT logs requires substantial domain knowledge to map low-level sensor readings to high-level activities. LLMs, trained on diverse datasets, can understand and abstract sensor data into meaningful events without deep domain-specific knowledge. This allows more users with limited background knowledge to generate valuable insights and make IoT data analytics more feasible.
- **Handling multi-modality.** IoT systems often collect data from multiple sources, resulting in mixed granularity and inconsistencies in data formats and characteristics. LLM-based models can understand and integrate heterogeneous data inputs, by handling this complexity. These models can align data from various sensors, ensuring a cohesive and consistent event log, particularly for scenarios involving time misalignment and varied data structures.
- **Real-time processing.** LLMs are capable of handling streams of data inputs, making them suitable for online applications. Initially, in our study, we triggered LLM calls whenever there was a change in sensor values. However, this approach can be computationally expensive and slow in certain applications. To address this limitation, we propose using batching—calling the LLM fewer times by grouping input data—similar to our approach for event log generation. This means that LLMs can process incoming data in real-time, handling sensor records one by one without relying on maintaining extensive change histories, and continuously updating the event logs with new information. This ensures that the event logs are always current, facilitating timely analysis and decision-making.
- **Performance optimisation with prompt engineering and user feedback.** Designing prompts that clearly describe the task, input data, and expected output, can significantly enhance the quality of the LLM's output. Moreover, users' and domain experts involvement in the process remains crucial for fine-tuning the LLM's outputs and improving accuracy over time. A continuous feedback loop between LLMs and users helps in refining the model, adapting to evolving data patterns, and ensuring that the system meets the specific needs of the application. This collaboration between humans and intelligent systems ensures that the benefits of LLMs are fully realised while maintaining high standards of accuracy and reliability.

In the end, LLMs offer a powerful solution to the challenges of event log abstraction and integration in IoT systems. By automating the manual, repetitive, and error-prone data preparation tasks, LLMs streamline the process, reduce the need for extensive domain knowledge, and handle the complexities of multi-modality and mixed granularity in data. Additionally, with proper prompt engineering and user feedback mechanisms, LLMs can provide accurate and up-to-date event logs, enhancing the overall efficiency and effectiveness of IoT data analytics. Hence, integrating LLMs with IoT systems can revolutionise the way we process and analyse sensor data, unlocking new potential for real-time insights and decision-making.

6   Conclusion and Future Works

The rapid growth of IoT technologies offers valuable data-driven insights, but a gap between raw and processed data needed for mining limits their full potential. This study explores using LLMs to pre-process raw IoT data into event logs for techniques like Process Mining. The results highlight LLMs' potential for event log abstraction and integration, though their application in this field is still in its early stages. Future research could focus on refining LLM performance, applying them to text-based data, and improving prompt engineering to enhance their usefulness in data preparation.

Acknowledgments. Mohsen Shirali, is now affiliated at UCLouvain in Belgium under grant Win4Collective number 2310088. The work of Zahra Ahmadi and Estefanía Serral was supported by the Flemish Fund for Scientific Research (FWO) with grant number G0B6922N.

References

1. Bernardi, M.L., Cimitile, M., Di Francescomarino, C., Maggi, F.M.: Do activity lifecycles affect the validity of a business rule in a business process? Inf. Syst. **62**, 42–59 (2016)
2. Cook, D.J., Crandall, A.S., Thomas, B.L., Krishnan, N.C.: Casas: a smart home in a box. Computer **46**(7), 62–69 (2013)
3. Di Federico, G., Burattin, A.: CvAMoS—event abstraction using contextual information. Future Internet **15**(3) (2023). https://doi.org/10.3390/fi15030113
4. Diba, K., Batoulis, K., Weidlich, M., Weske, M.: Extraction, correlation, and abstraction of event data for process mining. Wiley Interdisc. Rev. Data Min. Knowl. Discov. **10**(3), e1346 (2020)
5. van Eck, M.L., Lu, X., Leemans, S.J.J., van der Aalst, W.M.P.: PM^2: a process mining project methodology. In: Advanced Information Systems Engineering, pp. 297–313. Springer, Cham (2015)
6. Fahland, D.: Extracting and pre-processing event logs. CoRR abs/2211.04338 (2022). https://doi.org/10.48550/arXiv.2211.04338
7. Falah Rad, M., Shakeri, M., Khoshhal Roudposhti, K., Shakerinia, I.: Probabilistic elderly person's mood analysis based on its activities of daily living using smart facilities. Pattern Anal. Appl. 1–14 (2022)
8. Fani Sani, M., Sroka, M., Burattin, A.: LLMs and process mining: challenges in RPA. In: Process Mining Workshops, pp. 379–391. Springer (2024)
9. Fani Sani, M., van Zelst, S.J., van der Aalst, W.: Conformance checking approximation using subset selection and edit distance. In: 32nd International Conference on Advanced Information Systems Engineering (CAiSE), pp. 234–251. Springer (2020)
10. Janssen, D., Mannhardt, F., Koschmider, A., van Zelst, S.: Process model discovery from sensor event data. In: Process Mining Workshops, pp. 69–81. Springer (2021)

11. Karkouch, A., Mousannif, H., Al Moatassime, H., Noel, T.: Data quality in Internet of Things: a state-of-the-art survey. J. Netw. Comput. Appl. **73**, 57–81 (2016)
12. Li, C.Y., van Zelst, S., van der Aalst, W.: An activity instance based hierarchical framework for event abstraction. In: 3rd International Conference on Process Mining (ICPM), pp. 160–167. IEEE (2021)
13. Lull, J.J., Bayo-Monton, J.L., Shirali, M., Ghassemian, M., Fernandez-Llatas, C.: Interactive process mining in IoT and human behaviour modelling, pp. 217–231. Springer, Cham (2021)
14. Mangler, J., et al.: From Internet of Things data to business processes: challenges and a framework (2024). https://arxiv.org/abs/2405.08528
15. Rebmann, A., van der Aa, H.: Unsupervised task recognition from user interaction streams. In: International Conference on Advanced Information Systems Engineering, pp. 141–157. Springer (2023)
16. Rebmann, A., Pfeiffer, P., Fettke, P., van der Aa, H.: Multi-perspective identification of event groups for event abstraction. In: International Conference on Process Mining, pp. 31–43. Springer (2022)
17. Sani, M.F.: Preprocessing event data in process mining. In: CAiSE (Doctoral Consortium), pp. 1–10 (2020)
18. Sarno, R., Effendi, Y.A.: Hierarchy process mining from multi-source logs. Telecommun. Comput. Electron. Control **15**(4), 1955–1970 (2017)
19. Seiger, R., Franceschetti, M., Weber, B.: An interactive method for detection of process activity executions from IoT data. Future Internet **15**(2) (2023)
20. Senderovich, A., Rogge-Solti, A., Gal, A., Mendling, J., Mandelbaum, A.: The road from sensor data to process instances via interaction mining. In: 28th Conference on Advanced Information Systems Engineering, pp. 257–273. Springer (2016)
21. Shirali, M., Ahmadi, Z., Fernández-Llatas, C., Bayo-Monton, J.L.: Synergy of information in multimodal IoT systems–discovering the impact of daily behaviour routines on physical activity level. arXiv preprint arXiv:2403.14707 (2024)
22. Shirali, M., Bayo-Monton, J.L., Fernandez-Llatas, C., Ghassemian, M., Traver Salcedo, V.: Design and evaluation of a solo-resident smart home testbed for mobility pattern monitoring and behavioural assessment. Sensors **20**(24) (2020)
23. Tax, N., Sidorova, N., Haakma, R., van der Aalst, W.: Mining process model descriptions of daily life through event abstraction. In: Intelligent Systems and Applications (IntelliSys), pp. 83–104. Springer (2016)
24. Wei, J., et al.: Chain-of-thought prompting elicits reasoning in large language models. Adv. Neural. Inf. Process. Syst. **35**, 24824–24837 (2022)
25. Yong, G., Jeon, K., Gil, D., Lee, G.: Prompt engineering for zero-shot and few-shot defect detection and classification using a visual-language pretrained model. Comput.-Aided Civil Infrastruct. Eng. **38**(11), 1536–1554 (2023)
26. van Zelst, S., Mannhardt, F., de Leoni, M., Koschmider, A.: Event abstraction in process mining: literature review and taxonomy. Granular Comput. **6**, 719–736 (2021)

17th International Workshop on Social and Human Aspects of Business Process Management (BPMS2 2024)

17th International Workshop on Social and Human Aspects of Business Process Management (BPMS2 2024)

Human involvement in Business Process Management occurs at both social and individual levels. Social information systems [1], including social media, Enterprise 2.0, and social platforms [2], are rapidly growing in society, organizations, and economics. The integration of business process management with social information systems is becoming more common [3, 4], with novel approaches frequently emerging.

Merging business process management with social information systems facilitates the development of innovative business models using social platforms. TripAdvisor, Uber, and AirBnB exemplify this integration. The value-creating mechanisms of social information systems enable business models that were previously unattainable. The human aspect includes creating process models, communication, collaboration, coordination, and cooperation, as well as representing specific human-related aspects in models. With this background, the goal of the BPMS2 workshop [5] is to explore how social information systems integrate with business process management [6], and how business process management may profit from this integration [7]. Furthermore, the workshop investigates the human aspects of Business Process Management by involving human actors. Examples include using crowdsourced knowledge and tasks and the need for new user interfaces, e.g., augmented reality and voice bots.

Three papers were accepted for presentation at the BPMS2 2024 workshop:

Sylwia Białas and Piotr Wróbel examine DEI integration in HR processes within Poland's financial sector, analyzing 101 best practices across 20 commercial banks. The study focuses on recruitment, development, compensation, and communication, highlighting gender diversity most frequently, followed by age, disability, and LGBT+ inclusion. Internationally affiliated banks implement more DEI practices than local ones. The authors conclude that incorporating DEI into existing Business Process Management frameworks can promote sustained cultural change, stressing the importance of considering organizational context and DEI maturity when designing solutions.

Baumann et al. investigate the challenges visually impaired individuals encounter in business process modeling through qualitative interviews. The study assesses current non-visual process modeling methodologies, tools, and techniques, emphasizing issues like sequential data access, restricted editing capabilities, and the necessity for diverse representation formats. It examines factors influencing tool suitability and underscores the significance of inclusive collaborative modeling. The research provides insights for enhancing accessibility in business process modeling and promotes inclusive tool development and research practices.

The paper "Mining for Well-Being: The Potential of Process Mining for Evaluating Employee Well-Being" by Braakman et al. explores using process mining to assess employee well-being. A case study compares process mining variables with survey measures of work characteristics such as monotonous work, time pressure, workload, social support, and autonomy. While strong correlations between corresponding survey and process mining variables were not found, significant correlations emerged between process mining variables of workload, social support, and autonomy with survey measures of work engagement. The study suggests that combining process mining with traditional surveys could provide a more comprehensive understanding of work-related well-being, improving data collection efficiency and enabling continuous monitoring of work engagement.

We wish to thank all the people who submitted papers to BPMS2 2024 for sharing their work with us, the many participants who created fruitful discussions, and the BPMS2 2024 Program Committee members, who made a remarkable effort in reviewing the submissions. We also thank the organizers of BPM 2024 for their help with organizing the event.

References

1. Schmidt, R., Alt, R., Nurcan, S.: Social information systems. In: Proceedings of the 52nd Hawaii International Conference on System Sciences, pp. 2642–2646. Hawaii (2019)
2. Schmidt, R., Kirchner, K., Razmerita, L.: Understanding the business value of social information systems – a research agenda. In: Proceedings 53rd Hawaii International Conference on System Sciences (HICSS), pp. 2639–2648 (2020). https://doi.org/10.24251/HICSS.2020.321
3. Schmidt, R., Nurcan, S.: BPM and social software. In: Ardagna, D., et al. (eds.) Business Process Management Workshops, pp. 649–658. Springer, Berlin Heidelberg (2009)
4. Bruno, G., et al.: Key challenges for enabling agile BPM with social software. J. Softw. Maintenance Evol.: Res. Pract. **23**, 297–326 (2011). https://doi.org/10.1002/smr.523
5. Nurcan, S., Schmidt, R.: Introduction to the first international workshop on business process management and social software (BPMS2 2008). In: Business Process Management Workshops, pp. 647–648 (2009)
6. Schmidt, R., Nurcan, S.: Augmenting BPM with social software. In: Business Process Management Workshops, pp. 201–206. Ulm (2010)
7. Erol, S., et al.: Combining BPM and social software: contradiction or chance? J. Softw. Maintenance Evol.: Res. Pract. **22**, 449–476 (2010). https://doi.org/10.1002/smr.460

Organization

Organizing Committee

Selmin Nurcan — Université Paris 1 Panthéon-Sorbonne, France

Rainer Schmidt — Munich University of Applied Sciences, Germany

Program Committee

Jan Bosch — Chalmers University of Technology, Sweden

Jochen De Weerdt — KU Leuven, Belgium

Sebastian Dünnebeil — Munich University of Applied Sciences, Germany

Holger Günzel — Munich University of Applied Sciences, Germany

Marlen Jurisch — Munich University of Applied Sciences, Germany

Kathrin Kirchner — Technical University of Denmark, Denmark

Michael Möhring — Hochschule Reutlingen, Germany

Selmin Nurcan — Université Paris 1 Panthéon-Sorbonne, France

Luise Pufahl — TU Munich, Germany

Flavia Santoro — UERJ, Brazil

Rainer Schmidt — Munich University of Applied Sciences, Germany

Johannes Tenschert — Friedrich-Alexander-Universität Erlangen-Nürnberg, Germany

Irene Vanderfeesten — KU Leuven, Belgium

Alfred Zimmermann — Reutlingen University, Germany

Non-visual Process Models: How Do Blind and Low-Vision Users Model Business Processes?

Lisa Baumann[1]([⊠]), Anjo Seidel[2], and Mathias Weske[2]

[1] Technical University of Munich, Arcisstraße 21, 80333 Munich, Germany
lisa.baumann@tum.de
[2] Hasso Plattner Institute, University of Potsdam,
Prof.-Dr.-Helmert-Str. 2-3, 14482 Potsdam, Germany
{anjo.seidel,mathias.weske}@hpi.de

Abstract. Business processes are typically represented with graphical models. This can make engaging with them a challenge for people with visual disabilities. It is unclear whether existing tools are used by and meet the requirements of those users. In general, little research has focused on making business process models more accessible. Based on this observation, this study explores how blind and low-vision users currently understand and model business processes. Utilising a Grounded Theory methodology, we analyse qualitative data obtained from interviews at German educational institutions where business process modelling is taught to blind and low-vision students. The empirical results give insights into process modelling tools used in practice, their limitations, and how collaborative process modelling can become more inclusive. We find that textual, tactile but also visual models are in use. While there are common challenges, blind and low-vision users have individual preferences and strengths and, therefore, require a diversity of approaches beyond visual modelling to understand and create business process models.

Keywords: Business Process Management · Business Process Modelling · Accessibility · Inclusion · Visual Disabilities

1 Introduction

The accessibility of conceptual or business process models for users with disabilities is an understudied field of research [19]. Business process models often contain visual elements [17], which can pose a barrier to engaging in the creation or discussion of those models for blind and low-vision users [9]. Inclusive UML modelling for people with visual disabilities has been investigated [11,18]. However, their broader experience with business process modelling remains understudied. We conducted semi-structured interviews with blind, low-vision, and

© The Author(s), under exclusive license to Springer Nature Switzerland AG 2025
K. Gdowska et al. (Eds.): BPM 2024 Workshops, LNBIP 534, pp. 155–167, 2025.
https://doi.org/10.1007/978-3-031-78666-2_12

sighted staff members in German educational institutions where business process modelling is taught by and to people with visual disabilities. The qualitative data is analysed with the Grounded Theory methodology [4].

Our analysis sheds light on four important aspects: (1) What tools and assistive technologies people with visual disabilities use, (2) the challenges they face with existing tools, (3) what determines the suitability of different tools, and (4) how inclusive collaborative modelling can be facilitated. We discuss how future research into and development of business process modelling tools can improve the modelling experience of users with visual disabilities by taking their needs and perspectives into account.

Following the introduction, we situate our paper in the business process management research landscape in Sect. 2, before we explain and motivate our methodology in Sect. 3. Subsequently, we present our findings in Sect. 4 and use Sect. 5 for a discussion before concluding the paper in Sect. 6.

2 Related Work

In the following, we present existing research on the accessibility of business process models and highlight research gaps.

Sarioğlu et al. evaluate the accessibility of conceptual modelling tools for users with disabilities, showing that many tools pose barriers for users with visual disabilities due to incompatibility with text-to-speech tools and a lack of customisation options for font and color [19]. In the same paper, they conduct a literature review on this topic. The authors show the lack of research but note that visual disability is the most discussed type of disability with nine papers. However, they find a need for empirical research involving people with disabilities.

Few papers have taken up the use case of business process modelling specifically. Vaziri and DeOliveira, who is blind himself, investigate the accessibility of business process modelling by evaluating Event-Driven Process Chains and the modelling tool ARIS 7.1 [22]. While barriers for blind and low-vision users such as inadequate keyboard support are discussed, the disabilities of the participants are not specified. Following up on the initial research, they partially evaluate ARIS and the Signavio process editor with one blind and two low-vision participants based on the WCAG 2.0 (Web Content Accessibility Guidelines) [21,23]. Again, several accessibility issues are discovered.

Vaziri and Bossauer evaluate the usability and accessibility of the ARIS Business Architect 7.2 [2]. Again, accessibility is measured by conformity to the WCAG 2.0. They find that only 37.5% of the criteria are met. While the participants' visual abilities are not explicitly stated, the use of eye-tracking software during the evaluation suggests that they were not blind.

Due to accessibility issues, business process modelling using standard tools and languages poses a challenge for users with visual disabilities. However, user experiences have to be considered to fully determine a tool's accessibility [3].

For improving the accessibility of business process modelling, we can identify promising approaches in the existing research. Text-based examples are textual

notations [17] and tools that generate a business process model from a textual
process description [10] and vice versa [14], potentially by using Large Language
Models [24]. Such transformations have been proposed for the collaboration of
people with and without visual disabilities in modelling [15]. As drag and drop
motions that are widely used in business process modelling tools [10] can present
a challenge to some users with visual disabilities [16], tools that enable them to
create a business process model via alternative input formats such as voice [20]
would likely improve accessibility. Lastly, tactile representations of business process models [7] would allow users with visual disabilities to use their tactile sense
to explore a process. We present those approaches to our interview participants
to gain an understanding of their potential for users with visual disabilities.

While potential solutions exist, to our knowledge no existing research provides insights into the current business process modelling practices of individuals
with visual disabilities. Therefore, it remains unclear which solutions meet the
needs of those users and can be transferred to practice. We address those questions in the interviews.

3   Grounded Theory Methodology

In the following, we first provide an introduction to the Grounded Theory
methodology before reporting on our process of data collection and analysis.
For reproducibility, our data can be found online[1].

3.1   Grounded Theory

The Grounded Theory methodology was developed by Glaser and Strauss [8].
The aim is to develop a theory that is "'grounded' in the data", for example,
interview data [1,4].

Conducting Interviews. We conducted semi-structured interviews based on
a prepared interview guide. Following Charmaz, we encouraged participants to
share their perspectives with open-ended questions [4]. We also remained open
to deviating from the guide based on the responses [1].

Open Coding. In Grounded Theory, open coding is the first phase of data
analysis [1]. We traversed the transcript line for line, and assigned descriptive
terms, so-called codes, to its content to describe what was said [4]. To support
the coding process, MAXQDA[2] was used as a software tool. Codes are developed
from looking at the data and do not come from prior research [13]. To report the
participants' perspectives and not our interpretations, we looked for actions in
the data when coding and used gerunds in the assigned codes, as recommended

---

[1] https://github.com/libaumann/nonvisualmodels.
[2] https://www.maxqda.com.

by Charmaz [4]. For example, several participants told us that low-vision students can perceive a visual model when they magnify the graphic on their screen. However, they can only see one part of the model at a time. During open coding, we assigned the code "Not seeing everything at once when magnifying".

Axial Coding. In the subsequent axial coding phase, the codes are brought into relation with each other and assembled into categories [5]. The following scheme [4,5] was used to create categories: (1) A **Phenomenon** is identified, for which (2) **Conditions** can be determined, i.e., the reasons or circumstances that lead to the phenomenon. (3) **Strategies** describe behaviours in response to the phenomenon. (4) **Consequences** are the results of a strategy. And lastly, (5) the **Context** can give additional information on the phenomenon. Coming back to the example above, we created a broader category, "Having a limited perception of the model". A condition is the use of textual descriptions and tactile materials. Memorising different parts can help (Strategy). A consequence of sequential processing is needing additional time. The ability of sighted individuals to perceive a graphic at once made this category significant (Context).

Identifying Gaps and New Directions. Instead of collecting all data before the analysis, both happen simultaneously in Grounded Theory [1]. Based on the analysis of initial data, more participants are approached and more data is gathered to expand and strengthen those findings [1,4]. When we contacted institutions that offer apprenticeships, we were told that degree programs might use different approaches. Therefore, we extended our search to universities.

Memo Writing. Memo writing is an important activity during data collection and analysis [1]. Those short texts can include explanations of codes and thoughts on the relationship between codes as well as open questions [4].

Theorising. The last step was bringing the memos and the results of the coding sessions together into a theoretical understanding of the topic [4]. According to Charmaz, one approach to theorising is focusing on actions and processes and considering how different events are connected and experienced by individuals [4]. Using the category above, we built a theory of how business process models are processed differently by people with and without visual disabilities. We elaborate on this further in Sect. 4.2.

3.2 Interview Setup

Participants. We contacted twelve German educational institutions that teach blind and low-vision students and include business process modelling in their curricula. Eventually, we recruited eight interview participants who are presented in Table 1. Some have experience with teaching business process modelling to students with and without visual disabilities, and facilitating collaboration in

Table 1. Overview of interview participants by role and visual disability.

Number	Role	Visual disability
1	Adapting class materials	Sighted
2	Former student	Blind since birth
3	Teacher	Late-blind
4	Teacher	Sighted
5	Former student	Visual disability, not specified
6	Teacher	Sighted
7	Teacher	Sighted
8	Case Manager	Blind

the classroom. One participant (1) makes class materials accessible for students with visual disabilities. Another (8) supports students with visual disabilities throughout their education. As former students of such institutions, two participants could provide a perspective on the barriers they encountered and the tools they used. The others reported on the experiences of the students they had worked with.

Interviews. Three interviews were conducted via Zoom and recorded. We interviewed two participants on the phone and sent the notes we took for revisions. We interviewed three participants together in person and recorded the conversation. The recordings were then transcribed following rules by Dresing et al. [6].

4 Process Modelling for Blind and Low-Vision Users

We now present our insights into four areas as visualized in Fig. 1. We describe (1) different techniques and tools and (2) the challenges faced by blind and low-vision users as they use them to access business process models. Because not all tools are equally suitable, (3) we consider factors that determine whether a tool is a good match for a user. In the end, (4) we reflect on the challenges of inclusive collaborative modelling.

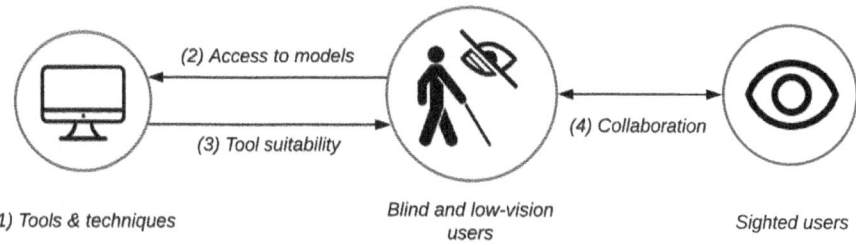

Fig. 1. A visual overview of the four areas of analysis mentioned above.

4.1   Tool Support for Blind and Low-Vision Users

Process models can be depicted in various ways. Typically, a graphical representation or textual description is used [17]. However, blind or low-vision users cannot perceive graphics or black letter text to the same extent as sighted users. Therefore, the participants shared specific tools that allowed them and their students to access digital and analogue process representations.

Textual Process Descriptions. To represent processes in a textual format, the participants mentioned full-text descriptions, the use of textual notations such as DOT[3] and PlantUML[4] or a custom textual syntax. Several participants noted that notations can take less time to read and provide a standardised representation, but one participant found that certain syntax elements like brackets can be harder to comprehend.

The participants shared different ways to access digital textual information. Using specific software, some users with low vision can magnify text on their screen. Alternatively, users can let screen readers read out digital text to them. Refreshable braille displays let them comprehend text displayed on a screen, one line of text at a time. Through openings, spikes are raised or lowered to automatically display the respective braille characters. Those solutions give users sequential text access.

Even though techniques exist to automatically translate graphical models into textual descriptions [14], none of the interview participants mentioned using them. Rather, transformations are done manually. Several participants said they provide more or less information depending on the needs of the student or leave out details such as the spatial location of elements. To make those decisions, they draw on their professional experience and feedback from students.

Digital Tool Support. The interview participants shared their perspectives on digital tools for accessing graphical models.

Low-vision users can use magnifying software. One participant also mentioned ongoing research into braille displays with multiple lines that would be more suited to displaying a graphic than refreshable braille displays with only one line.

In process modelling tools such as ARIS, the participants encountered accessibility issues. Instead, they use tools like Microsoft Word and Microsoft Excel. The mentioned advantages of these tools are their readability and access for users with and without visual disabilities.

Analogue Tool Support. Analogue process models were also in use. Many interview participants incorporate tactile models into their work, although several said that creating them takes time and effort. Tactile process representations

---

[3] https://graphviz.org/doc/info/lang.html.
[4] https://plantuml.com.

give students a spatial understanding of the process and the shape of visual elements. A blind person can trace a tactile model with their fingers and then translate what they perceive into the flow and elements of the model. Several participants mentioned that people who were born blind cannot draw on their prior experience of sight and can have problems with this. The successful use of tactile materials depends on the user's prior tactile experiences and tactile skills.

The interview participants find creative solutions to create tactile teaching materials such as using punched-out symbols from a flowchart drafting template or crafting magnetic elements out of foam rubber that can then be placed on a magnetic board. Besides 3D printing and braille printing, printing a graphic on swell paper was mentioned as a more technical approach: When the paper is heated, the outline of the graphic swells and can be perceived through touch.

One blind teacher includes textual information on tactile models in braille but notes that, when students are not proficient braille readers, auditive captions could be used as well. In Germany alone, multiple types of braille are in use of which some include contractions [12]. Eurobraille, a standard for digital texts, uses eight dots instead of the usual six for one symbol. The teacher noted that, as a result, braille texts take up more or less space. Readability can also be impacted: As a late-blind person, they mentioned that it takes them more time to read Eurobraille.

4.2   Access to Models

As they use the tools and techniques described above, blind and low-vision users can encounter challenges.

The Challenge of Sequential Data Access. Without visual abilities that allow quick processing of a complete graphic, data access is sequential. Several participants mentioned that, if using tools such as magnifying software or refreshable braille displays, users process one part of a model at a time and require more time than sighted people who look at a graphic and perceive it at once.

To cope with this issue, people with visual disabilities can memorize previous parts of the model to, in the end, have an understanding of the whole. Good memorisation skills make this task easier for students.

When exploring a model sequentially, it can be a challenge to keep an overview. However, understanding the position of elements within a model and knowing which elements are nearby are important pieces of information to some participants. To explore a tactile model, one former student explained that one hand can serve as an anchor and the other can be moved around.

Limited Editing Access. While several participants stressed that people with visual disabilities also have an interest in creating and editing visual models, the possibility to create or edit non-textual models is currently limited.

Using notations such as PlantUML and DOT, users can edit a textual process description and automatically transform it into the responding graphical model.

Yet, several people noted that they cannot check whether the graphic is correct or contains errors as, in a graphical format, the model becomes a "black box" to them. For this, they will need to consult a sighted person, which compromises their ability to work independently and costs time.

In an exploratory question, we asked about the potential of using one's voice to create or edit process models. One former student expressed doubts about whether voice input would always be processed correctly. As described above, resulting errors in the graphic would be difficult to spot. The participant would, for this reason, prefer textual input even though it can take longer. Other participants highlighted the potential of voice input for people who lack motor skills or vision to model graphically.

4.3   Suitability of Process Modelling Tools

A single tool is often not sufficient to meet all the needs of a user. Therefore, multiple tools are used alongside each other. For example, in exams that involved diagrams, a former student used both textual descriptions and the diagrams printed in braille to get a feeling for the original exam content. In the following, we reflect on factors affecting the suitability of tools as reported by our participants.

Suitability Is Determined by Individual Factors. The user's specific visual disability, skill levels, prior experiences, and personal preferences influence which approach would be suitable for a student.

The participants reported that some of their low-vision students could make use of and work with visual models. The same is not possible for blind students. A recurring theme in the interview data is the differences between people who have been blind from birth or a young age and those who became blind later in life. For example, in the experience of some participants, late-blind people can lack tactile skills. But it was also noted that, compared to people who have been blind from birth, it can be easier for them to develop an understanding of visual models by drawing on their prior visual experiences.

Users with visual disabilities often work with assistive technologies such as screen readers to access process models. A good command of those technologies is needed and has to be taught. Skills that are specific to the experience of users with visual disabilities such as braille literacy can also not be taken for granted but have to be developed.

Visual abilities can change. According to one participant, some users with low vision might perceive a graphic well on some days but not every day. A change in visual abilities can be accompanied by the need for different assistive technologies. Tools that are primarily accessible with one skill set or a specific visual disability can become unusable.

Cost Limits the Use of Tools. For the participants, affordability was one factor in the decision for or against a certain tool. Another consideration is whether the tool will be paid for by funding agencies. They explained that assistive technologies such as refreshable braille displays and tools that target users with disabilities are often expensive.

Availability and Accessibility Issues Limit the Use of Tools. Even if users find a tool they can work well with, issues can still arise. Some participants experienced that tools would be replaced and no longer be supported. This means additional work to find and adjust to new tools.

Participants recounted frustration when working with tools, stemming from accessibility issues such as limited keyboard support and inconvenient design decisions like elements that could only be differentiated from each other by colour.

4.4   Inclusion in Collaborative Process Modelling

Process models are often collaboratively created and then accessed by different stakeholders [14]. This can be difficult if the visual abilities of the involved stakeholders differ. Our participants described the challenges of collaboration and gave insights into overcoming them.

Graphical Models Are the Standard. The participants reported that they and their students work with the visual languages UML, Event-Driven Process Chains, and flow charts. Several participants feel that a visual representation is the dominant way to depict process models. In their experience, this is preferred by sighted people. As they want to make communication with those users easier, they see a need for users with visual disabilities to adapt and learn to work with graphical models. To do this, they employed different strategies. As a student, one participant had sighted classmates explain what different elements look like. One of the teachers used tactile diagrams.

The Value of Independence. From wanting to be as independent as possible to accepting help, the interview participants shared mixed feelings toward receiving support. While generally open to assistance, one participant stresses the importance of being independent in the workplace where having to rely on an assistant to translate diagrams could clash with time constraints.

Collaboration Requires Different Model Types. Noting the prevalence of graphical models, some participants found the need for a tool that allows blind and low-vision users to create such models. However, a teacher remarked that this does not have to be done using solely graphical modelling tools.

Instead, as proposed by some participants, a modelling tool could switch between alternative representations of the same model. For example, after a blind user has described a model textually, the tool could transform it into the corresponding visual model. The participants liked the idea of different output options, for example, printing a graphic on swell paper or in braille. Giving users the possibility to switch between model representations would enable collaboration while allowing all collaborators to choose the representation they work best with.

Beyond the Visual. The participants with visual disabilities noted the current need for sighted assistance to both understand and create graphical business process models in their roles as teachers and students. Yet, the interviews also revealed that others might not always be available to help, and blind or low-vision users are not always open to relying on assistance.

One former student highlighted that the successful inclusion of people with visual disabilities also requires a team that is open to adjusting how they work. We can conclude that the adaption of new ways of collaborating in modelling activities that go beyond visual models is necessary.

5   Discussion

In this section, we evaluate our findings and, while addressing their limitations, also highlight areas for future research to improve the modelling experience of blind and low-vision users.

Our research has several limitations. All interviews were conducted in the German educational context. The perspectives of students were underrepresented. Due to the qualitative research approach and the limited number of interview participants, the results are not representative of all users with visual disabilities. While the use of the Grounded Theory methodology limits our analysis, it has allowed us to set aside our assumptions and build an initial understanding of the topic from the perspective of those who experience barriers first-hand.

Despite their limitations, the results point towards relevant topics for future research. More studies should focus on investigating the adaptation of business process modelling practices by users with visual disabilities. A larger number of participants from different countries and outside of the educational contexts should be included. Based on more data, requirements for business process modelling tools and especially collaborative features can be formulated. Factors such as the onset of blindness, educational opportunities and additional disabilities should be taken into account when selecting participants. The interviews indicate that a broad range of tools and techniques are needed to satisfy the diverse and changing requirements of users with visual disabilities. Therefore, tools should offer different model representations that can be edited and accessed in line with the diverse preferences and skills of users with and without visual disabilities. The potential of Large Language Models for users with visual disabilities could be another interesting topic of research. By including people with visual disabilities throughout the development process, their needs can be met better.

6   Conclusion

The business process management research community has compared the benefits and drawbacks of visual and text-based models [17]. By focusing on the user group of people with visual disabilities, this paper adds a new perspective to the debate. For many users with visual disabilities, especially blind users, alternatives to visual models are not only a question of preference but a necessity for their participation in business process modelling. However, our interview participants show that such alternatives exist and are already used successfully to give blind and low-vision users access to process models.

With this paper, we study how blind and low-vision users understand and create business process models in an educational context. The analysis of semi-structured interviews using the Grounded Theory methodology provides insights into tools already in use and their limitations. Factors that influence the adaptation and suitability of modelling tools are provided. Additionally, we present requirements for collaborative modelling.

Accessibility has to be considered when developing and researching business process modelling tools and languages. People with visual disabilities should be included in this process. It is, in the end, not only an opportunity for business process management practitioners and researchers to reach more people with their ideas but, given the barriers to the equal participation of people with visual disabilities, to grow into a more diverse and inclusive community.

Disclosure of Interests. The authors have no competing interests to declare that are relevant to the content of this article.

References

1. Birks, M., Mills, J.: Grounded Theory: A Practical Guide. SAGE (2010)
2. Bossauer, P., Vaziri, D.: Evaluation of BPM-tools in terms of it-usability and it-accessibility. EMISA Forum **34**, 32–38 (2014)
3. Calvo, R., Seyedarabi, F., Savva, A.: Beyond web content accessibility guidelines: expert accessibility reviews. In: Proceedings of the 7th International Conference on Software Development and Technologies for Enhancing Accessibility and Fighting Info-exclusion, pp. 77–84. Association for Computing Machinery (2016). https://doi.org/10.1145/3019943.3019955
4. Charmaz, K.: Constructing Grounded Theory: A Practical Guide Through Qualitative Analysis. Sage Publications (2017)
5. Corbin, J.M., Strauss, A.: Grounded theory research: procedures, canons, and evaluative criteria. Qual. Sociol. **13**, 3–21 (1990)
6. Dresing, T., Pehl, T., Schmieder, C.: Manual (on) Transcription: Transcription Conventions, Software Guides and Practical Hints for Qualitative Researchers (2015). https://www.audiotranskription.de/wp-content/uploads/2020/11/manual-on-transcription.pdf
7. Edelman, J., Grosskopf, A., Weske, M., Leifer, L.: Tangible business process modeling: a new approach. In: Proceedings of ICED 09, the 17th International Conference on Engineering Design, pp. 485–494 (2009)

8. Glaser, B.G., Strauss, A.L.: The Discovery of Grounded Theory: Strategies for Qualitative Research. Aldine Transaction (1967)

9. Groenda, H., Seifermann, S., Mueller, K., Jaworek, G.: The cooperate assistive teamwork environment for software description languages. Stud. Health Technol. Inform. **217**, 111–8 (2015)

10. Ivanchikj, A., Serbout, S., Pautasso, C.: From text to visual BPMN process models: design and evaluation. In: Proceedings of the 23rd ACM/IEEE International Conference on Model Driven Engineering Languages and Systems, pp. 229–239. Association for Computing Machinery (2020). https://doi.org/10.1145/3365438.3410990

11. King, A., Blenkhorn, P., Crombie, D., Dijkstra, S., Evans, G., Wood, J.: Presenting UML software engineering diagrams to blind people. In: Miesenberger, K., Klaus, J., Zagler, W.L., Burger, D. (eds.) ICCHP 2004. LNCS, vol. 3118, pp. 522–529. Springer, Heidelberg (2004). https://doi.org/10.1007/978-3-540-27817-7_76

12. Lang, M., Hofer, U., Winter, F.: The braille reading skills of German-speaking students and young adults with visual impairments. Br. J. Vis. Impairment **39**, 6–19 (2021). https://doi.org/10.1177/0264619620967689

13. Lazar, J., Feng, J.H., Hochheiser, H.: Research Methods in Human-Computer Interaction. Morgan Kaufmann (2017)

14. Leopold, H., Mendling, J., Polyvyanyy, A.: Generating natural language texts from business process models. In: Ralyté, J., Franch, X., Brinkkemper, S., Wrycza, S. (eds.) CAiSE 2012. LNCS, vol. 7328, pp. 64–79. Springer, Heidelberg (2012). https://doi.org/10.1007/978-3-642-31095-9_5

15. Luque, L., Santos, C., Cruz, D., Brandão, L., Brandão, A.: Model2gether: a tool to support cooperative modeling involving blind people. In: CBSOFT 2016 (2016)

16. Milne, L., Ladner, R.: Blocks4all: Overcoming accessibility barriers to blocks programming for children with visual impairments. In: Proceedings of the 2018 CHI Conference on Human Factors in Computing Systems, pp. 1–10. Association for Computing Machinery (2018). https://doi.org/10.1145/3173574.3173643

17. Ottensooser, A., Fekete, A., Reijers, H., Mendling, J., Menictas, C.: Making sense of business process descriptions: an experimental comparison of graphical and textual notations. J. Syst. Softw. **85**(3), 596–606 (2012). https://doi.org/10.1016/j.jss.2011.09.023

18. Petrausch, V., Seifermann, S., Müller, K.: Guidelines for accessible textual UML modeling notations. In: Miesenberger, K., Bühler, C., Penaz, P. (eds.) ICCHP 2016. LNCS, vol. 9758, pp. 67–74. Springer, Cham (2016). https://doi.org/10.1007/978-3-319-41264-1_9

19. Sarioğlu, A., Metin, H., Bork, D.: How inclusive is conceptual modeling? A systematic review of literature and tools for disability-aware conceptual modeling. In: Almeida, J.P.A., Borbinha, J., Guizzardi, G., Link, S., Zdravkovic, J. (eds.) ER 2023. LNCS, vol. 14320, pp. 65–83. Springer, Cham (2023). https://doi.org/10.1007/978-3-031-47262-6_4

20. Signavio: Process on Voice: speech recognition for BPMN modeling. https://www.signavio.com/news/process-on-voice-speech-recognition-for-bpmn-modeling/. Accessed 15 Mar 2024

21. Thatcher, J., et al.: Web Accessibility: Web Standards and Regulatory Compliance. Apress Berkeley (2006). https://doi.org/10.1007/978-1-4302-0188-5

22. Vaziri, D., DeOliveira, D.: Accessible business process modeling. Int. J. Bus. Econ. Eng. **6**, 117–128 (2012). https://doi.org/10.5281/zenodo.1055950

23. Vaziri, D., DeOliveira, D.: Improving the Accessibility of Business Process Modelling Tools, pp. 383–411. IGI Global (2014). https://doi.org/10.4018/978-1-4666-4667-4.ch015

24. Vidgof, M., Bachhofner, S., Mendling, J.: Large language models for business process management: opportunities and challenges. In: Di Francescomarino, C., Burattin, A., Janiesch, C., Sadiq, S. (eds.) BPM 2023. LNBIP, vol. 490, pp. 453–465. Springer, Cham (2023). https://doi.org/10.1007/978-3-031-41623-1_7

Diversity and Inclusion in HR Processes in Financial Sector

Sylwia Białas[iD] and Piotr Wróbel[✉] [iD]

University of Gdansk, Armii Krajowej 101, 81-824 Sopot, Poland
`piotr.wrobel@ug.edu.pl`

Abstract. The aim of the study was to identify best practices for incorporating Diversity, Equity, and Inclusion (DEI) in HR processes within the financial sector. The study reviewed 101 best practices for integrating DEI into HR processes within 20 financial sector entities in Poland. Significant potential for DEI implementation was found in HR processes such as communication, employee development and succession, recruitment and selection, compensation and benefits, readaptation. Practices used by banks in Poland primarily target gender diversity, reflecting trends observed in other countries. Age diversity practices are gaining importance due to demographic changes, with future focus expected on intergenerational cooperation and meeting the needs of different generations. Other dimensions, such as disability, neurodiversity, and LGBT+ inclusion, are less addressed but are anticipated to grow in prominence due to demographic shifts and technological advancements. The results of the study may serve as an inspiration for redesigning HR processes in the financial sector and other industries, aiming to create an inclusive work environment.

Keywords: Diversity · Equity and Inclusion (DEI) · Diverse Workforce · Inclusive Workplaces

1 Introduction

In numerous organizations, the implementation of Diversity, Equity, and Inclusion (DEI) leads to a transformation of organizational culture (Jones, 2022). Such a process requires time and typically begins with activities aimed at raising DEI awareness among employees. However, this alone is insufficient. To ensure lasting changes, embedding DEI within a company requires integrating DEI principles into existing Business Process Management (BPM) frameworks, thereby ensuring the change's permanence. The implementation of DEI in processes directly related to people, particularly Human Resources (HR) processes, is particularly important (Jackson et al., 2017). This is corroborated by DEI maturity models, which indicate that changes in HR processes are fundamental prerequisites for more profound organizational changes (Wieczorek-Szymańska, 2017).

Integrating DEI into HR processes is crucial, especially in the context of dynamic changes within process taxonomies, such as the APQC framework. The "iteration of specific processes" metric underscores the need for flexibility and adaptation. When combined with DEI considerations, can lead to more responsive and sustainable human resource management.

In contemporary economies, the financial sector plays a pivotal role by providing enterprises with funding sources and investment opportunities. Furthermore, many entities within the financial sector operate on an international scale, facing the challenges of employee diversity and the need to build inclusive workplaces (Bank of England, 2021). Meanwhile, there is a notable lack of research analyzing the implementation of DEI into processes within the financial sector, particularly HR processes. This creates a research gap at the intersection of the general DEI concept and its operationalization into existing Business Process Management frameworks in the financial sector.

The aim of the study presented in this article was to identify best practices for incorporating Diversity, Equity, and Inclusion (DEI) in HR processes within the financial sector. This objective is highly significant, given research findings indicating that many DEI programs implemented in organizations fail (Dobbin, Kalev, 2016). Moreover, although the issue of gender equality has been raised for a long time, the share of women at managerial levels is growing very slowly.

The authors selected the financial sector due to its propensity for adopting innovative management practices, thereby creating experiences that can be implemented in other sectors (Fasnacht, 2018). Unlike operational processes, HR processes possess a more universal character, facilitating the transfer of experiences across industries. Therefore, this study aims to propose solutions to enhance HR processes to foster the development of inclusive workplaces across various sectors of the economy.

Research methodology: The study used literature review (Scopus database), a review of best practices from the Forum for Responsible Business database (a Polish organization promoting the idea of DEI in business), and an analysis of activities conducted by all commercial banks in Poland (20 entities) based on their websites, social media, and Environmental, Social and Governance (ESG) reports.

2 Literature Review

2.1 DEI and HR Processes

Diversity, Equity, and Inclusion (DEI) have become central ideas in modern discussions about human resource management (Khan, Khan 2023). Tung (2016) examines the changes and challenges in the international human resource management context. She emphasizes that global war for talent, brain circulation, and the growing use of multicultural teams render the need for diversity management.

However, the use of DEI has evolved over time. Initially, it focused on compliance with anti-discrimination laws, but gradually expanded its scope and importance. Nowadays, contemporary HRM recognizes DEI as a strategic imperative (Jackson et al., 2017). Shen et al. (2009) concludes that diversity management and human resource management overlap, as diversity management is an approach that revolves around employees and the HRM function is the custodian of the people management processes.

Researchers point to several HR processes in which DEI elements are or should be implemented: recruitment and selection, training and development, promotions and succession planning, remuneration, communication.

Companies aim to create a workforce that reflects the diversity of the communities they serve. This includes implementing unbiased hiring practices and actively seeking out diverse candidates (Gutterman, 2023). Such practices include hiring women, older people, and persons with disabilities (Erickson et al. 2014). Moreover, implementing DEI increases the attractiveness of the company as an employer and allows it to attract talents (Coles, 2021).

Regular training on topics such as unconscious bias, cultural competency, and inclusive leadership can help employees and managers understand the importance of DEI and how they can contribute (Gutterman, 2023, Johnson-Mallard et al., 2019). Training and development programs support women in their professional aspirations, contributing to greater equality in managerial positions (Russell et al., 2023, Khasawneh et al., 2023). These activities can be supported by succession plans that create appropriate share of seats is reserved for women.

Implementing the DEI concept in compensation process aims to achieve pay equity and eliminate wage disparities based on gender, race, ethnicity, sexual orientation, and other demographic factors. To this end, compensation processes incorporate pay audits and systematic analyses of wage gaps (Anderson et al., 2023). The value of salaries and raises is determined based on transparent criteria. Sometimes, pay transparency is introduced, although the effects of this action can vary. For instance, (Gulyas et al., 2023) indicate that pay transparency implemented in large Austrian companies did not significantly impact the wage gap. Nonetheless, raising awareness about wage disparities and conducting training on gender equality and bias elimination is emphasized (Kalev 2023).

An important aspect of HR departments' activities is supporting the internal communication process. Many researchers (e.g., Borito, 2023) emphasize that building an inclusive workplace should be accompanied by a change in internal communication, particularly the use of inclusive language in daily practice.

Researchers identify several factors supporting the implementation of DEI in organizations. In particular, the importance of Employee Resource Groups (ERGs) in this process is emphasized. ERGs, which are voluntary, employee-led groups, provide support, networking opportunities, and advocacy for underrepresented or marginalized employees. ERGs can play a significant role in promoting diversity and inclusion within the organization (Cenkci et al., 2019, Ramshankar, Thomas, 2023).

Some researchers (eg. Bruni et al., 2023) highlight the sometimes-occurring problem of lack of definitive or standardized corporate metrics for DEI. This includes measuring DEI activities, the sense of inclusion among employees, and the diversity of human resources.

The significance of HR processes in implementing DEI is recognized in DEI maturity models. For example, in the Diversity & Inclusion Maturity Model (Bourke, Dillon, 2018), the final level (Integrated) indicates that DEI is fully integrated into employee and other business processes. Similarly, the Model of Diversity Management (Wieczorek-Szymańska, 2017) emphasizes the introduction of the diversity concept into HRM.

Ultimately, HR processes may contribute to shaping organizational culture by fostering an environment where diversity and inclusion are valued, celebrated, and deeply integrated into the organization's fabric. This involves promoting open communication, respect for diverse perspectives, and cultivating a sense of belonging for all employees (Jones 2022).

2.2 Determinants of DEI Implementation in Banks in Poland

Currently, there are 20 commercial banks operating in Poland, ranging from small entities employing a few hundred people to large banks with tens of thousands of employees (ZBP, WIB, 2023). The adoption of the DEI concept by these banks is influenced by several factors, particularly the ownership structure and the national culture, social, demographic, and legal environment.

Of the 20 banks, 12 have a majority foreign owner, meaning they are part of international banking structures. The organizational cultures of these banks combine values derived from Polish culture with the values of the country of capital origin, creating a unique blend of values. Employees in banks headquartered in Western European countries (e.g., France, Spain, Germany) are influenced by cultural patterns in which tolerance for diversity plays a significant role. In such banks, the implementation of diversity management is therefore easier, and changes resulting from DEI find a receptive environment. However, acceptance and relevance of DEI initiatives depend from national context. The meaning and scope of DEI areas vary across cultures (Goodman, 2013). Poland is a country with a relatively homogeneous national and ethnic culture (Witkowska-Chrzczonowicz, 2021). This can influence attitudes toward diversity. Traditional values also play a significant role, affecting ethical issues such as gender and sexual diversity. Specifically, dimensions of national culture, such as uncertainty avoidance and the masculinity, impact the level of acceptance of diversity management among employees or customers in Poland (Hofstede, 2001). The high level of uncertainty avoidance characterizing Polish culture is associated with greater conservatism and resistance to change, posing a challenge for DEI implementation, which requires flexibility and openness to new ideas. Moreover, in cultures with a higher level of masculinity, such as Polish culture, traditional gender roles may still be significant. This influences perceptions of women's roles and career opportunities, presenting another challenge for DEI initiatives.

Additionally, the workforce structure in banks is predominantly female; however, senior management is male-dominated. There is a growing presence of Generation Z employees in banks, who bring specific cultural values to the organizational culture, particularly the need for gender equality and openness to cultural, ethnic, and sexual orientation diversity (Mărginean, 2021). For many young people, the changes resulting from DEI implementation seem natural and self-evident.

In recent years, the number of legal acts stimulating actions in selected dimensions of diversity has been rapidly increasing. The European Union's regulations increasingly focus on issues of equal treatment and anti-discrimination. Directives on parental leave and work-life balance, as well as the directive on gender balance among directors of listed companies, provide a clear impetus for enterprises to implement the DEI concept (Paolone et al., 2024).

3   Study Design

The research gap underlying this study pertains to the lack of research on the operational implementation of the general DEI concept into existing Business Process Management frameworks in financial sector entities. The aim of the study was to identify best practices for incorporating DEI into HR processes in the financial sector.

Research questions:

1. How can the DEI concept be implemented in financial sector entities through HR processes?
2. Which HR processes offer the greatest potential for DEI implementation?
3. For which dimensions of diversity do banks most frequently implement DEI practices in HR processes?

The study examined all 20 commercial banks in Poland. It is noteworthy that the financial sector leads the way in implementing DEI in Poland, with 11 out of 20 commercial banks being signatories of the Diversity Charter. The substantial involvement of banks in creating inclusive workplaces also stems from legal regulations, such as those related to the implementation of ESG.

Two primary sources of information were used in the study. The first source was the best practices database maintained by the Responsible Business Forum (FOB internet database: Dobre praktyki ESG w Polsce). Founded in 2000, FOB is the largest non-governmental organization in Poland dedicated to the concept of sustainable development. The database is populated by FOB members, including 11 commercial banks operating in Poland. It contains 8,564 best practices from the period 2016–2023 applied in various economic sectors covering a wide range of issues, from consumer matters to the natural environment, human rights, and work organization. For the analysis, 87 practices were selected based on three criteria: 1) applied by financial institutions, 2) addressing two of the Sustainable Development Goals: Gender Equality (Goal 5), Reduced Inequalities (Goal 10), 3) addressing HR processes; we excluded for example activities focused on charity activities and anti-discrimination practices resulting from legal regulations.

The second source of information were the websites, social media, and ESG reports of all 20 commercial banks in Poland, providing an additional 14 practices. We used the same criteria as in the case of the FOB database. As a result, a total of 101 best practices were analyzed.

In analyzing these sources, we focused on best practices across seven HR processes: recruitment and selection, readaptation, development and succession, compensation and benefits, functioning of Employee Resource Groups, communication during DEI implementation, and ongoing internal communication. Within the scope of DEI, we analyzed various dimensions of diversity: gender, age, disability, neurodiversity, sexual orientation, and caregiving responsibilities.

4   Results

Among the best practices shared by financial institutions (FOB database, banks' websites and social media) the largest portion (58 out of 101) consists of those that lay the groundwork for the implementation of DEI, particulary through communication process. These practices include a broad spectrum of activities, ranging from identifying employee perceptions of diversity to recruiting DEI allies within the organization, educational initiatives, and communication campaigns focusing on selected dimensions of diversity (Fig. 1).

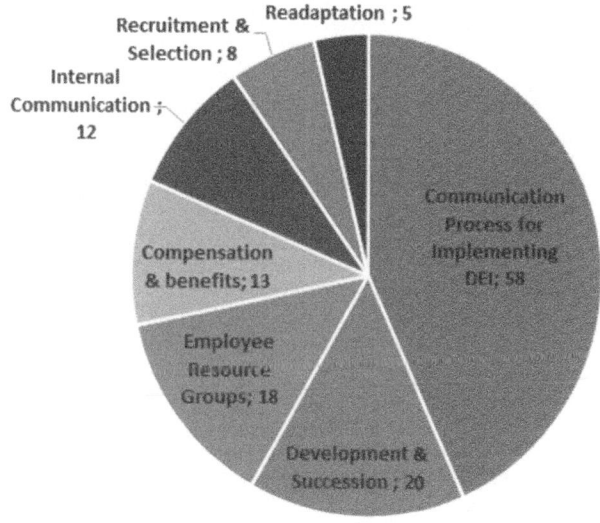

Fig. 1. FOB database, websites, social media, and ESG reports of banks

Many practices pertain to changes in existing HR processes. A relatively large portion is associated with development and succession processes (20%). This include initiatives such as coaching and mentoring programs primarily aim to support women, addressing the issue of the low representation of women among bank executives. Other development activities focus on reskilling older employees to adopt their skills to new reqiurements. Eighteen practices related to the functioning of Employee Resource Groups were identified across seven financial institutions. Some of the ERGs operate on a local level, while others were transnational. Their activities focuse on various employee groups, such as women, parents, and LGBT+. Practices related to compensation and benefits (eliminating the pay gap, inclusive benefits), internal communication (promoting inclusive language), recruitment and selection (minimizing biases and negative attitudes towards certain job candidates), and employee readaptation are less frequently presented.

Diversity management in financial sector addresses various dimensions, including gender, age, sexual orientation, disability, and caregiving responsibilities. The vast majority of the analyzed practices focus on gender, followed by age. Practices related to disability, neurodiversity, or sexual orientation are much less common. Communication activities, raising employee awareness about the DEI concept, are the most diverse in terms of dimensions they address. Recruitment and development activities also exhibit diversity, addressing gender, age, and disability dimensions. Other practices are predominantly focused on the gender dimension.

Not all best practices are process-oriented; some are one-time initiatives. However, these initiatives, when successful, can sometimes evolve into processes. For instance, in one bank, a single inclusive language workshop eventually led to the transformation of the entire internal communication system to use inclusive language (Table 1).

Table 1. Examples of Best Practices for Various HR Processes in the Financial Sector.

Process	Description of Best Practice	Comment
Recruitment and Selection Process	Inclusive job advertisements	Implementation of inclusive principles in drafting job advertisements
	Interview scenarios	Structuring interviews to avoid biases and unconscious negative attitudes toward candidates
	Metrics regarding the participation of underrepresented groups among new hires	Supporting the employment of people with disabilities
Employee Readaptation Process	Support for employees returning after a long absence	Additional training, buddy support, flexible working hours, and the possibility of reduced working hours
Development and Succession Process	Development programs for women	Individual potential diagnosis, training, inspirational meetings, coaching and mentoring, and networking
	Specialized competency development programs	Example: IT trainings for women
	Metrics regarding the participation of women at different management levels	Metrics may be accompanied by long-term goals
Compensation and Benefits Process	Incorporating goals related to minimizing the pay gap in compensation decisions	Example: annual salary increases
	Inclusive benefits	Benefit regulations consider the diversity of employees' close relations (e.g., informal and same-sex relations) and are available regardless of employment form. Additional leave related to the gender transition process
	Benefits for employees with caregiving responsibilities	Additional financial benefits, extra days off, and flexible work arrangements. Consultation with geriatric consultants

(continued)

Table 1. (*continued*)

Process	Description of Best Practice	Comment
Employee Resource Groups	ERGs supporting women, LGBT+, parents, etc.	Activities supporting chosen dimension of diversity
Communication Process for Implementing DEI	Employee opinion surveys on DEI	Periodic surveys of employee awareness followed by action planning based on the diagnosis
	Dedicating specific time periods to popularize important issues (thematic events)	One of the most frequently used practices. Examples: Diversity Days, Rainbow Culture Festival
	Recruiting allies for DEI within the organization	DEI ambassadors - from executive level to bank branches
	Recording videos with participation of employeees	Videos addressing selected dimensions and diversity issues
	Guides facilitating collaboration with persons with disabilities, LGBT+ community, etc.	Guides support internal collaboration and communication with clients
	Inclusive Leadership Trainings	Training for managers at all organizational levels
Ongoing Internal Communication Process	Introducing a standard of simple and inclusive communication (inclusive language)	The practice includes several activities - from developing the standard through workshops and training to modifying the website and bank documents regarding the language used
	Dialogue Skills Workshops	Supporting communication between different employee groups. Preventing hate speech and microaggressions

Sources: FOB database, websites, social media, and ESG reports of banks

5 Discussion and Conclusions

Goals

The vast majority of identified practices focus on managing existing employee diversity rather than expanding team diversity. Actions aimed at increasing the employment of underrepresented groups, such as individuals with disabilities or women in IT teams, are very rare.

In terms of diversity dimensions, the analyzed practices pertain to various aspects; however, the majority of them focus on gender. Polish banks do not differ significantly in this regard from companies in other countries. Several studies highlight the prevalence of diversity initiatives specifically targeting women (e.g., Cundiff et al., 2018; Gülsoy & Ustabaş, 2019). This enables banks to address the issue of gender disparities at different management levels and the gender pay gap. However, Polish banks are rather cautious in implementing these practices. Many bankers assert that changes require time and that a rapid shift in gender proportions at higher management levels could create unnecessary tensions among employees. There is an emphasis on providing women with greater career development opportunities while ensuring that final staffing decisions remain gender-neutral.

Practices targeting other dimensions of diversity, particularly age, are gaining importance. This is due to demographic changes that are becoming increasingly visible across the economy. Over time, the age dimension is expected to gain prominence in diversity management. As indicated by other researchers, such practices will focus on developing intergenerational cooperation (Baran & Kłos, 2014) and more effectively meeting the needs of employees from different generations (Li et al., 2022).

Other dimensions of diversity such as disability, neurodiversity, LGBT+ were rarely addressed and only recently. These areas are clearly neglected in the practices being developed. Previous studies emphasize that organizations' diversity and inclusion plans seldom include disability as a diversity dimension (Miethlich & Oldenburg, 2019). On the other hand, as indicated by Nweiser and Dajnoki (2024), demographic changes, along with substantial improvements in science, technology, architecture, and ergonomic design, will prompt more companies to seek employees among individuals with disabilities. Therefore, the dimensions of disability and neurodiversity are expected to gain importance in the studied entities. The small number of actions targeted at the LGBT+ community is not surprising in the context of previous studies on Polish enterprises (FOB, 2022).

Methods of Implementing DEI

The majority of the analyzed best practices (57%) focused on raising awareness among managers and employees about the DEI concept (communication process). This is likely because DEI is still a relatively new concept in Poland that requires popularization. On the other hand, companies that have long been implementing DEI also hire new employees, necessitating ongoing communication efforts in this area. This is particularly relevant in a country such Poland with a relatively homogeneous and conservative society. The importance of gaining support for DEI among managers and employees is emphasized by many researchers (e.g., CIPD, 2019).

Significant potential for DEI implementation was observed in HR processes such as employee development and succession (a considerable number of banks), ERG, compensation and benefits, internal communication, employee recruitment. Additionally, a few banks embed DEI in readaptation processes. Similar conclusions were reached by Daya (2014), who emphasized the importance of transparent recruitment, promotion, and development in her studies of companies implementing DEI in South Africa.

Virtually all the identified practices within specific HR processes have a universal character, allowing for their broad application in other sectors as well.

Involved Individuals and Entities

Typically the initiative to embed DEI in HR processes originates from HR departments, but DEI departments, board members, ESG committees, and employee resource groups also play important roles. Occasionally, the impetus for change comes from sales and marketing departments, which focus on communication with diverse customer groups, including disadvantaged groups such as people with disabilities, the LGBT+ community, and the elderly.

It is worth noting that external entities, such as experts, researchers, and NGOs, are often involved in the development and implementation of certain practices. Examples include the support of linguistic experts in implementing inclusive language or foundations supporting the LGBT+ community in developing a guide to facilitate collaboration with representatives of this group. This appears to be a factor increasing the likelihood of successful practice implementation.

Connections to BPM Frameworks

Integrating best DEI practices into existing Business Process Management frameworks, rather than implementing standalone projects, offers the potential for the lasting embedding of DEI values and the creation of an inclusive culture. However, the implementation of best practices must consider the organisation's DEI maturity level and its needs. The organizational context is crucial when designing DEI solutions to ensure the success of such changes.

The limitations of the study arise from the research method used, which relies on publicly available data provided by banks. The researchers did not have direct contact with the individuals responsible for DEI in the banks.

Disclosure of Interests. The authors have no competing interests to declare that are relevant to the content of this article.

References

Anderson, D., Bjarnadóttir, M.V., Ross, D.G.: Bridging the gap: applying analytics to address gender pay inequity. Prod. Oper. Manage. **32**(6), 1846–1864 (2023)

Bank of England: Diversity and inclusion in the financial sector – working together to drive change. Discussion Paper DP21/2 (2021)

Baran, M., Kłos, M.: Managing an intergenerational workforce as a factor of company competitiveness. J. Int. Stud. **7**(1), 94–101 (2014)

Borito, R.: Gender-inclusive language of office communications in an academic setting. Int. J. Innov. Res. Multidisc. Educ. **02**(09), 467–472 (2023)

Bourke, J., Dillon, B.: The diversity and inclusion revolution. Deloitte Rev. **22** (2018)

Bruni, S., Shenberger-Trujillo, J., Clark, T., Meyer, S.: Developing a human-centered corporate metric for inclusion, diversity, equity, & antiracism in a small business: a case study and lessons learned. In: 67th International Annual Meeting – Human Factors and Ergonomics Society, vol. 67, no. 1, pp. 1451–1456 (2023)

Cenkci, A.T., Zimmerman, J., Bircan, T.: The effects of employee resource groups on work engagement and workplace inclusion. Int. J. Organ. Divers. **19**(2), 1–19 (2019)

Chartered Institute of Personnel and Development (CIPD): Diversity management that works: an evidence-based view. London (2019)

Coles, A., Diversity and inclusion in talent acquisition. In: Golbeck, A.L. (ed.) Leadership in Statistics and Data Science: Planning for Inclusive Excellence, pp. 171–183. Springer (2021)

Cundiff, J.L., Ryuk, S., Cech, K.: Identity-safe or threatening? Perceptions of women-targeted diversity initiatives. Group Process. Intergroup Relat. **21**(5), 745–766 (2018)

Daya, P.: Diversity and inclusion in an emerging market context. Equal. Divers. Incl. Int. J. **33**(3) (2014)

Dobbin, F., Kalev, A.: Why diversity programs fail. Harv. Bus. Rev. **94**(7/8), 52–60 (2016)

Erickson, W., von Schrader, S., Bruyere, S.M., Vanlooy, S.: Disability-inclusive employer practices and hiring of individuals with disabilities. Rehabil. Res. Policy Educ. **28**(4) (2014)

Fasnacht, D.: Open Innovation in the Financial Services. Manage. Prof. **F627**, 97–130 (2018)

FOB: Dobre praktyki ESG w Polsce. https://odpowiedzialnybiznes.pl/dobre-praktyki/

Forum Odpowiedzialnego Biznesu: Diversity IN Check. Raport z badania. FOB, Warszawa (2022)

Gülsoy, T., Ustabaş, A.: Corporate sustainability initiatives in gender equality: organizational practices fostering inclusiveness at work in an emerging-market context. Int. J. Innov. Technol. Manage. **16**(4) (2019)

Godman, N.R.: Taking diversity and inclusion initiatives global. Ind. Commer. Train. **45**(3), 180–183 (2013)

Gulyas, A., Seitz, S., Sinha, S.: Does pay transparency affect the gender wage gap? Evidence from Austria. Am. Econ. J. Econ. Pol. **15**(2), 236–255 (2023)

Gutterman, A.S.: Embedding diversity, equity, and inclusion (2023)

Hofstede, G.: Culture's consequences. Comparing Values, Behaviors, Institutions and Organizations Across Nations, 2nd edn. SAGE (2001)

Jackson, S.E., Ruderman, M., McLester, D.: Diversity policies, diversity training, and organizational effort to support diversity and inclusion in public organizations. Public Pers. Manage. **46**(4), 407–431 (2017)

Johnson-Mallard, V., et al.: The robert wood johnson nurse faculty scholars diversity and inclusion research. Health Equity **3**(1), 297–303 (2019)

Jones, N.J.: Step 2: clarifying the role of diversity, equity, and inclusion in shaping culture. In: Rothwell, W.J., Ealy, P.L., Campbell, J. (eds.) Rethinking Organizational Diversity, Equity, and Inclusion: A Step-by-Step Guide for Facilitating Effective Change, pp. 45–55. Productivity Press (2022)

Kalev, A.: The impact of diversity training programs in the workplace and alternative bias reduction mechanisms. In: Oxford Research Encyclopedia of Business and Management. Oxford University Press (2023)

Khan, A.A., Khan, N.A.: Exploring diversity, equity, and inclusion (DEI) in the context of human resource management. ANVESHAK-Int. J. Manage. **53**(2) (2023)

Khasawneh, A., Akash, R.A., Al-Rawashdeh, A., Khatib, H.A.: Voice and agency: evaluating the "empowering women for leadership in administration roles" training program held at Yarmouk University. Dirasat. Hum. Soc. Sci. **50**(4), 170–184 (2023)

Li, Y., Kleshinski, C.E., Wilson, K.S., Zhang, K.: Age differences in affective responses to inclusion experience: a daily diary study. Pers. Psychol. **75**(4), 805–832 (2022)

Mărginean, A.E.: Gen Z perceptions and expectations upon entering the workforce. Eur. Rev. Appl. Sociol. **14**(22), 20–30 (2021)

Miethlich, B., Oldenburg, A.G.: Social inclusion drives business sales: a literature review on the case of the employment of persons with disabilities. In: 33rd International Business Information Management Association Conference (IBIMA), Education Excellence and Innovation Management through Vision 2020, Granada, Spain, pp. 6253–6267. IBIMA Publishing, King of Prussia (2019). http://hdl.handle.net/10419/200752

Nweiser, M.T., Dajnoki, K.: An overview insight into employment of disabilities at workplaces around the world: a review of the literature. Ann. Fac. Econ. Subotica (2024)

Paolone, F., Pozzoli, M., Chhabra, M., Di Vaio, A.: Cultural and gender diversity for ESG performance towards knowledge sharing: empirical evidence from European banks. J. Knowl. Manage. **28**(2) (2024)

Ramshankar, P., Thomas, M.R.: Role of employee resource groups (ERGs) in fostering workforce diversity in information technology (IT) organizations after COVID-19. In: Doğru, C. (ed.) Role of Human Resources for Inclusive Leadership, Workplace Diversity, and Equity in Organizations. IGI Global (2023)

Russell, M., Stewart, B., Brooks, L.: Advancing Gender Equity and Women's Leadership Capacity. Mentor. Netw. Train. Adv. Women Leadersh. J. **42**, 88–97 (2023)

Shen, J., Chanda, A., D'Netto, B., Monga, M.: Managing diversity through human resource management: an international perspective and conceptual framework. Int. J. Hum. Resour. Manage. **20**(2), 235–251 (2009)

Tung, R.L.: New perspectives on human resource management in a global context. J. World Bus. **51**(1), 142–152 (2016)

Wieczorek-Szymańska, A.: Organisational maturity in diversity management. J. Corp. Responsib. Leadersh. **4**(1), 79–91 (2017)

Witkowska-Chrzczonowicz K.: Interwar and contemporary Poland. A comparison of the protection of minority rights in the second and in the third Polish republics. Comparat. Law Rev. **27**, 407–422 (2021)

ZBP, WIB: Banki 2023. Raport o sytuacji ekonomicznej banków. Warszawa (2023)

Mining for Well-Being: The Potential of Process Mining for Evaluating Employee Well-Being

Mari A. J. Braakman[✉], Jos Zuijderwijk, Iris Beerepoot, Sven Lugtigheid,
Thomas Martens, Maria Peeters, Eva Knies, and Hajo A. Reijers

Utrecht University, Utrecht, The Netherlands
m.a.j.braakman@uu.nl

Abstract. Monitoring work-related well-being is crucial for organizational success and part of good employment practices. This paper explores how process mining can evaluate employee well-being by conceptualizing variables of various work characteristics using the Job Demands-Resources model (JD-R), which explains how work characteristics influence employee well-being. We explored how the process mining variables compare to validated survey measures. Data was collected in two ways: first, a survey was conducted to measure the work characteristics of monotonous work, time pressure, workload, social support, and autonomy and the well-being outcomes of burnout, boredom, and work engagement. Second, process mining was used to calculate scores for the same work characteristics so that the scores could be compared with the survey variables. No strong correlations were found between corresponding survey variables and process mining variables. However, results reveal strong correlations between process mining variables of workload, social support, and autonomy with the survey variable of work engagement. These findings suggest that process mining variables can be valuable for predicting work-related well-being, especially work engagement. The combination of process mining and survey research has the potential to increase our comprehension of work-related well-being, make data collection more efficient, and monitor work engagement continuously.

Keywords: Job demands-resources model · Work-related well-being · Work engagement · Process mining

1 Introduction

In addition to the ethical responsibility of creating a healthy work environment, numerous studies have demonstrated that higher employee well-being leads to increased productivity and greater business profitability [17,37]. Therefore, organizations need to monitor employee well-being and its causes to create a healthy and supportive work environment. Traditional methods of assessing employee well-being often rely on subjective self-report surveys [8,22]. Self-report surveys

K. Gdowska et al. (Eds.): BPM 2024 Workshops, LNBIP 534, pp. 180–191, 2025.
https://doi.org/10.1007/978-3-031-78666-2_14

are valuable in capturing employees' perceptions of various work characteristics that influence their well-being [22], such as work engagement, feelings of exhaustion, and boredom. However, they might not provide a complete comprehensive understanding of employee well-being as it misses the more objective as-is setting of the work environment [22]. Moreover, survey variables have some limitations [26]. For example, the format of the survey, including the formulation and ordering of questions, can influence how participants answer. Additionally, participants tend to follow their beliefs of social acceptability rather than answer true to their feelings and behavior [26].

Process mining can more objectively investigate work characteristics by analyzing data generated by employees' use of information systems. Although process mining has limitations of its own (e.g., possibility of incomplete data, scalability issues, lack of contextual understanding [1,2]), it could provide additional information and knowledge of the objective side of work-related well-being and its antecedents, which is an important factor of employee well-being [22]. Process mining can also help lessen the effects of survey biases, especially when survey variables are combined with process mining variables. Moreover, in the long term, process mining is more time-efficient and demands less active labor from employees.

This study will explore how process mining can be used in combination with survey variables to evaluate employee well-being by answering two questions:

1. *How do process mining variables compare to survey measures for predictors of employee well-being?*
2. *To what extent can process mining variables be used to evaluate employee well-being outcomes, such as work engagement, burnout (exhaustion), and boredom?*

To answer these questions, we performed a case study in which we formulated process mining variables that measure various work characteristics. We compared these results with those of their counterparts, as measured using a survey among the same employees. In addition, we explored how well process mining could explain work-related well-being *outcomes*, namely burnout, boredom, and work engagement. This study contributes to process mining research and work- and organizational psychology research by examining how work-related well-being and its antecedents can be measured using process mining. In the long term, this new way of evaluating work-related well-being can be used by organizations to monitor and improve their employees' well-being.

2   Related Work

In the next section, we will briefly discuss the related work on work-related well-being by explaining the Job Demands-Resources model and the resource perspective of process mining.

2.1 Work-Related Well-Being

Monitoring and evaluating employee well-being is part of good and ethical employment practices. In fact, in many countries (notably all countries in the European Union), organizations are obliged by law to protect employee well-being [29]. Apart from being a legal responsibility, it is also crucial to uphold employee well-being because it is related to higher performance [36]. One of the most used work- and organizational psychology models is the Job Demands-Resources model (JD-R model) [7,29]. The JD-R model categorizes work characterizes into job demands and job resources within the work environment. Job demands are those aspects of one's job that come with physical, psychological and physiological costs [15,30], for example, workload and time pressure. Job resources are those aspects of one's job that provide support to achieve one's goals and help cope with the existing job demands [15,30], for example, autonomy and support of colleagues. Both job demands and job resources considerably impact employee well-being [9]. First, the health impairment process explains that job demands are the most important antecedents for a decrease in well-being, such as sickness and burnout. Second, the motivational process explains that job resources are antecedents for organizational commitment and work engagement. The health impairment process decreases the motivational process and increases work performance [9].

2.2 The Resource Perspective of Process Mining

Employee well-being is crucial for organizations but it has not received much attention in the process mining literature. The majority of process mining studies are focused on the order of activities in a process, i.e., the control-flow perspective [6]. A different perspective that is sometimes taken, referred to as the resource or organizational perspective, is geared towards providing insights regarding the involvement of employees and their collaboration patterns [1]. Studies taking this perspective typically provide techniques for social network identification [4,25] or resource allocation [5,18]. However, some recent works are moving towards the direction of measuring well-being and work characteristics. For example, Tang and Matzner [35] and Burden et al. [13] discuss the potential use of process mining for analyzing aspects such as job satisfaction and workload. Other studies put some of these ideas into practice, for example, by analyzing the work-related preferences from event logs [12] or workloads and processing speeds of employees [24]. Although these studies provide interesting new methods and techniques that provide insight into aspects of employee well-being, they are typically (1) not applied to real-world data, (2) applied to isolated and individual well-being aspects such as workload, and (3) not compared to perceived well-being of employees. In this study, we explore the use of process mining to measure various job demands and resources and analyze their correlations with employee-perceived well-being. The ability to evaluate work-related well-being with process mining can help diminish survey bias and add an objective, real-life view of well-being and its predictors, creating a more complete view of the work context.

3   Conceptualization

This section discusses how work characteristics that influence work-related well-being may be measured using process mining and what attributes are needed to do so. We explore five work characteristics commonly discussed in the JD-R model [11]: monotonous work, time pressure, workload, social support, and autonomy.

Monotonous work refers to a work situation offering little variability in tasks. It typically involves repetitive and mundane activities that can lead to boredom [14]. Monotonous work can be found in numerous job positions. Examples are working at an assembly line, manual sorting in a postal distribution centre, or giving the same exact presentation five times a day. Process mining can investigate the monotony of employees' work by exploring the repetitiveness of activities that employees perform. The more similar the activities an employee performs, the more monotonous their work is. An attribute *activity type* is then needed in the event log to investigate monotonous work. There are two ways of measuring the repetitiveness based on the activity type attribute: (1) measuring the total number of tasks of the same activity type performed in succession and (2) measuring the average variance found in activity type per employee.

Time pressure refers to work done with limited time available to complete it [34]. This can be either in the form of a deadline on a fixed date or time or in the form of the available timeslot where work must be done in a specific amount of time after a trigger event. An attribute *deadline* or *due date* is needed in the event log to investigate if a deadline is connected to a particular activity and how far the deadline is in the future.

Workload can be described as the amount and difficulty of work. It can be measured using the number of tasks, hours of work, mental demands, or energy required to complete work [31]. A start and end timestamp are needed to examine the duration of tasks. The duration worked on tasks could be compared to the mean of the department or team to investigate what tasks are more complex or take longer than others. Another way of looking at workload could be to analyze the begin and end times of a workday per employee, assuming that employees with high workload are more likely to work longer hours.

Social support refers to the extent to which employees work together in a supportive work environment where employees are considerate and attentive to each other and resolve possible conflicts in a constructive manner [20]. Social support can be between coworkers, between supervisor and employee, and between organization and employees. The experienced support varies per employee and can even vary per contact moment. Analyzing the experienced social support using data collected from information systems is difficult because of its subjective nature. The closest thing to support that can be investigated is the amount of contact at work. An attribute *resource* is needed to analyze which employees work together on the same task. Moreover, tasks such as meetings could be labelled as contact moments within the attribute *activity type*. It should be noted that contact is different from support. Every measured contact moment could also be experienced as negative.

`Autonomy` refers to the degree to which employees have the freedom and capability to execute tasks independently [16]. This could be investigated in event logs by looking at deviations from the standard procedures, as autonomous people have the freedom to deviate. Additionally, the attribute *resource* could inform how many tasks are completed by a single employee. Although working alone is not the same as autonomy, it could indicate employees' experience that they can complete tasks independently without first passing the task on to a colleague or supervisor.

4   Research Methods

For the case study, we focused on two IT service departments at a large university in the Netherlands which work closely together. Both departments primarily deal with gathering and resolving IT-related incidents encountered by university employees and students. Data was collected using two methods. For process mining, one year worth of data was collected from the dominant information system to measure the five work characteristics mentioned in the previous section. A one-time survey was used to measure the same work characteristics as well as the employee well-being outcomes: burnout, boredom, and work engagement.

4.1   Data Collection: Survey

The survey was conducted in April 2022 and was filled in by 15 employees. It consisted of different scales to measure various job demands, job resources and well-being outcomes. To measure burnout the exhaustion scale of the Maslach Burnout Inventory was used [28,32], containing items such as 'I feel mentally exhausted by my work'. The Dutch Boredom Scale was used to measure boredom with items such as 'At work, times goes by very slowly' and 'I feel bored at my job' [27]. Work engagement was measured with the 3-item Utrecht Work Engagement Scale. One sample item is 'At work, I feel bursting with energy'. Both the job demands and job resources were measured using items from the Job-Content Questionnaire [19], outlined in Table 1.

4.2   Data Collection: Process Mining

Event data was collected from the records of TopDesk for a time period of one year. The employees were asked for consent to share their data. The TopDesk dataset consists of 105 attributes, of which 22 were domain independent and relevant for determining job demands and resources. We found that 20 attributes contained missing values, with 15 attributes sharing identical missing values across cases. Due to their minimal representation in the dataset and negligible impact on overall observations, these instances were omitted. Notably, the attribute `Activity End Date` exhibited missing values in over 50% of observations, rendering it impractical for analysis. Consequently, this lead to its exclusion from further consideration. Due to missing end dates, activity duration was calculated based on the difference between start times of activities within the same case, potentially incorporating waiting time.

The resulting cleaned dataset contains 17 attributes and 739,436 activities, filtered to include only employees from the selected departments that have worked at least 40 h between January 2020 and December 2022. Those not meeting these criteria were labeled as "unknown" to prevent incomplete case records. We created process mining variables for the work characteristics monotonous work, time pressure, workload, social support, and autonomy. Each variable is calculated as a weighted sum of its components. The process mining variables for the job demands and job resources can be found in Table 1.

4.3 Data Analysis

To explore the survey and process mining variables, bivariate correlational analyses were conducted to determine the correlations between the process mining and survey variables of the job demands and resources. Additionally, the correlations between the survey and PM variables and the well-being outcomes measured by the survey were compared. However, due to the small sample size the conclusions cannot be generalized to a larger population.

5   Results

5.1   Process Mining and Survey Variables

The correlations between the process mining variables and survey variables of the job demands and job resources can be found in Table 2. The results show that the survey variables with the corresponding process mining variables do not always strongly correlate. This indicates that the process mining variables did not measure the same concept as the survey variables. For example, the two time pressure variables had a small negative correlation ($r = -.11$), which might indicate that *perceived* time pressure (survey variable) is conceptually different from the time pressure measured with process mining.

5.2   Job Demands-Resources and Well-Being Outcomes

The correlations between the work characteristics and the employee well-being outcomes can be found in Table 3. A strong correlation was found between the process mining variable of monotonous work and boredom ($r = .55$). A similar strong correlation was found between the survey variable of monotonous work and boredom ($r = .57$). Together with the moderate correlation between the process mining and survey variable ($r = .31$), this could indicate that these two corresponding monotonous work variables relate to the same concept. For the survey variables, other strong correlations were between workload and boredom ($r = -.66$) and between boredom and work engagement ($r = -.52$).

When comparing the correlations of the PM variables and the survey variables, most correlations were found to have a relation in the same direction

Table 1. Job Demands and Resources measurements

Measurement	Weight
Monotonous work	
1. Variation coefficient of work type duration	1
Survey items	
(i) My work includes some repetitive tasks [19]	
(ii) My work includes many activities (reversed) [19]	
Time pressure	
1. Completion rate	1/4
2. Overdue rate	1/4
3. Median versus prescribed duration	1/4
4. Number of urgent cases	1/4
Survey items	
(i) My job requires me to work very quickly [19]	
Workload (work amount)	
1. Sum of average duration times frequency per activity type	1/2
Workload (work difficulty)	
2. Number of cases with a duration larger than 2 standard deviations	1/8
3. Number of activities with a duration larger than 2 standard deviations	1/8
4. Time worked via difficult channels	1/8
5. Number of cases that have been reopened	1/8
Survey items	
(i) I am required to do excessive work [19]	
Social support	
1. Number of handovers	1/4
2. Number of subcontracts	1/4
3. Number of people on the same case	1/4
4. Number of people doing the same activities	1/4
Survey items	
(i) If I want, I can get help from one or more colleagues [19]	
Autonomy	
1. Percentage of cases that the performer solves alone	1/3
2. Number of variants that a performer can solve alone	1/3
3. Number of deviations	1/3
Survey items	
(i) I can interrupt my work as I wish [19]	
(ii) I can determine my own work pace	
(iii) My job allows me to make many decisions	
(iv) I have much to say about what happens in my work	
(v) I can determine the order in which I perform my tasks	

Table 2. Correlations between process mining variables and survey job demands and resources

Variable	Monotonous work	Time pressure	Workload	Social support	Autonomy
Monotonous work	.31	−.31	−.54	.84*	−.72*
Time pressure	−.28	−.11	.25	−.19	.77*
Workload	−.13	−.10	.28	−.44	.24
Social support	−.08	−.08	.08	−.33	.29
Autonomy	−.04	.01	.06	−.05	−.12

Note. Vertically, the process mining variables are represented. Horizontally, the job demands and job resources measured using the survey are represented. N ranges between 10 and 13. * $p < .05$

(either positive or negative), however, there were quite some variations between the two types of measuring methods. For example, the PM variable of time pressure was negatively correlated with burnout ($r = -.24$) as expected from previous literature [15] but the survey variable of time pressure was positively correlated with burnout ($r = .36$). Either the different methods variable different concepts, or the objective work characteristic (PM variable) has a different effect on employee well-being than the subjective work characteristic (survey variable).

Table 3. Correlations between process mining and survey variables (vertical) and well-being outcomes (horizontal)

Variable	Burnout	Boredom	Work engagement
Monotonous work - PM	.13	.55	−.33
Monotonous work - Survey	.07	.57*	−.50
Time pressure - PM	−.24	−.30	.21
Time pressure - Survey	.36	−.49	−.10
Workload - PM	−.27	−.22	.61*
Workload - Survey	.06	−.66*	.15
Social support - PM	−.22	−.41	.61*
Social support - Survey	.26	.44	.11
Autonomy - PM	.26	−.07	.56*
Autonomy - Survey	−.28	−.41	.32

Note. N ranged between 10 and 13. For all survey variables the response options ranged from 1 to 5, except for work engagement which ranged between 1 and 7. * $p < .05$

6   Discussion

In this study, we formulated process mining variables measuring work characteristics that predict work-related well-being. We compared the process mining and survey variables in a case study. Although we did not find strong correlations between the process mining variables and their corresponding survey variables, our results suggest that process mining could provide valuable insights into employee well-being. The different correlations of the PM and survey variables to the well-being outcomes could indicate two things. First, the PM and survey variables measure two different concepts that both, to various extents, relate to employee well-being. Second, the process mining variables measure a different side of the same concept (objective rather than subjective), which shows that the perception of a work characteristic can have a different effect on well-being than the objective presence of the work characteristic. Both confirm our belief that the combination of process mining and survey variables provides a more comprehensive view of the effects of work characteristics on employee well-being.

Specifically, we found that the process mining variable of workload, social support, and autonomy were significantly positively correlated with work engagement. This finding supports previous research that suggests job resources such as social support and autonomy are important predictors of employee well-being [10], and that job demands such as workload can have a positive effect on work engagement when perceived as a challenge or if the workload is balanced [21,23]. Additionally, we found the process mining variable of monotonous work to have a moderate positive correlation with the survey variable of monotonous work, indicating a possible relation or similarity in construct for these two measurements. Further support for the potential similarity between the two variables can be found in the finding that both the process mining variable and survey variable are strongly correlated with the well-being outcome boredom.

Our study contributes to the broader discussion on survey bias, as process mining variables could diminish some of the limitations associated with self-report surveys. Process mining variables provide an objective and real-time assessment of job resources and job demands, reducing the potential for bias and improving accuracy. However, it is important to note that process mining variables have their own limitations. For instance, process mining assumes the researcher captures the complete data from the system, which may not always be the case [2]. This was also seen in our dataset, where the end time of activities was not always saved in the system. Moreover, the process mining approach does not consider the subjective aspects of job resources and demands, which is important for understanding the overall employee well-being [22]. For example, one employee might find a certain workload quite manageable and engaging, while another might find it too much. Therefore, it is essential to consider objective and subjective variables in conjunction when evaluating employee well-being. To accomplish this without overwhelming employees with frequent surveys, analysts can design a monitoring plan where surveys are administered at extended intervals, for example, annually. Process mining can then be used to monitor employee well-being in between these intervals and identify problems bottlenecks or possible risks for employee well-being. Additionally, process mining data can be collected from all employees, whereas surveys are more likely to have a smaller sample size due to response rate limitations [26]. The results of this study further support the argument for the use of both survey variables and process mining variables to evaluate employee well-being comprehensively (both the objective and perceived work characteristics) and more accurately (collecting real-time data from a larger sample).

This exploratory study is subject to a couple of limitations that should be considered. First, only a small group of participants filled in the survey. This decreases the generalizability of the findings. Future research should gather data from a more extensive and diverse sample to ensure more representative results, for example, by conducting similar research in other sectors. Second, some data, such as part of the activity's end times, were missing from the event data. Missing timestamps can impact the quality of the used data [3,33] and, thus, the accuracy of the process mining variables. Additionally, the data was collected from one system, which caused missing data from other systems used in the department. Accuracy can be improved in future studies by collecting data from all systems used within the sample. The process mining variables were formulated specifically for this context but could be calculated differently. Further research should consider what calculates scores would work best in the sample used.

7 Conclusion

The case study conducted in this research explored how process mining variables can be used to evaluate employee well-being. The results indicate that perceptions of job demands and job resources, such as time pressure, and autonomy do not strongly match the corresponding data extracted from information systems. However, interesting relations were found between the process mining variables of workload, social support, and autonomy and the well-being outcome of work engagement. This suggests that process mining may be a valuable tool for continuously monitoring the work-related well-being of employees, especially work engagement. Combining this with questionnaires that measure the corresponding employee perceptions will provide a more complete picture of work-related well-being.

Acknowledgements. We would like to thank Floris Smit, Jelmer Koorn, Maartje Orsel, and Feline Wafelaar for their contributions to this project.

References

1. van der Aalst, W.M.P.: Data Science in Action. Springer (2016)
2. van ver Aalst, W.M.P., et al.: Process mining manifesto. In: Business Process Management Workshops: BPM 2011 International Workshops, Clermont-Ferrand, France, 29 August 2011, Revised Selected Papers, Part I, pp. 169–194. Springer (2012)
3. van der Aalst, W.M.P., Santos, L.: May i take your order? On the interplay between time and order in process mining. In: International Conference on Business Process Management, pp. 99–110. Springer (2021)

4. van der Aalst, W.M.P., Song, M.: Mining social networks: uncovering interaction patterns in business processes. In: International Conference on Business Process Management, pp. 244–260. Springer (2004)

5. Arias, M., Rojas, E., Munoz-Gama, J., Sepúlveda, M.: A framework for recommending resource allocation based on process mining. In: Business Process Management Workshops: BPM 2015, 13th International Workshops, Innsbruck, Austria, 31 August–3 September 2015, Revised Papers 13, pp. 458–470. Springer (2016)

6. Augusto, A., et al.: Automated discovery of process models from event logs: Review and benchmark. IEEE Trans. Knowl. Data Eng. **31**(4), 686–705 (2018)

7. Bakker, A.B., Demerouti, E.: The job demands-resources model: state of the art. J. Manag. Psychol. **22**(3), 309–328 (2007)

8. Bakker, A.B., Demerouti, E.: Job demands-resources theory: taking stock and looking forward. J. Occup. Health Psychol. **22**(3), 273 (2017)

9. Bakker, A.B., Demerouti, E., Sanz-Vergel, A.I.: Burnout and work engagement: the JD-R approach. Annu. Rev. Organ. Psychol. Organ. Behav. **1**(1), 389–411 (2014)

10. Bakker, A.B., Demerouti, E., Verbeke, W.: Using the job demands-resources model to predict burnout and performance. Hum. Resour. Manage. **43**(1), 83–104 (2004)

11. Bauer, G.F., Hämmig, O., Schaufeli, W.B., Taris, T.W.: A critical review of the job demands-resources model: implications for improving work and health. In: Bridging Occupational, Organizational and Public Health: A Transdisciplinary Approach, pp. 43–68 (2014)

12. Bidar, R., ter Hofstede, A., Sindhgatta, R., Ouyang, C.: Preference-based resource and task allocation in business process automation. In: OTM Confederated International Conferences "On the Move to Meaningful Internet Systems", pp. 404–421. Springer (2019)

13. Burden, M., Keniston, A., Pell, J., Yu, A., Dyrbye, L., Kannampallil, T.: Unlocking inpatient workload insights with electronic health record event logs. J. Hospit. Med. 1–6 (2024)

14. Cummings, M.L., Gao, F., Thornburg, K.M.: Boredom in the workplace: a new look at an old problem. Hum. Factors **58**(2), 279–300 (2016)

15. Demerouti, E., Bakker, A.B., Nachreiner, F., Schaufeli, W.B.: The job demands-resources model of burnout. J. Appl. Psychol. **86**(3), 499–512 (2001)

16. Dworkin, G.: The Theory and Practice of Autonomy. Cambridge Studies in Philosophy. Cambridge University Press (1988)

17. García-Buades, M.E., Peiró, J.M., Montañez-Juan, M.I., Kozusznik, M.W., Ortiz-Bonnín, S.: Happy-productive teams and work units: a systematic review of the happy-productive worker thesis. Int. J. Environ. Res. Public Health **17**(1), 69 (2020)

18. Huang, Z., Lu, X., Duan, H.: Mining association rules to support resource allocation in business process management. Expert Syst. Appl. **38**(8), 9483–9490 (2011)

19. Karasek, R., Brisson, C., Kawakami, N., Houtman, I., Bongers, P., Amick, B.: The job content questionnaire (JCQ): an instrument for internationally comparative assessments of psychosocial job characteristics. J. Occup. Health Psychol. **3**(4), 322–355 (1998)

20. Kleine, A.K., Rudolph, C.W., Zacher, H.: Thriving at work: a meta-analysis. J. Organ. Behav. **40**(9–10), 973–999 (2019)

21. Li, P., Taris, T.W., Peeters, M.C.: Challenge and hindrance appraisals of job demands: one mans meat, another mans poison? Anxiety Stress Coping **33**(1), 31–46 (2020)

22. Li, Y., Tuckey, M.R., Bakker, A., Chen, P.Y., Dollard, M.F.: Linking objective and subjective job demands and resources in the JD-R model: a multilevel design. Work Stress **37**(1), 27–54 (2023)
23. Montani, F., Vandenberghe, C., Khedhaouria, A., Courcy, F.: Examining the inverted U-shaped relationship between workload and innovative work behavior: the role of work engagement and mindfulness. Hum. Relat. **73**(1), 59–93 (2020)
24. Ogunbiyi, N., Basukoski, A., Chaussalet, T.: Investigating the diffusion of workload-induced stress—a simulation approach. Information **12**(1), 11 (2020)
25. Panpanich, P., Porouhan, P., Premchaiswadi, W.: Analysis of handover of work in call center using social network process mining technique. In: 2015 13th International conference on ICT and knowledge engineering (ICT & knowledge engineering 2015), pp. 97–104. IEEE (2015)
26. Podsakoff, P.M., MacKenzie, S.B., Lee, J.Y., Podsakoff, N.P.: Common method biases in behavioral research: a critical review of the literature and recommended remedies. J. Appl. Psychol. **88**(5), 879 (2003)
27. Reijseger, G., Schaufeli, W.B., Peeters, M.C., Taris, T.W., van Beek, I., Ouweneel, E.: Watching the paint dry at work: psychometric examination of the Dutch boredom scale. Anxiety Stress Coping **26**, 508–525 (2013)
28. Schaufeli, W.B., Bakker, A.B., Hoogduin, K., Schaap, C., Kladler, A.: On the clinical validity of the maslach burnout inventory and the burnout measure. Psychol. Health **16**, 565–582 (2001)
29. Schaufeli, W.B.: Applying the job demands-resources model: a 'how to' guide to measuring and tackling work engagement and burnout. Organ. Dyn. **46**(2), 120–132 (2017)
30. Schaufeli, W.B., Bakker, A.B.: Job demands, job resources, and their relationship with burnout and engagement: a multi-sample study. J. Organ. Behav. **25**(3), 293–315 (2004)
31. Spector, P.E., Jex, S.M.: Development of four self-report measures of job stressors and strain: interpersonal conflict at work scale, organizational constraints scale, quantitative workload inventory, and physical symptoms inventory. J. Occup. Health Psychol. **3**(4), 356 (1998)
32. Steenbergen, E.F.V., van der Ven, C., Peeters, M.C., Taris, T.W.: Transitioning towards new ways of working: do job demands, job resources, burnout, and engagement change? Psychol. Rep. **121**(4), 736–766 (2018)
33. Suriadi, S., Andrews, R., ter Hofstede, A.H., Wynn, M.T.: Event log imperfection patterns for process mining: towards a systematic approach to cleaning event logs. Inf. Syst. **64**, 132–150 (2017)
34. Sussman, R.F., Sekuler, R.: Feeling rushed? perceived time pressure impacts executive function and stress. Acta Physiol. (Oxf) **229**, 103702 (2022)
35. Tang, W., Matzner, M.: Creating humanistic value with process mining for improving work conditions-a sociotechnical perspective. In: ICIS (2020)
36. Taris, T.W., Schaufeli, W.B.: Individual well-being and performance at work: a conceptual and theoretical overview. In: Cooper, C. (ed.) Current Issues in Work and Organizational Psychology, pp. 15–34. Routledge, London (2018)
37. van de Voorde, K., Paauwe, J., Van Veldhoven, M.: Employee well-being and the HRM-organizational performance relationship: a review of quantitative studies. Int. J. Manag. Rev. **14**(4), 391–407 (2012)

3rd International Workshop on Natural Language Processing for Business Process Management (NLP4BPM 2024)

3rd International Workshop on Natural Language Processing for Business Process Management (NLP4BPM 2024)

The NLP4BPM workshop aims to provide a forum for researchers and practitioners to present, discuss, and evaluate how the use of natural language processing (NLP) and large language models (LLMs) can be used to establish new or improve existing methods, techniques, tools, and systems that support the different phases of the BPM life-cycle. Furthermore, the workshop aims to promote an exchange on the advances, challenges, and barriers researchers encounter, and establish an environment where collaborations can naturally emerge.

In this context, NLP and LLMs can play a variety of roles. Among others, they can be used to describe processes in a comprehensible manner, define the meaning of events and activities, and provide support for the conduct of process analyses themselves, e.g., as an interface for process mining or modeling. In addition, it is possible to exploit the process, domain, and programming knowledge of LLMs to generate analysis or improvement suggestions.

The advent of LLMs has resulted in an exponential increase in the attention that NLP receives in the public domain, as well as in the possibilities of how NLP technology can be used. This also applies in the context of business process management, where we observe that the introduction of LLMs has enabled a broad range of new use cases and analytical tasks. Due to these developments, the third edition of our workshop attracted a record number of 15 submissions (in comparison to nine the year before). The submissions were each reviewed by three members of the Program Committee. From these, seven were accepted for the workshop and were presented at the BPM Conference in Krakow, Poland, attracting a large audience. After these research presentations, we also hosted a round table on *Guidelines for writing and reviewing LLM papers for BPM*, in which a large number of attendees discussed the future of paper writing and reviewing for research that focuses on the use of LLMs in business process management.

The seven accepted papers all focused on the use of LLMs, though for a variety of different tasks, ranging from conversational process modeling to causal reasoning and predictive monitoring:

Köpke and Safan focuses on conversational process modeling, in which a user interacts with an LLM to establish and refine a process model. Although such conversations can greatly improve the ease and efficiency of process modeling, it can also be costly, given that users are charged for each token in their prompts and the received output. Therefore, the work at hand proposes an approach that leads to token-optimized conversations, primarily thanks to a new meta-model for intermediate process models.

Voelter et al. investigates the vision capabilities of multimodal Generative Pre-trained Transformers (GPTs) to auto-generate structured process models from diagram- and text-based documents. Their experimental results reveal that generative vision models can be useful tools for semi-automated process modeling based on multimodal documents, whereas their paper also provides a new dataset for the research community.

Franzoi et al. reports on design science-based research in collaboration with a large multinational company to design a BPM system that leverages LLMs for process knowledge extraction from diverse enterprise content. Their prototype demonstrates that LLMs provide the means to organize and generate process knowledge independent of specific forms of representation, making it available in diverse output formats via prompting, resulting in representation-agnostic process knowledge.

Brzychczy et al. tackles the challenge of transforming low-level sensor data into high-level activities for effective process mining by using LLMs to automate the labeling process, a task that is traditionally reliant on domain experts. The paper demonstrates the accuracy and interpretability of LLM-generated labels compared to a Decision Tree Classifier, highlighting the potential of AI techniques to enhance process mining in Industry 4.0.

Buss et al. proposes ProcessLLM, a specialized LLM fine-tuned to support BPM tasks, aiming to address the inadequacies of general-purpose LLMs in complex BPM tasks. Evaluations using both artificial and real-life business settings demonstrate ProcessLLM's promising performance and a sophisticated understanding of BPM concepts and the BPMN language, paving the way for future research in LLM-enhanced BPM.

Brennig et al. studies the use of LLMs for predictive process monitoring (PPM) as a means to overcome the need for extensive preprocessing pipelines that are common practice in this task. Their results show that their LLM-based approach performs comparably to the state of the art in machine learning-based PPM when it comes to next activity prediction, while having a simplified prediction process with minimal preprocessing.

Fournier et al. introduces a benchmark to assess LLMs in reasoning about causal and process perspectives in business operations, referred to as Causally-augmented Business Processes (BPC). The benchmark consists of domain-specific situations, questions, and deductive rules, and can be used to test or train LLMs to improve their ability to reason in business contexts.

The organizers wish to thank all the people who submitted papers to the NLP4BPM 2024 workshop, the many participants who created fruitful discussions, and the NLP4BPM Program Committee members for their valuable work in reviewing the submissions. We look forward to future editions of the NLP4BPM workshop.

Organization

Organizing Committee

Manuel Resinas	University of Seville, Spain
Han van der Aa	University of Vienna, Austria
Adela del Río-Ortega	University of Seville, Spain
Henrik Leopold	Kühne Logistics University, Germany

Program Committee

Lars Ackermann	University of Bayreuth, Germany
Patrizio Bellan	Fondazione Bruno Kessler and Free University of Bozen-Bolzano, Italy
Josep Carmona	Universitat Politècnica de Catalunya, Spain
Irene Bedilia Estrada Torres	University of Seville, Spain
Mohammadreza Fani Sani	Microsoft, Denmark
Walid Gaaloul	Télécom SudParis, France
Chiara Ghidini	Free University of Bozen-Bolzano, Italy
Daniela Grigori	Paris Dauphine University, France
Christoph Kecht	Technical University of Munich, Germany
Wolfgang Kratsch	FIM, Germany
Hugo A. López	University of Copenhagen, Denmark
Fabrizio Maria Maggi	Free University of Bozen-Bolzano, Italy
Julian Neuberger	University of Bayreuth, Germany
Adrian Rebmann	SAP Signavio, Germany
Maxim Vidgof	Vienna University of Economics and Business, Austria
Sven Weinzierl	FAU, Germany
Karolin Winter	Eindhoven University of Technology, Netherlands

Straight Outta Logs: Can Large Language Models Overcome Preprocessing in Next Event Prediction?

Katharina Brennig$^{(\boxtimes)}$ ⓘ, Sascha Kaltenpoth ⓘ, and Oliver Müller ⓘ

Faculty of Business Administration and Economics, Paderborn University, Warburger Straße 100, 33098 Paderborn, Germany
{katharina.brennig,sascha.kaltenpoth,oliver.mueller}@uni-paderborn.de

Abstract. Predictive process monitoring (PPM) aims to predict the future behavior of process instances to mitigate process violations or take preventive measures. Current PPM methods for next event prediction (NEP) often utilize machine learning techniques, while first approaches also use deep learning techniques, especially natural language processing (NLP). Hence, these approaches often require extensive data preprocessing. To counteract this, we train and evaluate a fine-tuned large language model (LLM) to directly generate NEPs from XES-formatted event logs without any preprocessing. The results suggest that the proposed PPM approach performs comparably to the state-of-the-art in ML-based PPM, while contributing a simplified prediction process for NEP with minimal data preprocessing. Additionally, our LLM-driven approach produces valid XES outputs in nearly all cases, facilitating the direct export of predictions as event logs to be processed downstream (e.g., to employ process mining techniques or simulation). Further, our method offers easy integration into existing organizational infrastructures.

Keywords: Predictive Process Monitoring · Next Event Prediction · Natural Language Processing · Large Language Models

1 Introduction

Predictive process monitoring (PPM) aims to continuously anticipate the future behavior of ongoing process instances [15,16], such as the process outcome, sequences of future activities, or (remaining) lead time, to decrease the likelihood of process violations or unnecessary activities [1,15,27]. Recently, PPM methods to generate next event predictions (NEP) of a process instance have thoroughly been researched [15,16,45], which has involved the use of diverse machine learning methods (e.g., [10,23,39]). Additionally, deep learning methods (e.g., [9,35,40]) have been developed, especially approaches that incorporate natural language processing (NLP) [8,11,24,25,31–33,35,49]. All of these approaches provide promising results in NEP. However, they usually require preprocessing and encoding of the event log [15], which can often be a very extensive and time-consuming challenge.

K. Gdowska et al. (Eds.): BPM 2024 Workshops, LNBIP 534, pp. 197–208, 2025.
https://doi.org/10.1007/978-3-031-78666-2_15

To overcome this challenge, we argue that large language models (LLMs) with their capabilities in understanding and generating natural language, programming code, and markup languages [12,19,38] could be a promising approach to directly generate NEPs based on existing XES files without preprocessing the event log. Therefore, this study presents a new PPM approach using a fine-tuned LLM with the ability to directly generate NEPs in XES format. We leverage an existing state-of-the-art (SOTA) LLM, which we fine-tuned and trained separately on five benchmark event logs to generate NEPs. We evaluate our solution for every event log and compare our method to the SOTA *SNAP* method [31], as well as other recent studies (see Table 1). The results reveal that our method performs comparable but not better than the current SOTA, while drastically decreasing the preprocessing of an event log by mitigating the common prefix and event encoding steps [35]. In addition, we find that the LLM produces valid XES outputs in almost 100% of the cases, generating only four erroneous events while evaluating five folds of 74,142 events.

This study's contribution is manifold. From a research perspective, we show that LLMs are capable of identifying syntactical and semantical relationships between events and within event attributes. Based on that, they enable the generation of NEPs of a process instance using an event log with negligible preprocessing of the data. This not only offers researchers a less complex methodology for generating NEPs but also allows for a direct export of the results as an event log which can subsequently be processed downstream (e.g., to employ process mining techniques or simulation), due to the ability to generate NEPs based on XES formatting. Further, the generated NEPs in XES format allow us to predict more than just the next event by providing multi-task predictions including the name of the event, resource, timestamp, or other attributes. From a managerial perspective, our method enables an easy integration into an organization's infrastructure, because it could be directly connected to the existing process monitoring system. Further, it enables organizations to adapt the LLM-based approach to new domains (e.g., new processes) and tasks (e.g., new classes) through transfer learning without starting from scratch.

2   Background and Related Work

PPM is a process mining technique that has recently gained more attention [15,16,27]. It aims to continually anticipate future behavior of ongoing process instances (e.g., process outcome, sequences, or (remaining) lead time) [15,27,46]. During the training phase, a predictive model is learned from previously completed traces [15], in which the control flow of a process instance is considered, building the basis for PPM [43]. As additional input contextual or unstructured information can be used [15]. During the prediction phase, the predictive model is queried to forecast the future of a current instance [15]. Being able to predict the behavior of a running process instance allows organizations to capitalize on or adapt to desirable future developments, as well as respond and prevent an unwanted scenario by implementing suitable preventive countermeasures [15]. Nevertheless, due to the complexity and diversity of business processes various PPM approaches have been created [27], especially to generate NEPs of a process instance. Recurrent neural networks—more specifically, Long-Short-Term Memory (LSTM) architectures, build the foundation of the majority of the NEP PPM approaches [15]. However, existing benchmark studies [27,35,41,43,45] have provided comprehensive insights on the effectiveness of further developed PPM approaches.

With respect to approaches using NLP, there is also a recent rise in research noticable. Along with the first approaches for predicting the (remaining) lead time of a running process instance (e.g., [6,8,33]), there already exist approaches for generating NEPs (see Table 1). Further, Rama-Maneiro et al. [35] provide a comparison of NEP methodologies on an exhaustive benchmark revealing the current SOTA. Table 1 reveals the common preprocessing steps for prefix and event encoding. Embedding-based preprocessing methods reflect the prefix encoding (e.g., [25,32]) and additionally enable to encode events and their attributes (e.g., [8,24,49]). One of the most used embedding methods for preprocessing is the BERT model [8,11,14,24,25,49], which was first introduced by Devlin et al. [14], followed by Word2Vec [25,32] and Doc2Vec [33,35]. In addition, other preprocessing techniques include one-hot encoding [8,35], Bag-of-Words (BoW) [33], and time feature normalization [25]. Taking a closer look at the used prediction models of the studies presented in Table 1, it also appears that the majority of PPM approaches apply LSTMs to generate NEPs [8,24,25,32,33,35] whereas some approaches leveraged NLP techniques, especially BERT, as a prediction model [11,31,49].

With respect to approaches using LLMs, it appears that LLMs have scarcely been used in the field of PPM (see Table 1). LLMs typically comprise billions of parameters and are trained on large natural language datasets [7,29]. These generative models aim to generate meaningful and coherent text, given a sequence of tokens (word parts) as input, and are trained to autoregressively predict the next token t_{n+1} based on the conditional probability distribution $P(t_{n+1}|t_1...t_n)$ of a sequence of input tokens (context) [36]. This enables LLMs to answer questions, follow instructions, and also generate computer code and markup languages such as HTML and XML on a nearly human level [7,12,30,38]. Especially in BPM, these models enable new applications [19,44] as they have been shown to support process discovery by mining business process modeling notation (BPMN) and declarative process models from natural language descriptions [20]. They also enable the assessment of the suitability of process tasks from natural language descriptions [20]. Other approaches implement robotic process automation using LLMs [48]. Furthermore, LLMs can process directly-follows graphs, process variant information, and Petri net abstractions, and answer questions about them to support process mining [3].

Enabling a more in-depth investigation of existing LLMs, such as GPT-3 [7], GPT-4 [29], or Llama 3 [28], in the field of NEP generation could empower new applications. However, only one study (see Table 1) uses LLMs for event log preprocessing and as a prediction model to generate NEPs [31]. This LLM-driven method called SNAP [31] is based on stories comprising the foregone events in

Table 1. Related Studies

Preprocessing Strategies	Prediction Models				Bench-mark[a]	BPI Challenge				Other Benchmarks				Other Datasets[b]	
	LLM	LSTM	BERT	Other		2020	2016	2013	2012	Sepsis	EnvPermit	Helpdesk	NASA		
BERT, One-Hot	X					X									[8]
Llama 2, XGBoost	GPT-3	X		X		X				X	X		X	MIP	[31]
BERT, Masked Activity Model			X					X	X	X		X			[11]
BERT, Word2Vec, Time Feature Normalization		X												BPI19, private dataset	[25]
Word2Vec		X					X	X	X			X		BPI18	[32]
BERT, VGG, Layout XML, Convolution, LSTM		X												Claims	[24]
BERT			X		X	X			X			X		BPI17, Receipt	[49]
Doc2Vec, BoW, LDA		X					X							Hospital	[33]
Doc2Vec, One-Hot, Frequencies		X		DNN CNN GRU	X		X	X		X	X	X	X		[35]
not required	Llama 3				X		X	X	X		X				Ours

[a]Refers to studies comprehensively comparing existing studies.
[b]Event logs that appear only once.

a trace generated with Llama 2 [42] and employs GPT-3 to generate NEPs based on the generated stories [31]. However, this method requires exhaustive preprocessing, while our methodology directly processes XES-formatted data, as explained in the following.

3   LLM-Driven Next Event Prediction

3.1   NEP Pipeline

Our NEP pipeline comprises the fine-tuned Llama 3 model with 8 billion parameters, as shown in Fig. 1. An XES-formatted incomplete trace t is either read from a file or a string comprising the i completed events $(e_0, ..., e_i)$ in XES format. Based on this incomplete trace, Llama 3 generates the next event e_{i+1} in XES format. The prediction may be stored in an XES file for further analysis. This procedure can iteratively be repeated for every incomplete trace or XES file until reaching the end of the trace. As visible in the output frame of Fig. 1, the Llama 3 model generates all XES-formatted attributes of the event as text completion highlighted in green. The bold "concept:name" is our NEP. However, the pipeline generates all important event attributes such as the timestamp, the "org:resource", "org:role", or "lifecycle:transition". While the pipeline generates all values, enabling other tasks, we evaluate only the generated "concept:names" due to the study's NEP focus.

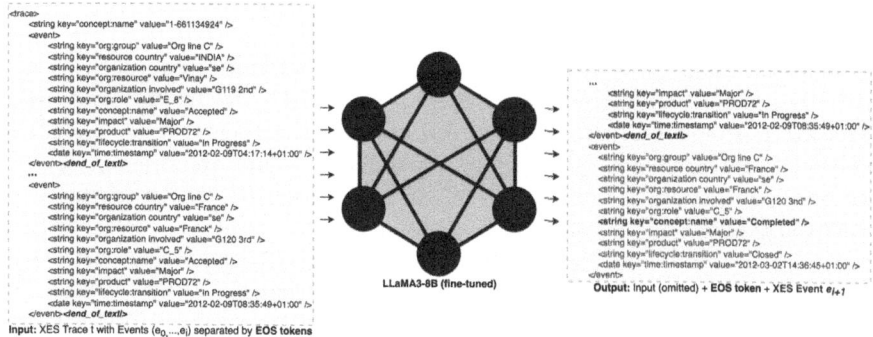

Fig. 1. NEP Inference Pipeline (Color figure online)

3.2   Benchmark Event Logs

For our comparison, we leverage five event logs including real-world processes, summarized in Table 2.

Table 2. Used Benchmark Event Logs

Dataset	Num. Cases	Num. Events	Num. Activities	∅ Case Length	Source
BPI13cp	1,487	6,660	7	4.48	[37]
BPI13in	7,554	65,533	13	8.68	[37]
Sepsis	1,049	15,214	16	14.48	[26]
BPI12	13,087	262,200	36	20.04	[18]
Helpdesk	4,580	21,348	14	4.66	[34]

The first two event logs were provided by the BPI Challenge 2013 [17]. The *BPI13cp* event log comprises closed problems from Volvo IT Belgium's incident and problem management system. The *BPI13in* event log comprises the corresponding incidents' traces. The *Sepsis* event log contains traces related to the sepsis case treatment procedures [26] in a hospital. Additionally, we consider the *BPI12* event log [18] which comprises traces of a loan application process used in the BPI Challenge 2012 as well as the *Helpdesk* event log which comprises events from a ticketing management process of the help desk of an Italian software company [34]. While the related work provides more benchmark event logs, we leverage only the five event logs stated in Table 2. We consider only event logs for our evaluation that occur three times or more within the related work.

3.3   Training and Evaluation

Training the Llama 3 model has been chosen for our initial approach as it is explicitly provided as an open-source LLM [28] and outperforms the significantly larger GPT-3 model [7], which has been applied in the current SOTA SNAP implementation [31]. For our training pipeline, we adapt the domain of XES files by further training our models on the next token prediction of the XES-formatted traces, applying the method of Gururangan et al. [21], referred to as fine-tuning in the following. Our prompt contains the maximum of past events that fit in the context length separated by end-of-sequence (EOS) tokens (see Fig. 1). Due to this strategy, the pipeline does not have a fixed prefix length but can process prefixes of up to 20 events. In contrast, we restrict the LLM's context length to 4096 tokens (word pieces) to run the experiments on a Titan RTX GPU with 24 GB. We additionally fine-tune the model using quantization-based parameter efficient fine-tuning (PEFT) to reduce GPU requirements [13]. We implement our pipeline using the Transformers library [47] and the pre-trained and instruction-tuned Llama 3 version provided on the HuggingFace hub *Meta-Llama-3-8B* [2]. To ensure reproducibility, we provide all required libraries, random states (seeds), and the program code in a GitHub-repository[1].

Evaluating our approach comprises 5-fold cross-validation, fine-tuning the Llama 3 LLM five times on every event log using a different random split of 64% training, 16% validation, and 20% test traces. Although some studies incorporate a time-based split due to their simultaneous NEP and remaining time prediction, we follow the two identified benchmark studies, which also leverage 5-fold cross-validation. Following the trace-based 5-fold cross-validation split strategy leads to the same number of fold traces but to different event counts in the fold test sets. For the calculation of the accuracy and F1 score, we convert the ground truth XES files and the Llama 3-generated XES files into tabular data using the *pm4py*-library [4]. We subsequently compare the generated "concept:names" to their true values per fold and calculate an average accuracy. We compare our results to Oved et al. [31] as well as to the approaches evaluated in the benchmark study by Rama-Maneiro et al. [35], as they both incorporate 5-fold cross-validation. Other recent studies do not apply 5-fold cross-validation, which disables a direct comparison. We calculate the accuracy, as it is the only measure reported by both studies [31,35]. For future comparability with other studies, we also report the weighted and macro average F1-Score. Finally, we track the syntax errors, hallucinations, and the training and evaluation times.

4   Results

Table 3 summarizes the average training and evaluation time required for every fold of our comparison. Overall, including all folds, our comparison needed 616 h, including around 75 h for training and 541 h of evaluation. As visible in Table 3, the evaluation took much more time than the training. This is caused by the

---

[1] https://github.com/skaltenp/straight_outta_logs.

difference between the evaluation and training process. During training, a complete trace is generated, whereas every event iteratively needs to be generated based on the prefix events during evaluation. However, our pipeline only needs 3 to 11 s to generate a complete event.

Table 3. Training and Evaluation Time

| Dataset | Average Fold Train-/Val- | | | Average Fold Evaluation | | | | |
| | Traces | Events | Time | Traces | Events | Time (hh:mm:ss) | | |
	Abs.	Abs.	(hh:mm:ss)	Abs.	Abs.	Overall	per Trace	per Event
BPI13cp	1,189	5,352	00:20:01	298	1,308	02:22:37	00:00:29	00:00:07
BPI13in	6,043	52,384	03:32:28	1,511	13,149	28:42:36	00:01:08	00:00:08
Sepsis	839	12,181	00:29:27	210	3,033	03:47:34	00:01:05	00:00:03
Helpdesk	3,664	17,085	01:32:54	916	4,263	08:52:02	00:00:35	00:00:11
BPI12	10,469	209,811	09:00:45	2,618	52,389	64:32:11	00:01:29	00:00:04

Table 4 summarizes the accuracies of the related benchmark studies compared to our approach. The colorization denotes the performance ranking of the NEP approach on the specific event log. The *Average* column denotes the average over the different event logs along with the average *Rank* of an approach. For the calculation of the average, we assume that all approaches not evaluated on an event log show median performance (written in red), as this does obtain the overall median and, thus, fairness. While our method shows lower performance than the SOTA models, it is still equal to or above the median in all event log evaluations. Considering the average of all event logs, our method still performs above the median and shows only 3.0% lower performance than the best model on average.

The syntax errors tracked during the evaluation are stated in Table 5. In addition to the errors stated in Table 5, we found that our fine-tuned Llama 3 produced trace closing tags on events that usually occur at the end of a trace. Although this may compromise the integrity of an XES file, we omit the error during evaluation as it would serve as a stop sign for the process in a real-world environment. As Table 5 shows, our approach generates nearly zero syntax errors. Llama 3 only generated two additional event tags that cause syntax errors in all five folds of the *BPI13in* event log evaluation and one falsely generated quotation mark, leading to an average error rate of 0.0046% across the five folds. The last error is one missing value in a string XES tag that causes an error rate of 0.0066% for the *Sepsis* event log. These are the only syntax errors within all datasets. Overall, nearly zero syntax errors are produced during generation. In addition to 0.0583% (six) missing concept names, we found 21 generated concept names in all 5 folds that do not exist in the event logs, so-called "hallucinations" [22]. On average these are 4.2 hallucinated events per fold of 74,142 events (0.1314%).

Table 4. Accuracy Comparison

	BPl13cp		BPI13in		Sepsis		Helpdesk		BPI12		Average		Total
	Accuracy	Rank	Accuracy	Rank	Accuracy	Rank	Accuracy	Rank	Accuracy	Rank	Accuracy	Rank	Rank
Camargo [35]	0.547	8	0.667	7	0.610[a]	5	0.829	5	0.833	6	0.697	6	6
Evermann [35]	0.588	7	0.668	6	0.400	11	0.336	11	0.593	10	0.517	9	10
Hinnka [35]	0.635	5	0.747	3	0.635	3	0.831	4	0.867	2	0.743	3.4	4
Khan [35]	0.436	11	0.519	9	0.210	12	0.800	8	0.429	12	0.479	10.4	12
Mauro [35]	0.249	12	0.367	12	0.615	4	0.318	12	0.847	5	0.479	9	10
Pasquadibisceglie [35]	0.475	10	0.469	11	0.562	9	0.840	3	0.833	6	0.634	7.8	7
Tax [35]	0.640	4	0.701	5	0.642	2	0.842	2	0.855	3	0.736	3.2	2
Theis [35]	0.595	6	0.594	8	0.557	10	0.788	10	0.829	9	0.673	8.6	8
Venugopal [31]	0.484	9	0.496	10	0.610[a]	5	0.797	9	0.547	11	0.587	8.8	9
Rama-Maneiro [31]	0.675	2	0.777	1	0.610[a]	5	0.852	1	0.874	1	0.757	2	1
Oved [31]	0.696	1	0.774	2	0.655	1	0.816[a]	6	0.833[a]	6	0.755	3.2	2
Ours	0.645	3	0.724	4	0.610	5	0.816	6	0.848	4	0.729	4.4	5
Median	0.592		0.668		0.610		0.816		0.833		0.685		
Deviation from max	0.051		0.053		0.045		0.036		0.023		0.028		
Ours[b] (Weighted F1)	0,637		0,701		0,599		0,803		0,845				
Ours[b] (Macro F1)	0,377		0,449		0,382		0,413		0,675				

[a] We assume the median on missing values to provide a fair comparison.
[b] We report the weighted and macro average F1-Score for future comparability.

These negligible number of hallucinations are mostly similar but not the same as existing events, such as the hallucinated "IV Antifungal" and the existing event "IV Antibiotics" or the hallucinated "ER" in "ER Admission NC".

Table 5. Errors

Dataset	Average Fold Testset		Average Fold Syntax Errors		Average Fold Missing Concept Names		Average Fold Hallucinations	
	Traces	Events	Abs.	%	Abs.	%	Abs.	%
BPI13cp	298	1,308	0	0.0000	0.0	0.0000	0.2	0.0153
BPI13in	1,511	13,149	0.6	0.0046	0.0	0.0000	0.0	0.0000
Sepsis Cases	916	3,033	0.2	0.0066	1.8	0.0593	2.8	0.0923
Helpdesk	210	4,263	0	0.0000	0.0	0.0000	1.0	0.0235
BPI12	2,618	52,389	0	0.0000	0.0	0.0000	0.2	0.0004
Overall	5,553	74,142	0.80	0.0112	1.80	0.0593	4.20	0.1314

5 Discussion

Based on the complex preprocessing required in text-aware and NLP-based NEP approaches that we identified in related studies, we aimed to evaluate the use of the Llama 3 8 billion parameter LLM for NEP directly using XES-formatted event logs. While our approach could not outperform the SOTA SNAP method [31], it still performs above the median for most of our evaluation benchmark event logs. Nevertheless, Table 4 shows a varying accuracy in NEPs for each event log. Taking into account the statistics in Table 2, a relation is visible between an event logs performance and its statistics. Thus, the accuracy for NEP is lower for event logs that contain a high number of different activities, resulting in a higher average case length and longer prefixes exceeding the LLM's prefix context of 20 events, but only a small number of training data. The more training data the LLM can be trained on, the better the results can be [5,7]. Furthermore, our approach directly generates XES-formatted event predictions with nearly zero syntactic errors per event on average. The above-median performance and syntactically correct output highlight that the fine-tuned Llama 3 model captures the relations between events and attributes in events not only syntactically but also in a semantic way. This agrees with the results of Oved et al. [31], who found that their fine-tuned GPT-3 model captures the relations in their generated stories based on previous events. Furthermore, our approach using XES formatting shows that LLMs such as Llama 3 or GPT-3 can identify semantical relationships in structured languages, as proposed by Sui et al. [38].

Besides this important implication for researchers working on NEP, our approach is simply applicable to an existing process monitoring system for practitioners. It requires only an XES file, which is necessary in most process monitoring cases, directly generating XES formatting for further analysis in other process mining software. Whereas other approaches may need process adaptations and interruptions (e.g., extracting process information) [24,31], our approach directly generates NEPs based on an XES file. Besides the simple implementation, our LLM-driven approach allows the generation of NEPs, including all attributes, possibly allowing process simulations. Along with this, our approach allows the generation of additional information and attributes such as timestamps, possibly enabling remaining time prediction [6,8,35]. Further, our LLM-based approach can enable organizations to adapt new domains (e.g., new processes) and tasks (e.g., new classes). The feasibility of transfer learning in NLP-based PPM has already been demonstrated by Liessmann et al. [25] by transferring knowledge of one organization's process to a similar process in another organization, enabling PPM in the target organization. As an LLM learns the trace's structure and formatting, it enables transfer learning without training the LLM from scratch.

6   Limitations and Outlook

Naturally, our approach has limitations. First, the training and evaluation time takes several hundred hours. However, with a generation time of 3 to 11 s per event, our approach is still fast and valid, considering that it can generate additional information and attributes of an event. Second, our model can generate hallucinated concept names. Nevertheless, this only occurs in very rare cases. Third, Our approach has a lower accuracy than the SOTA. While additional preprocessing would possibly increase the prediction accuracy, it would lead to a less simple usage of the approach. This leads to a trade-off between simplicity and accuracy, where the extent of process simplicity vs. increasing the accuracy by additional preprocessing needs to be decided. Nevertheless, a higher accuracy is critical if applied in contexts such as medicine [26]. Therefore, an investigation of larger LLMs, such as Llama 3 with 70 billion parameters [28], could improve the accuracy. Additionally, leveraging a global training strategy on a dataset combining all existing open event logs in future research could improve the models performance as shown in prominent examples such as BERT [14], Llama 2 [42], and the GPT variants [7,29]. This may lead to improvements in LLM-driven NEP approaches and enable the implementation of a real-world organizational use case. Further, as we only reported the accuracy of the prediction for the attribute "concept:name", similar investigations for the other predicted attributes (e.g., resources, timestamps) could be performed. In addition, it could be interesting to consider the prediction of the whole process sequence. Lastly, methods of AI alignment that strive to align LLMs to the task [5] could not only help to mitigate the rare but existing hallucinations but to improve the accuracy.

References

1. van der Aalst, W.M.P.: Process mining: a 360 degree overview, pp. 3–34. Springer, Cham (2022)
2. AI@Meta: Llama 3 model card (2024). https://github.com/meta-llama/llama3/blob/main/MODEL_CARD.md
3. Berti, A., Schuster, D., van der Aalst, W.M.P.: Abstractions, scenarios, and prompt definitions for process mining with LLMS: a case study. In: De Weerdt, J., Pufahl, L. (eds.) Business Process Management Workshops, pp. 427–439. Springer, Cham (2024)
4. Berti, A., van Zelst, S., Schuster, D.: PM4Py: a process mining library for Python. Softw. Impacts **17**, 100556 (2023)
5. Bommasani, R., et al.: On the opportunities and risks of foundation models. arXiv (2021)
6. Brennig, K., Benkert, K., Löhr, B., Müller, O.: Text-aware predictive process monitoring of knowledge-intensive processes: does control flow matter? In: De Weerdt, J., Pufahl, L. (eds.) Business Process Management Workshops, pp. 440–452. Springer, Cham (2024)
7. Brown, T., et al.: Language models are few-shot learners. In: Larochelle, H., Ranzato, M., Hadsell, R., Balcan, M., Lin, H. (eds.) Advances in Neural Information Processing Systems, vol. 33, pp. 1877–1901. Curran Associates, Inc. (2020)
8. Cabrera, L., Weinzierl, S., Zilker, S., Matzner, M.: Text-aware predictive process monitoring with contextualized word embeddings. In: Cabanillas, C., Garmann-Johnsen, N.F., Koschmider, A. (eds.) Business Process Management Workshops, pp. 303–314. Springer, Cham (2023)
9. Camargo, M., Dumas, M., González-Rojas, O.: Learning accurate LSTM models of business processes. In: Hildebrandt, T., van Dongen, B.F., Röglinger, M., Mendling, J. (eds.) Business Process Management, pp. 286–302. Springer, Cham (2019)
10. Ceci, M., Lanotte, P.F., Fumarola, F., Cavallo, D.P., Malerba, D.: Completion time and next activity prediction of processes using sequential pattern mining. In: Džeroski, S., Panov, P., Kocev, D., Todorovski, L. (eds.) Discovery Science, pp. 49–61. Springer, Cham (2014)

11. Chen, H., Fang, X., Fang, H.: Multi-task prediction method of business process based on BERT and transfer learning. Knowl.-Based Syst. **254**, 109603 (2022)
12. Chen, M., et al.: Evaluating large language models trained on code (2021)
13. Dettmers, T., Pagnoni, A., Holtzman, A., Zettlemoyer, L.: QLoRA: efficient fine-tuning of quantized LLMs (2023)
14. Devlin, J., Chang, M.W., Lee, K., Toutanova, K.: BERT: pre-training of deep bidirectional transformers for language understanding. In: Burstein, J., Doran, C., Solorio, T. (eds.) Proceedings of the 2019 Conference of the North American Chapter of the Association for Computational Linguistics: Human Language Technologies, Volume 1 (Long and Short Papers), pp. 4171–4186. Association for Computational Linguistics, Minneapolis (2019)
15. Di Francescomarino, C., Ghidini, C.: Predictive process monitoring, pp. 320–346. Springer, Cham (2022)
16. Di Francescomarino, C., Ghidini, C., Maggi, F.M., Milani, F.: Predictive process monitoring methods: which one suits me best? In: Weske, M., Montali, M., Weber, I., vom Brocke, J. (eds.) BPM 2018. LNCS, vol. 11080, pp. 462–479. Springer, Cham (2018). https://doi.org/10.1007/978-3-319-98648-7_27
17. van Dongen, B.F., Weber, B., Ferreira, D.R., De Weerdt, J.: Report: business process intelligence challenge 2013. In: Lohmann, N., Song, M., Wohed, P. (eds.) Business Process Management Workshops, pp. 79–87. Springer, Cham (2014)
18. van Dongen, B.: BPI challenge 2012 (2012). https://doi.org/10.4121/UUID: 3926DB30-F712-4394-AEBC-75976070E91F
19. Feuerriegel, S., Hartmann, J., Janiesch, C., Zschech, P.: Generative AI. Bus. Inf. Syst. Eng. **66**(1), 111–126 (2023)
20. Grohs, M., Abb, L., Elsayed, N., Rehse, J.R.: Large language models can accomplish business process management tasks, pp. 453–465. Springer (2024)
21. Gururangan, S., et al.: Don't stop pretraining: adapt language models to domains and tasks. In: Jurafsky, D., Chai, J., Schluter, N., Tetreault, J. (eds.) Proceedings of the 58th Annual Meeting of the Association for Computational Linguistics, pp. 8342–8360. Association for Computational Linguistics (2020)
22. Ji, Z., et al.: Survey of hallucination in natural language generation. ACM Comput. Surv. **55**(12) (2023)
23. Lakshmanan, G.T., Shamsi, D., Doganata, Y.N., Unuvar, M., Khalaf, R.: A Markov prediction model for data-driven semi-structured business processes. Knowl. Inf. Syst. **42**(1), 97–126 (2015)
24. Levich, S., Lutz, B., Neumann, D.: Utilizing the omnipresent: incorporating digital documents into predictive process monitoring using deep neural networks. Decis. Support Syst. **175**, 114043 (2023)
25. Liessmann, A., Wang, W., Weinzierl, S., Zilker, S., Matzner, M.: Transfer learning for predictive process monitoring. In: ECIS 2024 Proceedings (2024)
26. Mannhardt, F.: Sepsis cases - event log (2016). https://doi.org/10.4121/UUID: 915D2BFB-7E84-49AD-A286-DC35F063A460
27. Marquez-Chamorro, A.E., Resinas, M., Ruiz-Cortes, A.: Predictive monitoring of business processes: a survey. IEEE Trans. Serv. Comput. **11**(6), 962–977 (2018)
28. Meta: Introducing meta llama 3: the most capable openly available LLM to date (2024). https://ai.meta.com/blog/meta-llama-3/
29. OpenAI: GPT-4 technical report (2024)
30. Ouyang, L., et al.: Training language models to follow instructions with human feedback. In: Koyejo, S., Mohamed, S., Agarwal, A., Belgrave, D., Cho, K., Oh, A. (eds.) Advances in Neural Information Processing Systems, vol. 35, pp. 27730–27744. Curran Associates, Inc. (2022)

31. Oved, A., Shlomov, S., Zeltyn, S., Mashkif, N., Yaeli, A.: SNAP: semantic stories for next activity prediction (2024)
32. Pasquadibisceglie, V., Appice, A., Castellano, G., Malerba, D.: Darwin: an online deep learning approach to handle concept drifts in predictive process monitoring. Eng. Appl. Artif. Intell. **123**, 106461 (2023)
33. Pegoraro, M., Uysal, M.S., Georgi, D.B., Van der Aalst, W.M.: Text-aware predictive monitoring of business processes. In: Business Information Systems, pp. 221–232 (2021)
34. Polato, M.: Dataset belonging to the help desk log of an Italian company (2017). https://doi.org/10.4121/UUID:0C60EDF1-6F83-4E75-9367-4C63B3E9D5BB
35. Rama-Maneiro, E., Vidal, J.C., Lama, M.: Deep learning for predictive business process monitoring: review and benchmark. IEEE Trans. Serv. Comput. **16**(1), 739–756 (2023)
36. Shanahan, M., McDonell, K., Reynolds, L.: Role play with large language models. Nature (2023)
37. Steeman, W.: BPI challenge 2013, closed problems (2013). https://doi.org/10.4121/UUID:C2C3B154-AB26-4B31-A0E8-8F2350DDAC11
38. Sui, Y., Zhou, M., Zhou, M., Han, S., Zhang, D.: Table meets LLM: can large language models understand structured table data? A benchmark and empirical study. In: Proceedings of the 17th ACM International Conference on Web Search and Data Mining, WSDM '24, pp. 645–654. Association for Computing Machinery, New York (2024)
39. Tama, B.A., Comuzzi, M.: An empirical comparison of classification techniques for next event prediction using business process event logs. Expert Syst. Appl. **129**, 233–245 (2019)
40. Tax, N., Verenich, I., La Rosa, M., Dumas, M.: Predictive business process monitoring with LSTM neural networks. In: Dubois, E., Pohl, K. (eds.) Advanced Information Systems Engineering, pp. 477–492. Springer, Cham (2017)
41. Teinemaa, I., Dumas, M., Rosa, M.L., Maggi, F.M.: Outcome-oriented predictive process monitoring: review and benchmark. ACM Trans. Knowl. Discov. Data **13**(2) (2019)
42. Touvron, H., et al.: Llama 2: open foundation and fine-tuned chat models (2023)
43. Verenich, I., Dumas, M., Rosa, M.L., Maggi, F.M., Teinemaa, I.: Survey and cross-benchmark comparison of remaining time prediction methods in business process monitoring. ACM Trans. Intell. Syst. Technol. **10**(4), 1–34 (2019)
44. Vidgof, M., Bachhofner, S., Mendling, J.: Large language models for business process management: opportunities and challenges, pp. 107–123. Springer (2023)
45. Weinzierl, S., Zilker, S., Dunzer, S., Matzner, M.: Machine learning in business process management: a systematic literature review. Expert Syst. Appl. 124181 (2024)
46. Weske, M.: Business Process Management: Concepts, Languages, Architectures, 3rd edn. Springer, Heidelberg (2019)
47. Wolf, T., et al.: Transformers: state-of-the-art natural language processing. In: Liu, Q., Schlangen, D. (eds.) Proceedings of the 2020 Conference on Empirical Methods in Natural Language Processing: System Demonstrations, pp. 38–45. Association for Computational Linguistics (2020)
48. Ye, Y., et al.: ProAgent: from robotic process automation to agentic process automation (2023)
49. Yuan, Y., Liu, X., Lu, K.: Multi-perspective data fusion framework based on hierarchical BERT: provide visual predictions of business processes. Comput. Mater. Continua **78**(1), 1227–1252 (2024)

Enhancement of Low-Level Event Abstraction with Large Language Models (LLMs)

Edyta Brzychczy[1] , Krzysztof Kluza[1(✉)] , and Leszek Szała[2]

[1] AGH University of Krakow, Mickiewicza Av.30, 30-059 Krakow, Poland
{brzych3,kluza}@agh.edu.pl

[2] Department of Mathematics, Informatics and Cybernetics, Faculty of Chemical Engineering, University of Chemistry and Technology, Technicka 5, 166 28 Prague 6, Prague, Czech Republic
Leszek.Szala@vscht.cz

Abstract. Event abstraction enables transformation of low level events into higher level events making process mining (PM) on sensor data available. There are many approaches to event abstraction described in literature, however the main approaches include supervised or unsupervised techniques. We address the challenge of transforming low-level sensor data into high-level activities required for effective process mining, a task traditionally reliant on domain experts. By leveraging LLMs to automate the labelling process of sensor data clusters, we bridge the gap between raw data and process models. Motivated by a mining industry use case, we validated the effectiveness of LLMs in accurately labelling operational phases. Our LLM-generated labelling rules demonstrated high accuracy and interpretability, simplifying the understanding for domain experts. Additionally, we compared our LLM-based approach with a Decision Tree Classifier, highlighting the advantages of LLMs in generating simpler, more understandable labelling functions. Our work underscores the potential of advanced AI techniques to enhance the efficiency and accuracy of PM, contributing to the Industry 4.0 initiative.

Keywords: Event abstraction · Sensor data · LLMs · Process mining

1 Introduction

Process mining (PM) is nowadays a state-of-the-art approach to process analysis in the Business Process Management (BPM) field [25]. Its potential is recognized by many industrial branches, driven by Industry 4.0 solutions, aiming to achieve more effective processes and competitive advantage. The Industry 4.0 concept assumes the wide usage of heterogeneous data sources to discover knowledge about processes to analyse and improve them [2]. One of the possible data sources is sensor data, which requires preprocessing before their direct usage, in the form of an event log, for process analysis [17,21].

K. Gdowska et al. (Eds.): BPM 2024 Workshops, LNBIP 534, pp. 209–220, 2025.
https://doi.org/10.1007/978-3-031-78666-2_16

The creation of event logs from low-level event data is currently a challenging task in the PM domain. It results mainly from a gap between the abstraction level between the raw event data (e.g., in the form of sensor readings) and higher-level activities expressed by process models [13].

The structure of a typical event log comprises: *Case ID*, *Activity* and *Timestamp* [23]; however, in the raw sensor data, very often, Case ID and Activity are not present [17]. Hence, the creation of an event log requires preprocessing methods related to so-called event abstraction and event correlation [9].

Event abstraction enables the transformation of low-level events into higher-level events (or activities). The literature describes many approaches to event abstraction (discussed in Sect. 2); however, the main ones include supervised or unsupervised techniques. One of the main unsupervised techniques is clustering. What is important, clustering results require additional transformation to be useful in event log creation, known as event labelling. In the literature devoted to unsupervised event abstraction for PM purposes, cluster labelling requires domain knowledge, which is very often not available at the moment [14]. Also, labelling itself is a challenging task for many domain experts. In this scope, Large Language Models (LLMs) can find application and automate tasks traditionally carried out by humans [15].

In our paper, we would like to present a real-life example of how the LLMs can support the labelling of the data and clustering results for creating event logs with satisfying quality/reliability. Our work was motivated by an industrial use case from the mining domain, namely the longwall shearer operation process.

The paper is structured as follows: Sect. 2 describes the low-level event abstraction in the PM domain. Section 3 presents related work on LLMs' usage in PM and event abstraction. Section 4 presents our use case and results of experiments. Section 5 evaluates and discusses our results. Section 6 concludes the paper and highlights future work.

2 Low-Level Event Abstraction in PM

The basic artefact needed for PM is the event log [1]. The main requirements for the event log are [23]:

- each event should be linked to a case or process instance (Case ID),
- each event should correspond to an activity executed in the process (Activity),
- each event should be ordered in the sequence (e.g., Timestamp).

In sensor data, process activity is not present, and very often also Case ID is not straightforwardly available [23]. These two issues are sources of challenges to applying the PM to sensor data to improve industrial processes. In the PM domain, these issues are addressed by event correlation and abstraction techniques used for creating event logs.

Our work focuses on event abstraction, which is mapping event data to events representing activity executions [9]. This task is one of the most important challenges of BPM for IoT implementations, defined as bridging the gap between sensor data and event logs for process mining [13,19].

Two main strategies for event abstraction in process mining are supervised and unsupervised. In a supervised approach, the event abstraction technique expects some form of additional input, e.g., a reference model or expert rules, that guides the translation of the low-level events into higher-level activities. In the case of an unsupervised approach, no form of additional input is required [24].

In terms of the richness of the information that is considered as a starting point, the event abstraction approaches can be ordered as follows [9]: unsupervised learning (e.g., clustering), supervised learning, behavioural patterns, process models, and further domain knowledge.

From the point of view of event abstraction tasks in the industrial domain, two main conditions occur: (1) we have vast low-level data from monitoring systems (e.g., SCADA) to be analysed, (2) process models or behavioural patterns for low-level data are not available. Therefore, clustering is a popular approach in the industrial domain to discover high-level activities from low-level data [6]. The important advantage of clustering over supervised techniques is that some actual non-typical behaviour (states or activities) can be identified [8].

One of the first works using clustering to abstract sensor data is [10]. Van Eck et al. presented the discovery of process models of smart bottle usage based on sensor data, including temperature, a 3D accelerometer, a light intensity sensor, and a sound level sensor. In this use case, clustering was used to support the generation of implicit labels for data segments (parts of time-series sensor data in time windows). Obtained clusters had been annotated by a domain expert.

Mayr et al. in the [20] focused on process-state and product-state data and proposed an approach to event abstraction based on selected features and their segmentation. For each segment, statistics are calculated, which are clustered to identify similar statistical behavior. In that use case also, domain knowledge is used for cluster labelling.

The example of clustering in the heavy industry domain – mining – was presented in [7]. The authors used clustering to abstract sensor data from a long-wall shearer monitoring system to discover higher-level activities. They used also domain knowledge to understand the meaning of discovered clusters.

Despite its popularity, clustering has limitations [5,21], especially the need for cluster meaning explanation by domain experts. This fact motivated us to undertake the research in support of unsupervised techniques used in the event abstraction of sensor data with LLMs.

3   Related Works

LLMs are nowadays a very popular technology, which offers various benefits in many industries and tasks. Possibilities and challenges of LLMs in the BPM domain in reference to the BPM lifecycle were presented in [22]. Authors pointed out the capabilities of LLMs to support organisation in process improvement.

In the PM domain LLMs can be used in various tasks [3]. In [4], authors present usage of textual abstractions of PM artefacts, e.g., event logs and process models, to leverage the application of LLMs in the PM domain. They argue that because of size limitations, original event logs or process models can not be directly used. They propose a set of abstraction of PM object and different prompt generation strategies to obtain process description, conformance suggestions or process improvement ideas.

Kourani et al. propose in [18] the framework that integrates LLMs into process modelling to enhance flexibility, efficiency, and accessibility for expert and non-expert users. Its main components include advanced techniques in prompt engineering, model generation, error handling and interactive feedback loop to model refinement. The framework uses Partially Ordered Workflow Language (POWL) as an intermediate representation of the process, enabling transformations to BPMN and Petri nets notations.

In the paper [16], the authors introduced the concept of conversational process modelling, describing the process of creation and process model improvement based on the interaction between domain experts and chatbots. They identified application scenarios for conversational process modelling (e.g., "gather information", "process modelling", "compare and access") and evaluated them with existing chatbots.

Other examples of LLMs in process mining tasks are presented in [12]. Authors investigated the mining of imperative and declarative models from textual descriptions and assessing the suitability of process tasks for RPA.

Extensive example of prompt engineering in PM proposed Jessen et al. [15]. Authors developed a framework using LLMs to mimic the task and skills of various process mining roles. They used conversational agent architecture, employing GPT-3.5-turbo and GPT4. Evaluation of the created framework showed that the LLMs were able to fully or partially understand the question and propose a suitable solution in 77% of cases.

All mentioned papers present the usage of LLMs on a higher level of model abstraction with well-established artefacts in the form of an event log or/and process model (or its textual description). There is no example of usage of LLMs to abstract low-level data to create an event log, however their capabilities to group and label tasks are presented in [11]. We would like to present our idea of how the LLMs can support log engineering based on sensor data starting from the use case described in the next section.

4   Experiments

In our work we use a dataset from longwall shearer which operates in the underground coal mine. Its basic process operation phases are (1) stoppage, (2) moving, (3) cutting, and (4) working without moving (idle). Sensors installed on machinery monitor the shearer's operation. The main features that can be used in event abstraction collected by the monitoring system include currents on drums, currents on tractors (haulages), movement directions, shearer speed, and location. Table 1 presents an example of collected data.

Table 1. Exemplary sensor data

Timestamp	OL [A]	OR [A]	HL [A]	HR [A]	Move Left	Move Right	Location [m]	Speed [m/s]
03:43:17	0	0	53	54	1	0	71	4.2
03:43:18	0	0	55	57	1	0	71	4.2
03:43:19	0	0	59	59	1	0	71	4.3
03:43:20	0	0	59	59	1	0	71	4.4
03:43:21	0	0	60	60	1	0	72	4.4

4.1   Experiment 1: Unsupervised Data Labelling

For unsupervised data labelling, we used ChatGPT 4o with the following prompt: *I try to understand the process phases of the longwall shearer, which operates in the underground coal mine. HL and HR denotes variables relating to haluages work, OL and OR denotes variables relating to drum work. Label each row in the data with a single label from the four possible process operation phases: 1. Stoppage, 2. Move, 3. Cutting, 4. Idle.*

Apart from the prompt, a CSV file with balanced sample of unlabelled data (1000 random examples of each state) was given to the chat. As the file is uploaded to a temporary workspace, it does not count to the token limit. To conduct the experiment, we prompted the chat with the same prompt 31 times (the sample size was chosen arbitrarily to provide a robust sample size for evaluation), which allowed us to gather and compare the responses obtained. In each iteration[1], ChatGPT 4o generated a different Python labelling function, e.g.

```
1  # Experiment 1 iteration 1
2  def label_phase1(row):
3      if row['Speed'] == 0 and row['Move_left'] == 0 and row['Move_right'] ==
           0 and row['OL'] == 0 and row['OR'] == 0:
4          return 'Stoppage'
5      elif (row['Move_left'] == 1 or row['Move_right'] == 1)  and (...):
6          return 'Move'
7      elif (...):
8          if row['Speed'] == 0:
9              return 'Idle'
10         else:
11             return 'Cutting'
12     else:
13         return 'Unknown'
```

Listing 1.1. Excerpt of the result of the Experiment 1 iteration 1

Our full experimental data is unbalanced (Cutting: 106951, Stoppage: 101743, Idle: 32146, Move: 2444). The performance metrics (Accuracy, Precision, Recall, F1-Score, Cohen's Kappa) for all 31 labelling iterations for the whole dataset against the ground truth (labelled by experts) are presented in Fig. 1.

---

[1] The term iteration refers to repeated trials under identical experimental conditions to assess the consistency and range of the LLM's outputs.

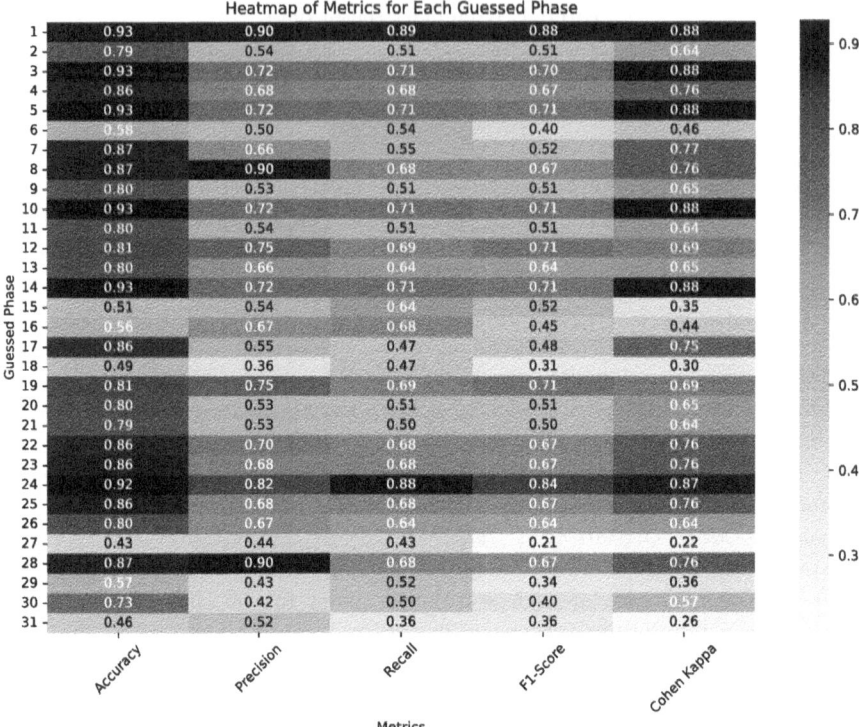

Fig. 1. The heatmap presenting the evaluation metrics for each iteration of the experiment (guessed phase)

The heatmap for Experiment 1 demonstrates varying performance across different guessed phases, with certain phases showing high metrics in accuracy, precision, recall, F1-Score, and Cohen's Kappa, indicating strong model performance. Due to the non-deterministic nature of LLMs, it demonstrates varying performance across different experiment instances, with certain instances like 1, 8, 14, and 24 showing high metrics indicating strong model performance. It is important to mention that this is the unsupervised solution, as we did not provide labels for generating the rules. The varying performance across iterations might point to the LLMs' limitation, but also showcases the potential of LLMs. For the well-performed instance (as presented in Listing 1.1), the rules are rather easy to understand, interpretable and self-explanatory.

To assess the possibility of getting high metrics in the rule-based approach, we also performed supervised training using the classical machine learning approach—Decision Tree Classifier. We used decision trees as these also provide, in theory, self-explainable rules. The confusion matrix for the classifier is presented in Fig. 2. Although such a classifier achieved quite good results (precision, recall and F1-score 0.97), because of complexity and very specific threshold values, such rules are not really generalized and comprehensible, see the example rule for Cutting phase in Listing 1.2.

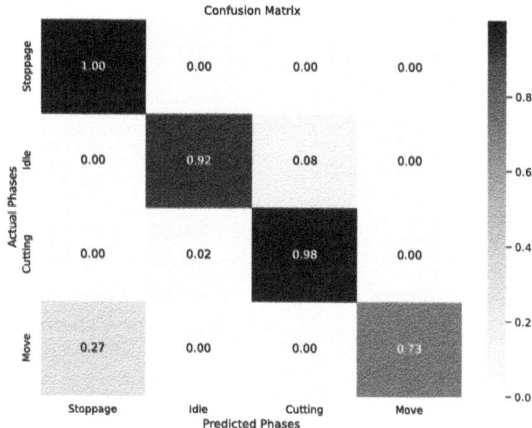

Fig. 2. The confusion matrix for Decision Tree Classifier

```
1    if (OL <= 17.50 and Speed > 0.50 and OR > 19.50) or \
2        (OL > 17.50 and HR <= 37.50 and Speed > 0.50) or \
3        (OL > 17.50 and HR > 37.50 and HL <= 149.50):
4        return Cutting
```

Listing 1.2. Example rule for Cutting phase

To assess the overall performance of the LLM in this task, for each iteration of the experiment for each labelling, we counted the most common guessing by LLM. The confusion matrix in Fig. 3 and 4 compares the actual phases to the most commonly guessed phases. Each cell in the matrix indicates the number of instances of sensor data where a specific actual phase was guessed as another phase. The rows represent the actual phases, the columns represent the guessed phases, while the diagonal elements represent the number of correctly guessed phases.

We can see that the guess's quality differs among the phases. The Stoppage phase was guessed correctly in almost 100% of cases. The Idle state was very often indicated as a stoppage (80%); however, it is a very specific state in longwall operation, and only two variable values differ the idle stage from full stoppage. Cutting phase was indicated correctly in 84% and Move in almost 73% of cases. What is important, longwall shearer is a heavy machine and it is not possible to move it in one second; thus we have a mixture of predicted stoppages and moves indicated as the Move phase.

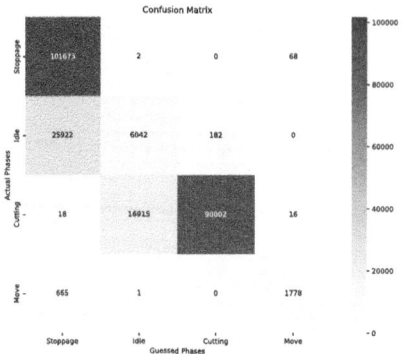

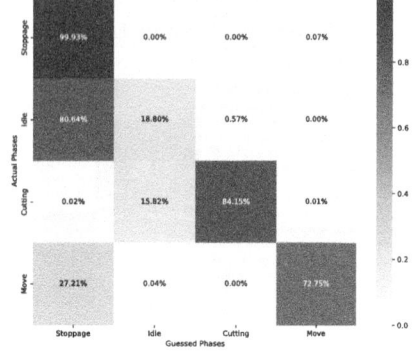

Fig. 3. The confusion matrix for Experiment 1

Fig. 4. The normalized confusion matrix for Experiment 1

4.2 Experiment 2: Unsupervised Label Matching

In Experiment 2, we provided to the ChatGPT 4 phases encoded by numbers, and wanted the LLM to match the encoded phase with one of the possible operation phases. We used the following prompt: *I try to understand the process phases of longwall shearer which operates in the underground coal mine. In the file there is a sample of data. HL and HR denotes variables relating to haluages work, OL and OR denotes variables relating to drum work. Each row in the data is labeled with the phase number in the column 'Phases_encoded'. Match each phase from the 'Phases_encoded' with one of the possible process operation phases: Stoppage, Move, Cutting, Idle. Provide one to one matching.*

Note that 'Phases_encoded' can be understood as label from a clustering algorithm or other source.

Experiment 2 was designed to validate the LLM's capability to match encoded labels with meaningful process phases, even when labels are available. Such validation step can be relevant for ensuring that the LLM's understanding aligns with domain knowledge, e.g. for scenarios where labels require validation or refinement.

In Fig. 5, we present a distribution of guess types for each phase number in the actual phase of the process. We can observe that for all four phases, most of the LLM's guesses are correct. The main issues are related to the Idle phase, indicated as Stoppage or Move. The best guessing of the phase is for the Cutting phase indicated marginally as Move (twice) or Idle (once) per all 31 experiments.

5 Statistical Evaluation and Discussion

In the evaluation of results, we used shearer statuses from the monitoring system, including (1) Stoppage, (2) Cutting, (3) Move, and (4) Idle. We cross-checked

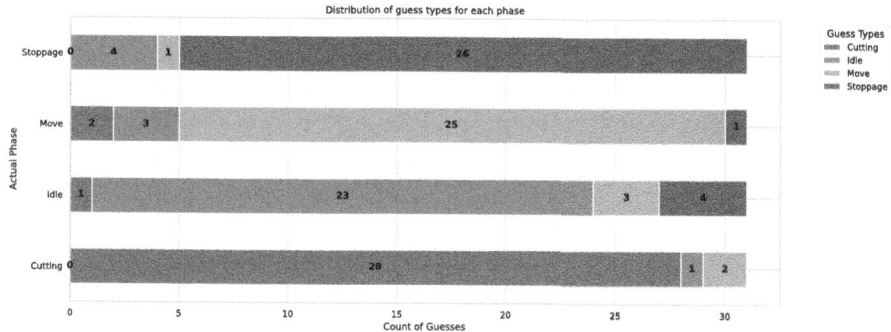

Fig. 5. The distribution of guess types for each phase

labels of the obtained phases from LLMs with statuses received from domain experts.

5.1   Experiment 1

We consider a pair of random variables (ξ, ζ), whose values are members of set $\{cutting, idle, move, stoppage\}$. The realizations of ξ are experts' judgements and the realizations of ζ are the most frequently occurring answers given by ChatGPT in 31 independent queries for each dataset. We verify independence of ξ and ζ using the χ^2-test. Typically, the null hypothesis assumes that ξ and ζ are independent, and the alternative hypothesis assumes that ξ and ζ are dependent. The obtained test statistic equals 403105 and the p-value is $2.2 \cdot 10^{-16}$. At significance level of $\alpha = 0.05$ the test rejects the null hypothesis in favour of the alternative hypothesis. There is evidence that (ξ, ζ) are dependent. Therefore this suggests that there is a significant connection between the experts' opinions and ChatGPT's answers, showing that they are not independent and their responses are related.

5.2   Experiment 2

We verify whether the probability of a mistake in a single query does not depend on i, where $i = 1, 2, 3, 4$ correspond to

$$cutting, idle, move, stoppage$$

respectively. We use the χ^2-test. Typically the null hypothesis assumes that the probabilities of a mistake are equal for all i and the alternative hypothesis claims the opposite. We define random variables $N_{i,j}$, where $j = 1, 2, \ldots, 31$ correspond to the answers given by ChatGPT in 31 independent queries with:

$$N_{i,j} = \begin{cases} 0 & \text{if the correct answer was returned,} \\ 1 & \text{if an error was made.} \end{cases}$$

Let $N_i = \sum_{j=1}^{31} N_{i,j}$ is the number of mistakes occurred for i given, $N = \sum_{i=1}^{4} N_i$ is the total number of mistakes and $\hat{N}_i = \frac{N}{4}$ means the expected number of mistakes for each i with the assumption of independence. The test statistic

$$T = \sum_{i=1}^{4} \frac{\left(N_i - \hat{N}_i\right)^2}{\hat{N}_i}$$

was equal to 2.3636 and the p-value 0.5004. At significance level of $\alpha = 0.05$ the test does not reject the null hypothesis of the equality of the probabilities considered. It is necessary to remember that statistical testing could never prove the null hypothesis. Either it rejects the null hypothesis, or there is no reason to reject it (as it is in this situation). This means there is no evidence that the model makes a mistake more frequently for any of the phases of the operational process (*cutting, idle, move, stoppage*).

6   Conclusions

In this paper, we demonstrated the significant potential of using Large Language Models (LLMs) to support the labelling of data for creating event logs, focusing on a real-life industrial use case from the mining domain.

We introduced the novel application of LLMs to automate the labelling process of sensor data, a task traditionally reliant on domain experts. This approach bridges the gap between low-level sensor data and high-level activities required for effective process mining.

Through our experiments, we validated the effectiveness of LLMs in accurately labelling phases of a longwall shearer operation. We showed that the LLM-generated labelling rules might be not only highly accurate but also interpretable, facilitating easier understanding and validation by domain experts. We also compared our LLM-based approach with a supervised machine learning technique, specifically a Decision Tree Classifier. While the classifier achieved high accuracy, the complexity of its rules underscored the advantage of LLMs in generating simpler and more understandable labelling functions.

Two experimental setups – unsupervised data labelling and unsupervised label matching – highlighted the versatility of LLMs in handling different aspects of the labelling process. The results showed that LLMs could successfully match encoded phases to actual process phases, further demonstrating their utility in scenarios where domain knowledge is limited.

Our work contributes to the broader Industry 4.0 initiative by showcasing how advanced AI techniques can enhance the efficiency and accuracy of process mining in industrial settings. The ability to generate reliable event logs from raw sensor data supports the goal of achieving more effective processes and competitive advantage. Future work will focus on refining the LLM-based approach, optimizing its parameters such as temperature and seed, exploring its applicability to other industrial domains, and integrating it with real-time data processing systems to further enhance its practical utility.

References

1. van der Aalst, W.M.P.: Process Mining - Data Science in Action, 2nd edn. Springer (2016). https://doi.org/10.1007/978-3-662-49851-4
2. Alcácer, V., Cruz-Machado, V.: Scanning the industry 4.0: a literature review on technologies for manufacturing systems. Eng. Sci. Technol. Int. J. **22**(3), 899–919 (2019). https://doi.org/10.1016/j.jestch.2019.01.006
3. Berti, A., Qafari, M.S.: Leveraging large language models (LLMs) for process mining. Technical report (2023)
4. Berti, A., Schuster, D., van der Aalst, W.M.P.: Abstractions, scenarios, and prompt definitions for process mining with LLMs: a case study. In: De Weerdt, J., Pufahl, L. (eds.) Business Process Management Workshops, pp. 427–439. Springer, Cham (2024)
5. Bertrand, Y., Van den Abbeele, B., Veneruso, S., Leotta, F., Mecella, M., Serral, E.: A survey on the application of process discovery techniques to smart spaces data. Eng. Appl. Artif. Intell. **126**, 106748 (2023). https://doi.org/10.1016/j.engappai.2023.106748
6. Brock, J., Rempe, N., von Enzberg, S., Kühn, A., Dumitrescu, R.: A framework for the domain-driven utilization of manufacturing sensor data in process mining: an action design approach, pp. 771–781 (2023). ESSN: 2701-6277. https://doi.org/10.1016/j.engappai.2023.106748
7. Brzychczy, E., Trzcionkowska, A.: Process-oriented approach for analysis of sensor data from longwall monitoring system. In: Burduk, A., Chlebus, E., Nowakowski, T., Tubis, A. (eds.) Intelligent Systems in Production Engineering and Maintenance, pp. 611–621. Springer, Cham (2019)
8. Brzychczy, E., Żuber, A., van der Aalst, W.: Process mining of mining processes: analyzing longwall coal excavation using event data. IEEE Trans. Syst. Man Cybern. Syst. **54**(5), 2723–2734 (2024). https://doi.org/10.1109/TSMC.2023.3348496
9. Diba, K., Batoulis, K., Weidlich, M., Weske, M.: Extraction, correlation, and abstraction of event data for process mining. WIREs Data Min. Knowl. Discov. **10**(3) (2020). https://doi.org/10.1002/widm.1346
10. van Eck, M.L., Sidorova, N., van der Aalst, W.M.P.: Enabling process mining on sensor data from smart products. In: Tenth IEEE International Conference on Research Challenges in Information Science, RCIS 2016, Grenoble, France, 1–3 June 2016, pp. 1–12. IEEE (2016). https://doi.org/10.1109/RCIS.2016.7549355
11. Fani Sani, M., Sroka, M., Burattin, A.: LLMs and process mining: challenges in RPA: task grouping, labelling and connector recommendation. In: International Conference on Process Mining, pp. 379–391. Springer (2023)
12. Grohs, M., Abb, L., Elsayed, N., Rehse, J.R.: Large language models can accomplish business process management tasks. In: De Weerdt, J., Pufahl, L. (eds.) Business Process Management Workshops, pp. 453–465. Springer, Cham (2024)
13. Janiesch, C., et al.: The Internet of Things meets business process management: a manifesto. IEEE Syst. Man Cybern. Mag. **6**(4), 34–44 (2020). https://doi.org/10.1109/MSMC.2020.3003135
14. Janssen, D., Mannhardt, F., Koschmider, A., van Zelst, S.J.: Process model discovery from sensor event data. In: Leemans, S., Leopold, H. (eds.) ICPM 2020. LNBIP, vol. 406, pp. 69–81. Springer, Cham (2021). https://doi.org/10.1007/978-3-030-72693-5_6

15. Jessen, U., Sroka, M., Fahland, D.: Chit-chat or deep talk: prompt engineering for process mining. Working paper. arXiv.org (2023). https://doi.org/10.48550/arXiv.2307.09909

16. Klievtsova, N., Benzin, J.V., Kampik, T., Mangler, J., Rinderle-Ma, S.: Conversational process modelling: state of the art, applications, and implications in practice. In: Di Francescomarino, C., Burattin, A., Janiesch, C., Sadiq, S. (eds.) Business Process Management Forum, pp. 319–336. Springer, Cham (2023)

17. Koschmider, A., et al.: Process mining for unstructured data: challenges and research directions. arXiv preprint arXiv:2401.13677 (2023)

18. Kourani, H., Berti, A., Schuster, D., van der Aalst, W.M.: Process modeling with large language models. In: International Conference on Business Process Modeling, Development and Support, pp. 229–244. Springer (2024)

19. Mangler, J., et al.: From Internet of Things data to business processes: challenges and a framework (2024)

20. Mayr, M., Luftensteiner, S., Chasparis, G.C.: Abstracting process mining event logs from process-state data to monitor control-flow of industrial manufacturing processes. In: Longo, F., Affenzeller, M., Padovano, A. (eds.) Proceedings of the 3rd International Conference on Industry 4.0 and Smart Manufacturing (ISM 2022) (2021). Procedia Comput. Sci. **200**, 1442–1450. Elsevier. https://doi.org/10.1016/j.procs.2022.01.345

21. Seiger, R., Franceschetti, M., Weber, B.: An interactive method for detection of process activity executions from IoT data. Future Internet **15**(2) (2023). https://doi.org/10.3390/fi15020077. https://www.mdpi.com/1999-5903/15/2/77

22. Vidgof, M., Bachhofner, S., Mendling, J.: Large language models for business process management: opportunities and challenges. In: Di Francescomarino, C., Burattin, A., Janiesch, C., Sadiq, S. (eds.) Business Process Management Forum, pp. 107–123. Springer, Cham (2023)

23. Weerdt, J.D., Wynn, M.T.: Foundations of process event data. In: van der Aalst, W.M.P., Carmona, J. (eds.) Process Mining Handbook. Lecture Notes in Business Information Processing, vol. 448, pp. 193–211. Springer (2022). https://doi.org/10.1007/978-3-031-08848-3_6

24. van Zelst, S.J., Mannhardt, F., de Leoni, M., Koschmider, A.: Event abstraction in process mining: literature review and taxonomy. Granular Comput. **6**(3), 719–736 (2021). https://doi.org/10.1007/s41066-020-00226-2

25. Zerbino, P., Stefanini, A., Aloini, D.: Process science in action: a literature review on process mining in business management. Technol. Forecast. Soc. Change **172**, 121021 (2021). https://doi.org/10.1016/j.techfore.2021.121021

ProcessLLM: A Large Language Model Specialized in the Interpretation, Analysis, and Optimization of Business Processes

Alina Buss[1], Wolfgang Kratsch[2,4,5], Sebastian Johannes Schmid[3,4,5]($\boxtimes$), and Hongyang Wang[1]

[1] TUM School of Management, Technical University of Munich, Munich, Germany
{alina.buss,hongy.wang}@tum.de
[2] Technical University of Applied Sciences Augsburg, Augsburg, Germany
[3] University of Bayreuth, Bayreuth, Germany
[4] Branch Business and Information Systems Engineering of the Fraunhofer FIT, Augsburg and Bayreuth, Germany
[5] FIM Research Center for Information Management, Augsburg and Bayreuth, Germany
{wolfgang.kratsch,sebastian.schmid}@fim-rc.de

Abstract. Generative AI, notably embodied by Large Language Models (LLMs), has the potential to revolutionize key aspects of Business Process Management (BPM). Addressing the inadequacies of general-purpose LLMs in complex BPM tasks, we follow the Design Science Research paradigm and propose ProcessLLM, a specialized LLM fine-tuned to support BPM tasks. Evaluated in both artificial and real-life business settings, ProcessLLM demonstrates promising performance and a sophisticated understanding of BPM concepts and the BPMN language. Our research serves as a foundational exploration at the intersection of BPM and LLMs, paving the way for future research in LLM-enhanced BPM. Validated by business experts, our research also demonstrates the real-world utility and applicability of LLMs in BPM, thus setting the stage for their broader adoption in operational BPM settings.

Keywords: Business Process Management · Large Language Models · Generative Artificial Intelligence · Natural Language Processing

1 Introduction

Business processes are crucial for enterprises, serving as the mechanisms that convert inputs into valuable outputs for customers and stakeholders [1]. With the ever-increasing pace of global business and the constant emergence of disruptive technologies, organizations face rapidly changing customer expectations, heightened competition, and the need for agility in their processes [2]. In this setting, Business Process Management (BPM) serves as the backbone of modern businesses, helping to stay adaptable and efficient in volatile business environments [3].

K. Gdowska et al. (Eds.): BPM 2024 Workshops, LNBIP 534, pp. 221–232, 2025.
https://doi.org/10.1007/978-3-031-78666-2_17

Meeting the complex requirements of contemporary BPM demands innovative strategies and tools that can address today's volatile business environments [4]. Specifically, combining BPM with technological advancements can lead to more efficient, scalable, and adaptable processes [2]. The introduction of Artificial Intelligence (AI), for example, has enabled BPM capabilities that encompass predictive process monitoring, real-time decision-making, or process optimization [5]. Building on these advancements, the emergence of generative AI represents a significant paradigm shift at the intersection of BPM and AI [6]. Known for its ability to create novel data instances, Generative AI has already shown its efficacy, for example, in automated process improvement [7] or event log quality enhancement [8]. A key development in generative AI is the advent of Large Language Models (LLMs) like ChatGPT, which achieved rapid popularity within a short time frame. Thereby, LLMs are recognized for their reasoning and creative capabilities, evident across a variety of domains and tasks [6]. As BPM is an integral information systems discipline, it therefore has the potential to realize considerable value by integrating LLMs in its activities [9]. Consequently, researchers are currently investigating the notion of using LLMs in BPM. Preliminary studies discuss opportunities and challenges [6] or suggest conceptual frameworks for designing such artifacts [10]. Furthermore, general-purpose LLMs are successfully applied to selected BPM tasks such as extracting process models from textual process descriptions [11].

While general-purpose models have extensive knowledge due to their training on large and diverse corpora of text, they are proven to have limitations when they are used to conduct domain specific tasks [12]. Meanwhile, fine-tuned, domain-specific LLMs have demonstrated better performance in specific domains such as finance [13] or software engineering [14]. Recognizing the potential of fine-tuned LLMs, it becomes imperative to explore the capabilities of more specialized, domain-tailored models in the BPM context. To date, research on fine-tuned LLMs for BPM is absent, and general-purpose LLMs have been limited to selected applications such as process discovery. Consequently, we follow calls in the literature [12] and pose the following research question: How can domain-specific LLMs be designed to support BPM activities?

To answer this question, we follow the Design Science Research (DSR) paradigm proposed by Peffers et al. [15] and design ProcessLLM, an LLM fine-tuned on BPM knowledge and the BPMN language. The artifact is extensively evaluated following the DSR evaluation framework by Sonnenberg and vom Brocke [16] that consist of four evaluation stages. As the evaluation yields satisfactory results, our research contributes theoretically by providing a proof-of-concept at the intersection of BPM and LLMs that should stimulate discussion and further research. Practically, we provide an open-source prototype[1] to support practitioners in various BPM activities as well a detailed code repository[2] to assist developers in the design of LLMs.

---

[1] Hugging Face repository: https://huggingface.co/ProcessLLM-developers/ProcessLLM.

[2] GitHub repository: https://github.com/ProcessLLM-developers/ProcessLLM.

2 Theoretical Background

2.1 Business Process Management

BPM is the systematic approach of optimizing and managing a company's business processes to improve efficiency, effectiveness, and adaptability [17]. Therefore, BPM activities typically follow a lifecycle beginning with process identification followed by discovery, analysis, redesign, implementation, and monitoring [1].

Given the growing significance of BPM, there has been an evolution of process modelling tools. Organizations primarily engage in process modelling for purposes encompassing documentation, organizational redesign, and knowledge management [18]. Thereby, an array of modelling languages has emerged, including but not limited to BPMN, POWL, Petri Nets, UML, EPC, and YAWL. Among all process modelling languages, BPMN stands out as one of the most popular ones [19]. BPMN is used for the graphical representation and modelling of business processes in a flowchart-like manner that enhances user comprehension. Nonetheless, the increasing complexity of contemporary process models poses a challenge for non-expert users [18].

2.2 Large Language Models in Natural Language Processing

Natural Language Processing (NLP) intersects computational science and linguistics, effectively equipping machines with the ability to comprehend, interpret, and produce human language [20]. LLMs represent a significant milestone in the evolution of NLP. These models, characterized by their vast number of parameters, are trained on extensive datasets to achieve proficiency in both natural language understanding and generation [21]. The introduction of the Transformer architecture [22] marked a paradigm shift in the design of LLMs. Leveraging attention mechanisms, transformer-based models, such as GPT, have achieved state-of-the-art performance in diverse NLP tasks. The LLM landscape is currently undergoing constant change and innovation. Meta's LLaMA series, for example, has demonstrated remarkable efficacy even with fewer parameters compared to other models. As LLaMA is available open source, it brought forth many follow-up models, for example Vicuña, Alpaca and Guanaco.

The initial training of an LLM enables a broad understanding of language as well as the concepts and knowledge embedded in language. Fine-tuning allows to refine this generalized knowledge for specialized tasks. This process involves adjusting the model based on a task-specific dataset, thereby tailoring its performance to a particular domain or application. For instance, fine-tuned on financial data, BloombergGPT showed significantly enhanced performance in financial tasks without compromising general abilities [13]. It demonstrates that fine-tuning enables transferring knowledge to niche areas, optimizing the model's performance without exhaustive retraining.

Fine-tuning LLMs requires methods to adjust the parameters of pre-trained models effectively and efficiently. One common strategy is Parameter Efficient Fine-Tuning (PEFT), which introduces new modules between layers of an existing network [23]. This strategy retains the original weights of the pre-trained model and modifies only the new adapter layer, whose parameters are initially randomized [23]. Diverse techniques have emerged within the PEFT domain among which Low-Rank Adaptation (LoRA) is of particular interest as it allows a significant reduction in GPU demands - up to threefold - compared to conventional methods [24].

3   Design Objectives

Stemming from the problem definition in Sect. 1, motivated by the theoretical foundations outlined in Sect. 2, and validated by a survey conducted with business experts, we derive four design objectives (DOs) that will guide the design of our artifact.

(DO1) Domain-specific knowledge in BPM and BPMN. Domain-specific data can enable LLMs to outperform its baseline model in specific tasks [13]. As the research question revolves around supporting BPM activities, the artifact should be provided with knowledge regarding the principles of BPM. Furthermore, process modeling is used to document processes. As BPMN is the most popular process modeling language [19], an in-depth understanding of BPMN ensures the artifact's practical applicability in real-world scenarios. Moreover, combining BPM and BPMN knowledge equips the artifact with the ability to adjust to specific business contexts.

(DO2) Interpretation of BPMN diagrams in natural language. The complexity of BPMN diagrams and its various elements poses a significant challenge to individuals without knowledge about these modelling languages [18]. Without an advanced understanding of the underlying process, BPM activities such as analysis and redesign become impossible. Since our artifact aims to support BPM activities, it should be able to interpret the XML code associated with BPMN models [6].

(DO3): Support for BPM activities. Within the BPM lifecycle, various activities are crucial [9]. However, these activities are not independent of each other: for instance, building upon process analysis results, business process redesign brings significant value to an organization as it enables competitive advantages such as customer satisfaction [1]. Thus, the proposed solution should support all BPM activities.

(DO4) Intuitive interaction with users. Non-expert users have difficulties understanding complex BPMN models [18]. To enable our artifact to support BPM activities, the artifact should enable an intuitive and user-friendly interaction. Thereby, chatbots are promising as they provide convenient support for users' inquiries. Thus, the design should allow for intuitive user interaction with reasonable waiting times.

4   Artifact Design

We now present our LLM-artifact - the ProcessLLM – that is designed to support BPM activities. Such a domain-specific LLM can be achieved by fine-tuning a base model with topic-specific data [25]. Therefore, we start the development of our artifact with the data preparation and base model selection stage. Afterwards, we conduct two fine-tuning steps, namely unsupervised fine-tuning, and supervised fine-tuning. The unsupervised fine-tuning step enhances the artifact's general understanding of BPM and BPMN concepts as defined in DO1. The supervised fine-tuning step focuses on the artifact's ability to correctly interpret the XML code of BPMN models as defined in DO2. Consequently, this allows ProcessLLM to effectively support BPM activities as defined in DO3. Due to differences in the underlying datasets and purposes, both fine-tuning stages are conducted sequentially in two distinct phases. The fine-tuning itself, is carried out using the PEFT fine-tuning method LoRA [24] in both cases. Lastly, we initialize ProcessLLM within a Gradio app, providing the user with a chat interface.

We incorporate multiple evaluation metrics into the training pipeline as continuous monitoring of the model's performance is essential in the development of machine learning models. This ensures that weaknesses and inaccuracies of the model are identified quickly so that refinements can be incorporated into subsequent training iterations. The focus of the different iterations was mainly on the optimization of the underlying datasets and hyperparameters. The following sections provide a detailed description of the artifact after the final design iteration.

4.1 Data Preparation

The training pipeline of ProcessLLM begins with data preparation. This step focuses on the collection and preprocessing of topic-specific data that will be used in the different fine-tuning steps. Data preparation is crucial in fine-tuning as it ensures that the model is exposed to patterns that align with its intended application, hence improving its performance in that specific domain. Therefore, we focus on collecting data that revolves around BPM and BPMN. Furthermore, data preparation involves providing the correct format of data as different fine-tuning methods require different formats of datasets. For example, unsupervised fine-tuning requires unstructured data while supervised fine-tuning requires structured data in a tabular format. Lastly, data preparation encompasses methods to assure sufficient data quality by cleaning the data.

Unstructured Data
First, we collect and preprocess the unstructured data required for the unsupervised fine-tuning. The primary objective of the unstructured dataset is to provide ProcessLLM with knowledge in the field of BPM and BPMN. This includes, for example, detailed information about the BPM lifecycle and elements of the BPMN language such as activities and gateways. To initiate this process, we conduct targeted keyword searches for "Business Process Management" and "Business Process Modelling and Notation" across search engines including Google Scholar, and different online libraries. Subsequent analysis of their helpfulness within the quantitative evaluation allows us to select a total of six documents, with three focused on BPM knowledge and three focused on the BPMN language[3]. These documents are retrieved in PDF format, converted to TXT files, and finally cleaned. The cleaning process involves the removal of empty lines, table of contents, and references that hold no relevance to our designated use case. Afterwards, the data is split into unstructured training and evaluation datasets, with the former used for unsupervised fine-tuning and the latter for efficacy assessment through a quantitative

---

[3] *The complete business process handbook* by Rosing et al. (2015); *Handbook on business process management 2* by vom Brocke and Rosemann (2010); *Business Process Management* by Gadatsch (2023); *Business Process Model and Notation (BPMN)* by Object Management Group (2011); *Real-Life BPMN: With introductions to CMMN and DMN* by Freund and Rücker (2012); *BPMN 2.0 Symbols - A complete guide with examples* by Camunda.

evaluation. One evaluation document each is selected for BPM and BPMN concepts, while the remaining four documents are used for training.

Structured Data

Next, we create the structured dataset that is required for the supervised fine-tuning. The primary objective of this dataset is to enhance the model's ability to understand processes in the BPMN notation from their XML-code and to explain them in natural language as defined in DO2. This covers the comprehension of individual symbols, their combinations, as well as complete processes.

To achieve this, we design an instruction dataset in the alpaca-format. The alpaca-format is the standard instruction dataset format used for supervised fine-tuning of LLMs. It is organized into three columns: *instruction, input*, and *output*. For our use case, the *instruction* column employs 140 different prompt commands, such as *"What does this symbol mean?"* or *"Please summarize the following process"*. The *input* column contains the XML code of the symbols or processes in BPMN notation. A manually created answer is then provided to the model over the *output* column. The manual creation of the instruction dataset ensures high quality output; however, we acknowledge that it might also introduce the authors' bias into the model.

Additionally, we define varying degrees of BPMN complexity in our dataset. Level 1 complexity ensures a basic understanding of individual BPMN symbols in XML code by only presenting a single symbol to the model and asking for its meaning. Level 2 complexity presents combinations of BPMN symbols with instructions to either explain what their combination means or to answer specific questions about it. Lastly, level 3 complexity ensures an advanced understanding of end-to-end processes. The model is thereby asked to either summarize the process or to answer specific questions about it.

The instruction dataset is afterwards split into 80% training data and 20% evaluation data, ensuring that the original distribution of the three BPMN complexity levels is maintained within both datasets. Particular attention is paid that, especially for complexity levels 2 and 3, the LLM encounters combinations of symbols and processes in the evaluation dataset that it has not previously encountered in the training dataset. The train-test-split results in 226 records in the training dataset that later builds the foundation for the supervised fine-tuning and 47 records in the evaluation dataset that is used to assess the effectiveness of the fine-tuning through a final, quantitative evaluation.

4.2 Base Model Selection

The base model selection involves choosing a pre-trained model that aligns best with the target task or domain, considering factors such as effectiveness and computational efficiency. This decision is crucial as it determines the starting point for fine-tuning. For ProcessLLM, we decide on the LLaMA-2 series developed by Meta. This series represents the best performing open-source models available at the time of development, producing similar results as OpenAI's commercial GPT 3.5 [26]. We specifically opt for the LLaMA-2 chat series, developed for chatbot applications, to ensure that Process-LLM can interact with the user as defined in DO4. Within this series, we had to decide between the 7, 13, and 70 billion parameter variants. In general, variants within a model series with more parameters perform better or at least as good as variants with fewer parameters. As part of several iterations of our quantitative evaluation, we recognize that this observation also applies to our case. However, we also observe that the increase in performance between the 7 and 13 billion parameter variants is more pronounced than the improvement observed between the 13 and 70 billion parameter version. Here, it is important to acknowledge the inherent challenges associated with training and operating a model of such proportions as the 70 billion parameter variant. Consequently, to balance model capacity, practical manageability, and effectiveness while fulfilling DO4 that includes reasonable waiting times, we opted for the LLaMA-2 13 billion chat series as the most suitable base model for the subsequent fine-tuning steps.

4.3 Unsupervised Fine-Tuning

Unsupervised fine-tuning involves exposing the base model to unstructured data, allowing it to capture underlying patterns and structures related to the targeted task or domain. This procedure enhances the model's domain-specific understanding and promotes adaptability to diverse input variations. For ProcessLLM, the primary goal of unsupervised fine-tuning is to enhance the model's knowledge of BPM and BPMN as defined in DO1. Therefore, we expose our base model to the unstructured training dataset created in Sect. 4.1. The fine-tuning is conducted with the LoRA [24] method. Within multiple quantitative evaluation iterations, we choose a hyperparameter combination[4] for the LoRA fine-tuning that maximizes performance while avoiding underfitting and overfitting effects on the training data. The resulting LoRA-adapter is then integrated into the base model, forming an intermediate model that becomes the new base model for the supervised fine-tuning.

4.4 Supervised Fine-Tuning

In supervised fine-tuning of LLMs, the base model is trained on structured data similar to future usage, using explicit task-related annotations to guide its learning process. This process differs from unsupervised fine-tuning, where the model learns from unstructured data to freely capture underlying patterns. In supervised fine-tuning, the model refines its parameters based on structured examples, allowing it to generate accurate outputs for the designated task. For ProcessLLM, the primary objective of this supervised fine-tuning step is to enhance the model's ability to understand BPMN diagrams using natural language, as defined in DO2. Therefore, we fine-tune the intermediate model on the structured training dataset created in Sect. 4.1. Analogous to the unsupervised fine-tuning, the LoRA method is employed using a hyperparameter combination[5] that maximizes performance while avoiding under and overfitting. By integrating the resulting LoRA-adapter into the intermediate model, we obtain the final version of ProcessLLM.

---

[4] Training-epochs: 3; Learning-rate: 0,0003; LoRA Rank: 8; LoRA Alpha: 16; Batch Size: 128; Micro Batch Size: 4; Cutoff Length: 256; LR Scheduler: linear; Overlap Length: 128; Prefer Newline Cut Length: 128.

[5] Training-epochs: 20; Learning-rate: 0.0003; LoRA Rank: 8; LoRA Alpha: 16; Batch Size: 128; Micro Batch Size: 4; Cutoff Length: 256; LR Scheduler: linear.

4.5 Initialization

The initialization of ProcessLLM prepares the model for interacting with the end user. Typically, machine learning models are developed by experts, necessitating specialized hardware, software, and expertise for validation. Consequently, non-technical collaborators and end-users encounter obstacles in effectively using the technology. Addressing this challenge, we use Gradio as an open-source software to integrate ProcessLLM in a chatbot frontend. To guide ProcessLLM in generating helpful responses, we initialize it with the following prompting template:

"You are a helpful chatbot. Your answers are clear and concise. If you summarize a process, the summary should include how the process starts, which activities happen afterwards and how it ends. Thereby, different XOR- and AND-Gateways as well as the underlying conditions should be described before the subsequent activities are described. Do not mention the gateways themselves. Only describe the consequences of the gateways. If you start to describe one path, describe it completely before you go on to the next path. The summary should not contain any IDs or meta-information."

Once initialized, ProcessLLM becomes accessible to users through a chat interface. Users can interact with the model by inputting the XML code of a BPMN-process into the prompt and posing specific questions about the underlying process.

5 Evaluation

5.1 EVAL1 – Problem Identification

In EVAL1, we focus on validating the significance of the defined DSR problem by analyzing the literature and conducting a subsequent survey with four business experts – three senior consultants and one manager - from a German insurance company to demonstrate that the design problem represents a research gap and is important to practice. As outlined in Sect. 1, current literature revealed potential for the utilization of fine-tuned LLMs in BPM activities. For example, Vidgof, Bachhofner, Mendling [6] identify numerous high-potential use cases such as automated process improvement or issue discovery followed by root cause analysis. While these and other use cases have conceptually been described, their feasibility in terms of a real-world instantiation has not been demonstrated. As such, current literature in the field underlines the novelty and importance of our DSR problem. From a practical perspective, the survey provided deeper insights into real-world BPM challenges. The participants mentioned the obscurity or complexity of process documentation, which hinders targeted process analysis. Additionally, there is uncertainty regarding the appropriate methodology for process analysis. Furthermore, a lack of understanding of the processes and insufficient resources were mentioned. These findings, along with current literature, justify our engagement in this DSR project whereby the addressed problem is academically significant and practically relevant.

5.2 EVAL2 – Design Evaluation

In EVAL2, we conduct a second survey with the same business experts to assess the correctness and completeness of our design specification. In this survey, we first introduced the general idea of the artifact. Afterwards, we asked the experts for feedback regarding the use of such a tool in their work environments. Key requirements identified include precise BPMN model comprehension, efficient response times, and user-friendly implementation. Concerns raised involve potential model inaccuracies and operational costs. Notably, one expert highlighted the need for human collaboration in BPM processes, prompting us to position our artifact as a human-in-the-loop approach rather than a full replacement for human labor. After collecting the experts' feedback, we introduced our DOs and asked the participants to rate them. The feedback was positive, with experts confirming the completeness and relevance of these DOs to practical BPM challenges. Therefore, EVAL2 confirmed that our design specification aligns closely with practical requirements of BPM practitioners.

5.3 EVAL3 – Quantitative Evaluation

EVAL3 is crucial for assessing the feasibility and effectiveness of our artifact. Therefore, we instantiate our artifact as a software prototype, conduct a quantitative evaluation, and benchmark its performance against ten other open-source LLMs in an artificial setting. EVAL3 was conducted using two different metrics and datasets. First, we evaluated the models' performance using perplexity scores. Perplexity, a metric in the domain of NLP, measures how well a model can predict a sample of data by quantifying the level of uncertainty with which it predicts the next word in a sequence of text. Thereby, a lower perplexity score indicates that the model is better at calculating the next word in the given sequence of text [27]. Perplexity scores were calculated on two unstructured BPM and BPMN evaluation documents (cf. Sect. 4.1) that ProcessLLM had not seen during training. This method helps us understand to which extent fine-tuning the LLM leads to a better comprehension of complex BPM and BPMN scenarios as defined in DO1 compared to non-fine-tuned, general-purpose LLMs. Additionally, we employed a loss evaluation on the structured evaluation dataset (cf. Sect. 4.1), that links the XML code of processes in BPMN-Notation to their translation in natural language. The loss values indicate the disparity between the model's predicted output and the actual output provided in the evaluation dataset. Therefore, a lower loss value is indicative of higher accuracy. By doing so, we can measure the model's ability to process BPMN diagrams in natural language as defined in DO2. The results of EVAL3 are summarized in Table 1. Detailed information about the change of metrics over the course of the development pipeline can be found in the GitHub-repository.

Upon analyzing the results, ProcessLLM demonstrates state-of-the-art performance, whereby it outperforms general-purpose LLMs across all evaluation metrics. Notably, its advancement is especially pronounced when contrasted with its base model, the LLaMA-2 13 billion chat model.

5.4 EVAL4 – Qualitative Evaluation

EVAL4 focuses on demonstrating the applicability and usefulness of ProcessLLM in real organizational contexts. Therefore, we conduct a two-staged qualitative evaluation to assess the fulfillment of DO1–4, considering both internal performance and external expert feedback. In the first stage, we utilized an exemplary Order-to-Cash process

Table 1. Result of the EVAL3 activities

Model	BPM-Perplexity-Value	BPMN-Perplexity-Value	Evaluation-Loss
LLaMA-2-7b-chat	8.5547	8.9524	2.1972
LLaMA-2-13b-chat	7.7176	8.0994	1.9421
LLaMA-2-70b-chat	7.1510	7.5020	1.9496
Alpaca-7b	9.3602	9.9803	2.4056
Alpaca-13b	8.9466	9.4420	4.9621
Vicuña-7b	8.0262	9.1286	1.8285
Vicuña-13b	7.7157	8.2962	1.7524
Guanaco-13b	7.8065	8.3673	1.9095
MPT-7b-chat	11.3516	13.9375	8.6159
MPT-30b-chat	10.3828	12.0625	1.9518
ProcessLLM (13b)	**7.1250**	**7.1856**	**0.7851**

featuring five activities and different XOR and AND gateways as input for ProcessLLM. This allowed us to internally evaluate the model's proficiency in handling general BPM and BPMN inquiries (DO1), processing BPMN diagrams in natural language (DO2), and providing support in BPM activities (DO3). Despite its ability to fulfill the DOs, we noted inconsistencies and occasional misinterpretations, particularly with complex process models, suggesting room for further refinement. The model's response times (DO4), were found to be reasonable, ranging from 30 s to 3 min. Selected excerpts from this internal evaluation phase can be found below.

[DO2] Capability to process BPMN diagrams in natural language
User: *Please summarize the following process in natural language: [XML-Code]*
ProcessLLM: *The process starts when an order is received and checked for stock availability. If there is no stock, the order is rejected. If there is stock, the order is confirmed, and an invoice is issued. The order is then shipped and fulfilled.*
[DO3] Capability to provide support for BPM activities
User: *What potential weaknesses do you see in the process and what are the consequences of these?*
ProcessLLM: *Here are some potential weaknesses and their potential consequences: 1. Lack of automation: The process is largely manual, which can lead to delays, errors, and increased operational costs. For example, the process of checking stock availability and confirming orders could be automated, freeing up the sales representative to focus on other tasks. [9 more weaknesses identified]*
User: *How would you redesign the process to tackle these weaknesses?*
ProcessLLM: *To address these weaknesses, I would suggest the following redesign: 1. Automate the order processing: Implement an order management system that can handle the entire process, from receiving the order to fulfilling it, and provide visibility into the status of orders. [9 more redesign suggestions identified]*

In the second stage, we used two real-life processes – an authority delegation and an underwriting referral process featuring multiple activities and gateways - that the business experts from EVAL1 and EVAL2 use in their daily work as input for ProcessLLM and analyzed them. The results were afterwards presented to the business experts within a workshop. The experts confirmed the fulfilment of all DOs, especially appreciating the model's rapid response times and user-friendly interaction. They expressed interest in employing the model as a preliminary analysis tool for processes, valuing its unique perspective and creative problem-solving capabilities. Specifically, they emphasized that this approach offered a new angle on problems, enhancing the ideation phase and fostering innovative solutions. However, they echoed our concerns regarding reliability, emphasizing its importance for widespread corporate use. Furthermore, the experts saw potential in process implementation, imagining the artifact as a tool for cross-verifying process steps, enhancing process execution accuracy. Overall, while ProcessLLM shows promising results in organizational contexts, the evaluation highlights areas needing attention, particularly in reliability and handling complex scenarios.

6 Conclusion

The integration of LLMs presents a promising avenue for enhancing the entire BPM lifecycle. While existing literature primarily takes a conceptual perspective, this study provides initial groundwork in the practical application of LLMs in BPM. The evaluation shows that ProcessLLM performs well, demonstrating a good understanding of BPM concepts and the ability to process BPMN diagrams, thus supporting BPM activities. This proof-of-concept highlights the feasibility and real-world applicability of using LLMs in BPM. Future research should refine ProcessLLM for better accuracy, explore advanced machine learning techniques, integrate diverse data sources, and apply LLMs across different BPM stages.

References

1. Dumas, M., La Rosa, M., Mendling, J., et al.: Fundamentals of Business Process Management. Springer, Berlin (2018)
2. Kerpedzhiev, G.D., König, U.M., Röglinger, M., et al.: An exploration into future business process management capabilities in view of digitalization. Bus. Inf. Syst. Eng. (2021)
3. Badakhshan, P., Wurm, B., Grisold, T., et al.: Creating business value with process mining. J. Strateg. Inf. Syst. (2022)
4. Röglinger, M., Plattfaut, R., Borghoff, V., et al.: Exogenous shocks and business process management. Bus. Inf. Syst. Eng. (2022)
5. Dumas, M., Fournier, F., Limonad, L., et al.: AI-augmented business process management systems: a research manifesto. ACM Trans. Manag. Inf. Syst. (2023)
6. Vidgof, M., Bachhofner, S., Mendling, J.: Large language models for business process management: opportunities and challenges. In: BPM 2023 Forum (2023)
7. van Dun, C., Moder, L., et al.: ProcessGAN: supporting the creation of business process improvement ideas through generative machine learning. Decis. Support Syst. (2023)
8. Schmid, S.J., Moder, L., et al.: Everything at the proper time: repairing identical timestamp errors in event logs with Generative Adversarial Networks. Inf. Syst. (2023)

 9. Klievtsova, N., Benzin, J.-V., Kampik, T., et al.: Conversational process modelling: state of the art, applications, and implications in practice. In: BPM 2023 Forum (2023)
10. Kampik, T., Warmuth, C., Rebmann, A., et al.: Large process models: business process management in the age of generative AI. arxiv.org/abs/2309.00900 (2023)
11. Grohs, M., Abb, L., Elsayed, N., et al.: Large language models can accomplish business process management tasks. In: BPM 2023 Workshops (2023)
12. Baldazzi, T., Bellomarini, L., Ceri, S., et al.: Fine-tuning large enterprise language models via ontological reasoning. In: Rules and Reasoning (2023)
13. Wu, S., Irsoy, O., Lu, S., et al.: BloombergGPT: a large language model for finance. arxiv.org/abs/2303.17564 (2023)
14. Nguyen, N., Nadi, S.: An empirical evaluation of GitHub copilot's code suggestions. In: 19th International Conference on Mining Software Repositories (2022)
15. Peffers, K., Tuunanen, T., Rothenberger, M.A., et al.: A design science research methodology for information systems research. J. Manag. Inf. Syst. (2007)
16. Sonnenberg, C., vom Brocke, J.: Evaluations in the Science of the Artificial-Reconsidering the Build-Evaluate Pattern in Design Science Research. Springer, Heidelberg (2012)
17. Beerepoot, I., Di Ciccio, C., Reijers, H.A., et al.: The biggest business process management problems to solve before we die. Comput. Ind. (2023)
18. Recker, J.: Opportunities and constraints: the current struggle with BPMN. Bus. Process Manag. J. (2010)
19. Farshidi, S., Kwantes, I.B., Jansen, S.: Business process modeling language selection for research modelers. Softw. Syst. Model. (2023)
20. Khurana, D., Koli, A., Khatter, K., et al.: Natural language processing: state of the art, current trends and challenges. Multimed. Tools Appl. (2023)
21. Brown, T., Mann, B., Ryder, N., et al.: Language models are few-shot learners. In: Advances in Neural Information Processing Systems (2020)
22. Vaswani, A., Shazeer, N., Parmar, N., et al.: Attention is all you need. In: Advances in Neural Information Processing Systems (2017)
23. Houlsby, N., Giurgiu, A., Jastrzebski, S., et al.: Parameter-efficient transfer learning for NLP. In: International Conference on Machine Learning (2019)
24. Hu, E.J., Shen, Y., Wallis, P., et al.: LoRA: low-rank adaptation of large language models. arxiv.org/abs/2106.09685 (2021)
25. Hu, Z., Wang, L., Lan, Y., et al.: LLM-Adapters: an adapter family for parameter-efficient fine-tuning of large language models. arxiv.org/abs/2304.01933 (2023)
26. Touvron, H., Martin, L., Stone, K., et al.: Llama 2: open foundation and fine-tuned chat models. arxiv.org/abs/2307.09288 (2023)
27. Colla, D., Delsanto, M., Agosto, M., et al.: Semantic coherence markers: the contribution of perplexity metrics. Artif. Intell. Med. (2022)

Towards a Benchmark for Causal Business Process Reasoning with LLMs

Fabiana Fournier, Lior Limonad$^{(\boxtimes)}$, and Inna Skarbovsky

IBM Research, Haifa, Israel
{fabiana,liorli,inna}@il.ibm.com

Abstract. Large Language Models (LLMs) are increasingly used for boosting organizational efficiency and automating tasks. While not originally designed for complex cognitive processes, recent efforts have further extended to employ LLMs in activities such as reasoning, planning, and decision-making. In business processes, such abilities could be invaluable for leveraging on the massive corpora LLMs have been trained on for gaining deep understanding of such processes. In this work, we plant the seeds for the development of a benchmark to assess the ability of LLMs to reason about causal and process perspectives of business operations. We refer to this view as Causally-augmented Business Processes (BP^C). The core of the benchmark comprises a set of BP^C related situations, a set of questions about these situations, and a set of deductive rules employed to systematically resolve the ground truth answers to these questions. Also with the power of LLMs, the seed is then instantiated into a larger-scale set of domain-specific situations and questions. Reasoning on BP^C is of crucial importance for process interventions and process improvement. Our benchmark, accessible at https://huggingface.co/datasets/ibm/BPC, can be used in one of two possible modalities: testing the performance of any target LLM and training an LLM to advance its capability to reason about BP^C.

Keywords: Large Language Models · Business Processes · Causally-augmented Business Processes · Reasoning · Benchmark

1 Introduction

Large Language Models (LLM) refers to statistical models of natural language that, based on the large corpora of text data they have been trained on, predict next plausible tokens (basic units of text) given an input string [22]. LLMs have been successfully applied to a wide range of applications including: Chatbots and virtual assistants (e.g., automated customer support); content generation and automation (e.g., articles and blogs generation); language translation; text

This project has received funding from the European Union's Horizon research and innovation programme under grant agreements no 101094905 (AI4GOV), 101092021 (AutoTwin), and 101092639 (FAME).

summarization and document analysis; and question answering. The interaction with LLMs is conventionally attained via a textual prompt in which the content and the instructions to the LLM are being constructed, also known as *prompt engineering*. The process is iterative, with the model's output being analyzed and the prompt adjusted accordingly. Prompt engineering is key to the efficient use of LLMs [18]. However, LLM models can only answer prompts accurately if they have been fed the right training data as they lack planning and reasoning capabilities (e.g., [11,21]). As stated by Yann LeCun[1], chief AI scientist of Facebook and Instagram, LLMs have "very limited understanding of logic...do not understand the physical world, do not have persistent memory, cannot reason in any reasonable definition of the term...". In fact, ChatGPT has a causal hallucination issue when tackling causal relationships which cannot be overcome relying solely on prompts [8].

While reasoning may not be inherent in the architecture of LLMs, critics do argue that the ability to reason does manifest itself as an emergent property in LLMs [20]. That is, once trained on sufficiently large corpora of examples, LLMs may become capable of statistically deriving statements that are also coincidentally sound with the given set of arguments in the input. Similar emergent abilities may also include planning, decision-making, in-context learning, and answering in zero-shot settings [15]. Whether reasoning is an inherent capability or an emergent property of future LLMs, particularly for interpreting business process models, is likely to remain a heated debate. This debate emphasizes the importance of establishing benchmarks for evaluating LLM performance. Our focus is on driving such enablement by using LLMs to analyze textual descriptions of Causally-augmented Business Processes (BP^Cs). Such descriptions differ from conventional process descriptions by also including statements about the causal execution dependencies among the activities as first-class citizens. Teaching LLMs to reason about BP^Cs could enhance the analysis and improvement of such processes.

To standardize and measure the ability of LLMs to facilitate such tasks, a designated benchmark is being developed. The purpose of the developed novel benchmark is twofold: as a testing dataset that can be employed to quantify the ability of an LLM to reason about BP^Cs, and as a training dataset that can be used to adapt the ability of an LLM for this task. We evaluated two open-source and three commercial LLMs on a small subset of the situations and questions. Our results highlight the importance of producing an objective numeric scale partitioned by different perspectives to compare and assess various LLM performances. They also indicate that there is room for further improvement of LLMs in the task of reasoning about BP^Cs. The evolving benchmark and corresponding prompts are available here: https://github.com/IBM/SAX/tree/main/NLP4BPM2024.

Given these observations, we are cautious in claiming that LLMs possess a genuine inherent ability to reason. We approach this by acknowledging that while LLMs may not inherently reason, they can achieve predictive accuracy

---

[1] https://www.ft.com/content/23fab126-f1d3-4add-a457-207a25730ad9.

through training on a large set of deductive textual statements. This accuracy results in syntactic output that reliably aligns with what would be produced by genuine logical reasoning. Whether reaching such a level of predictive accuracy is philosophically equivalent to actually having the ability to reason is something that we leave beyond the scope of our study.

2   Background

Business process management is the discipline that combines approaches for the design, execution, control, measurement, and optimization of business processes [1]. With the penetration of AI applications into organizations, the concept of AI-Augmented Business Process Management Systems (ABPMSs) was coined in [5]. ABPMSs are process-aware information systems that rely on trustworthy AI technology to continuously adapt and improve a set of business processes with respect to one or more performance indicators. In such systems, subsymbolic AI methods are not used to replace human or symbolic reasoning in crucial tasks, but rather to support human and machine decisions and actions [12]. A natural way of driving this interaction between humans and AI is through LLMs. To enable this, it is important for LLMs to support reasoning about business processes. In analogy to [10], by reasoning we refer here to the process of thinking about business processes in a logical and systematic way, using evidence and past experiences to reach a conclusion or make a decision, for the sake of process improvement. As shown in [7], understanding the temporal dependencies among the tasks in the process is not enough for reasoning about the consequences of interventions underlying process improvement decisions. To this aim, we specify here the Causally-augmented Business Process (BP^C) formalism as the concrete flavor for process descriptions in which not only the temporal flow but also the causal relations among the tasks are inherently captured.

A BP^C is a business process extended with inter-activity relations that reflect causal execution dependencies. A `causal execution dependence`, denoted as $A \xrightarrow{c} B$, implies that the time task B executes is determined by the time task A executes in a given process as defined in [7]. From a process perspective, we consider `followed-by` as a general form of a temporal relation among activities, hereafter denoted as $A \rightarrow B$, implying that according to most process observations (frequently evidenced in many process execution logs), the execution of activity B occurs either immediately after the execution of A or sometime after. Considering the causal process perspective, a BP^C may be represented in the basic form of a graph, with nodes designating activities and edges as causal execution dependencies among these activities. Three fundamental patterns, or "junctions" [17], can be composed to characterize any causal network: $A \xleftarrow{c} B \xrightarrow{c} C$ (confounder), $A \xrightarrow{c} B \xleftarrow{c} C$ (collider), and $A \xrightarrow{c} B \xrightarrow{c} C$ (mediator).

The potential of leveraging LLMs in the BPM field has been recently researched, for example, by analyzing which opportunities and challenges LLMs pose for the individual stages of the BPM lifecycle [19], in [4], where prompt engineering techniques are discussed as an alternative to fine-tuning a specific LLM, and in [12] where the authors introduce the notion of a Large Process Model (LPM), an envisioned neuro-symbolic software system that integrates process management knowledge in organizations with LLMs and statistical and inference methods for the automated inference of insights and actions. In addition, LLMs have been researched in a wide range of tasks related to BPM, including process mining [3,9], automation of portions of complex tasks[2], conversational process modelling [13], and explainability of business process outcomes [6]. To date, neither there is work on leveraging LLMs for causal reasoning about business processes, nor a relevant benchmark for testing LLMs. As an early member of this family of solutions, the PET dataset [2] presents a benchmark for question answering, featuring an initial corpus of business process descriptions annotated with activities, gateways, actors, and flow information. The dataset contains 45 documents with narrative descriptions of business processes and their annotations. However, the PET descriptions do not include causal execution dependencies among the activities. In [16] the TORQUE benchmark for temporal ordering has been investigated. The dataset encompasses 21k user-generated and fully answered temporal relation questions. Concerning both benchmarks, ours differs in tackling the task of causal reasoning about business processes.

3   Approach

Our goal is to develop a question-and-answer benchmark dataset to test the capacity of an LLM to produce sound answers to questions about textual narratives describing BP^Cs. Regardless of the debate about LLMs' suitability for reasoning tasks, we assume that with "sufficient" training and exposure to massive amounts of data, these models can become proficient. Yet, determining sufficiency requires some objective performance measurement. Thus, a benchmark is necessary. We believe the developed instrumentation and methodology, shown in Fig. 1, are suitable for model testing and can also serve as a dataset to train LLMs to reason about BP^C. The benchmark may also be further extended to more concrete domains and particular aspects of reasoning as the example given in this paper. We do acknowledge that the realization presented here is the first step in our longer journey. Hence, in addition to presenting our results so far, we also elaborate on our future evaluation intentions once the benchmark is instantiated at a greater scale.

Our approach is to define a core set of template questions and situations for basic reasoning about BP^C textual narratives. These descriptions combine statements about process activities, time precedence relations, and causal execution dependencies. At the root of such descriptions, we anticipate statements of the form "activity A occurs before activity B", and "the execution of A causes the execution of B". From a process perspective, textual descriptions typically include manifestations of the basic temporal relation followed-by. While there is a formal distinction between directly-follows and eventually-follows [1],

---

[2] https://www.infoworld.com/article/3714621/how-llms-can-help-streamline-busine ss-processes.html.

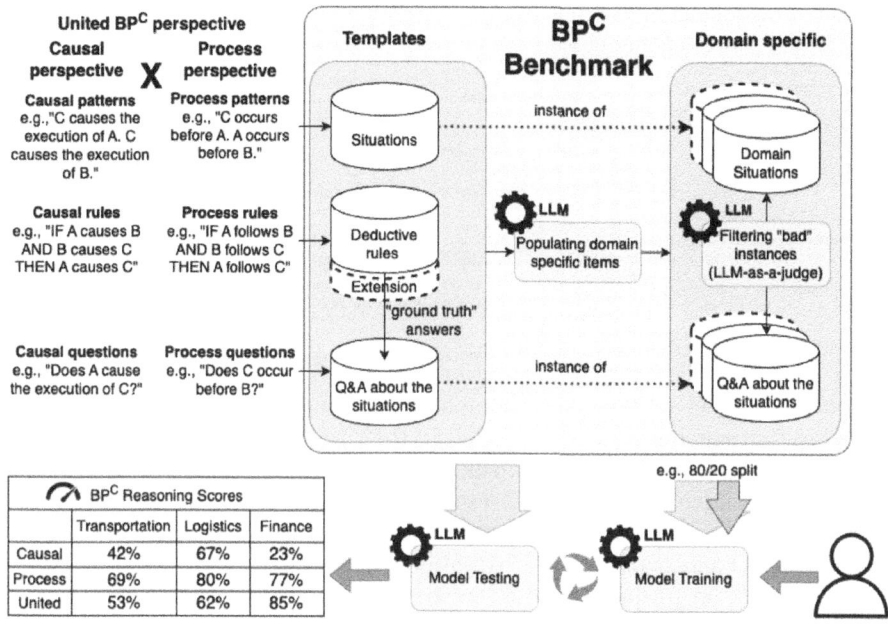

Fig. 1. BP^C benchmark for testing and training of LLMs

natural language often does not strictly differentiate between them. Hence, our work assumes a variety of manifestations of the core phrase 'occurs before/after' to be expressed in the populated text. Similarly, from a causal process perspective, the descriptions in the text are likely to include manifestations of the causal relation of `causal-execution-dependence`. Respectively, manifestations of the core phrase of "causes the execution of" are populated in the text.

Therefore, to account for complete coverage of textual narratives that are descriptive of any structural form of a BP^C, we partition the set of template situations into three subsets corresponding to each of the three fundamental causal patterns. Respectively, from a process perspective, each pattern is associated with either a congruent process structure or a simple execution sequence of the form $A \to B \to C$ (as shown in [7]). The core set of situations that can be composed to describe any real-world BP^C situation is shown in Table 1. A situation refers to the unified temporal and causal relations among any subset of three activities A, B, and C. These conditions provide the core content for generating domain-specific text based on the respective template phrases listed.

Respectively, we associate each situation with a set of deductive rules to facilitate the reasoning about each situation. The rules are split between causal, process, and the combination of the two perspectives, capturing the meaning of the relations in each situation, unfolding from fundamental properties of the relations in discrete mathematics, jointly with a closed world assumption according to which any unknown premise is deduced to be `False`. For each situation,

Table 1. The domain of five different BP^C situations as spanned by the three causal patterns and corresponding process structures

	Confounder		Collider		Mediator
Situation#	1	2	3	4	5
Causal structure	$A \xleftarrow{c} B \xrightarrow{c} C$		$A \xrightarrow{c} B \xleftarrow{c} C$		$A \xrightarrow{c} B \xrightarrow{c} C$
Process structure	$A \leftarrow B \rightarrow C$ (split)	$B \rightarrow A \rightarrow C$ (or $B \rightarrow C \rightarrow A$)	$A \rightarrow B \leftarrow C$ (join)	$A \rightarrow C \rightarrow B$ (or $C \rightarrow A \rightarrow B$)	$A \rightarrow B \rightarrow C$
Causal phrase template	B causes the execution of A, B causes the execution of C		A causes the execution of B, C causes the execution of B		A causes the execution of B, B causes the execution of C
Process phrase template	B occurs before A, B occurs before C	B occurs before A, A occurs before C	A occurs before B, C occurs before B	A occurs before C, C occurs before B	A occurs before B, B occurs before C

we generated a set of "Yes"/"No" template questions answerable by deductive reasoning that combines the facts stated in the situation with at least one associated rule. For example, the basic rule of transitivity on the relation of causal execution dependence, i.e., $A \xrightarrow{c} C \Leftarrow (A \xrightarrow{c} B) \wedge (B \xrightarrow{c} C)$, may be employed to resolve the answer to the question "Does A cause the execution of C?" when applied in the context of Mediator situation #5 (Table 1).

For brevity, we include here as an example the set of rules (see Table 2) and corresponding template questions (see Table 3) created for situation #2. This situation addresses a BP^C condition in which the confounder pattern in the causal perspective is associated with a sequence pattern of the same three activities in the process perspective. Concerning symmetry and reflexivity rules (Table 2), it is assumed that any cycle in the BP^C structure can be entangled with a method such as k-loop unrolling [14].

Table 2. Deductive rules associated with the confounder situation #2, considering the manifestation of $\rightarrow$ as "occurs before" and of $\xrightarrow{c}$ as "causes the execution of".

	Rules	Comments
Process related	PR1: $A \rightarrow C \Leftarrow (A \rightarrow B) \wedge (B \rightarrow C)$	Transitivity of the $\rightarrow$ relation
	PR2: $B \nrightarrow A \Leftarrow A \rightarrow B$	Asymmetry of the $\rightarrow$ relation
	PR3: $A \leftarrow B \Leftarrow B \rightarrow A$	Where $\leftarrow$ is an antonym manifestation of $\rightarrow$, e.g., "occurs after"
	PR4: $A \nrightarrow A$	No reflexivity
	PR5: $B \Leftarrow (A \rightarrow B) \wedge A$	Entailed from the meaning of $\rightarrow$: i.e., if A executes then B will execute at some later time
Causal related	CR1: $B \xcancel{\xrightarrow{c}} A \Leftarrow A \xrightarrow{c} B$	Asymmetry of the $\xrightarrow{c}$ relation
	CR2: $A \xrightarrow{c} C \Leftarrow (A \xrightarrow{c} B) \wedge (B \xrightarrow{c} C)$	Transitivity of the $\xrightarrow{c}$ relation. Relevant for situation #5
	CR3: $A \xleftarrow{c} B \Leftarrow B \xrightarrow{c} A$	Where $\xleftarrow{c}$ is an antonym manifestation of $\xrightarrow{c}$, e.g., "because the execution of"
	CR4: $\neg B \Leftarrow (A \xrightarrow{c} B) \wedge \neg A$ and $A \Leftarrow (A \xrightarrow{c} B) \wedge B$	Entailed from the meaning of $\xrightarrow{c}$: i.e., if A doesn't execute then B doesn't execute (and vice versa)
Process and causal structure	PCR1: $A \rightarrow B \Leftarrow A \xrightarrow{c} B$	Causal execution dependence implies time precedence in the process
	PCR2: $B \nrightarrow A \Leftarrow A \rightarrow B$	Time precedence in the process implies no causal execution dependence on the opposite direction

Table 3 shows for each of the populated template questions, which of the rules in Table 2 were involved in deducing its answer w.r.t the condition articulated in situation #2. These answers to the template questions remain as the "ground truth" for their corresponding instantiated domain-specific versions.

Table 3. A core set of template questions with answers deduced by the rules for the confounder situation #2

	Template question	Answer	Related rule(s)
Process related	QP1: Does C occur before B?	Yes	PR1
	QP2: Does B occur before C?	No	PR2 and QP1
	QP3: Does A occur after C?	Yes	PR3
	QP4: Does B occur after C?	Yes	PR3 and QP1
	QP5: Does C occur after A?	No	PR2
	QP6: Does C occur after B?	No	PR2 and QP1
	QP7: Does A occur after B?	No	PR3 and PR2
Causal related	QC1: Does A cause the execution of C?	No	CR1
	QC2: Does B cause the execution of C?	No	CR1
	QC3: Does A execute because of C?	Yes	CR3
	QC4: Does B execute because of C?	Yes	CR3
	QC5: If C doesn't execute, will A ever execute?	No	CR4
	QC6: If C doesn't execute, will B ever execute?	No	CR4
Process and causal related	QPC1: Does A cause the execution of B?	No	Close worldassumption
	QPC2: Does B cause the execution of A?	No	PCR2

Similar to the underlying set of situations, domain-specific questions were methodologically instantiated from these questions. While the core set of template questions is extensible (see Sect. 3.1), any specific set uniquely tags the benchmark version and also sets a bound to its expressiveness, that is, the grammatical richness the benchmark can accommodate. Pragmatically, this means that any two target LLMs can be compared only when assessed according to the same benchmark version.

3.1   Extending the Benchmark

As noted, the core set of questions may be extended with additional template questions to assess the capacity of an LLM to reason about any additional aspect of interest. For example, we may define the boolean function is_shortened(A) to denote that the execution time of activity A was expedited to finish its execution earlier. Respectively, the following rule can be added:

$$CR5 : is\_shortened(B) \Leftarrow A \xrightarrow{c} B \wedge is\_shortened(A)$$

Similarly, other functions can be added to denote other forms of temporal intervention, such as `is_extended`(A), `is_halted`(A), and `is_delayed`(A).

Respective to adding this rule, the causal set of questions for situation #2 above can be extended with the additional template questions as listed in Table 4.

Another form of extension can be achieved by extending the grammar underlying the process situations to express more complex structures (e.g., gateways).

Table 4. An extension to the core set of template questions for situation #2.

	Template question	Answer	Related rule(s)
Causal related	QC7: If we shorten A, will B be shortened?	No	CR5
	QC8: If we shorten B, will A be shortened?	No	CR5
	QC9: If we shorten C, will A be shortened?	Yes	CR5
	QC10: If we shorten C, will B be shortened?	Yes	CR5

3.2 Populating Domain-Specific Questions

We used the open-source Mixtral-instruct-8x-7b[3] LLM to instantiate the template questions for situation #2 to concrete problem domain statements for each of the three perspectives. For this work, we restricted the instantiation of the 15 questions in Table 3 and the additional 4 in Table 4, to having each template question populated with one corresponding domain-specific question. The domains were arbitrarily selected as one of the following: transportation, manufacturing, logistics, retail, finance, insurance, and medical. An example of a prompt employed for such instantiation attending to the process perspective in situation #2 (QP1) is illustrated as prompt #1 in Table 5.

---

[3] https://huggingface.co/mistralai/Mixtral-8x7B-Instruct-v0.1#model-card-for-mixtral-8x7b.

Table 5. Series of LLM prompts used during benchmark development

#	LLM prompt
1	**INPUT:** **Phrase:** A, B, and C are activities in some process. C occurs before A. A occurs before B. **Question:** Does C occur before B? **Instruction:** Considering the above question, choose any relevant concrete activities A, B, and C in the domain of ≪transportation≫ that retain the truth of the phrase statements. Using these terms, instantiate corresponding phrase statements and a question statement in a form that matches the above statements and question. **OUTPUT:** **Concrete Activities:** A: Boarding the plane B: Takeoff of the plane C: Check-in at the airport **Instantiated Phrase Statements:** Check-in at the airport occurs before boarding the plane. Boarding the plane occurs before takeoff of the plane. **Instantiated Question Statement:** Does check-in at the airport occur before takeoff of the plane?
2	**INPUT:** **Template question:** If we shorten C, will B be shortened? **Instantiated question:** If we shorten *Damage assessment team inspects the damage and estimates the cost of repair* in duration, will *Insurance policyholder receives the payout* be shortened? **Instruction:** Considering the instantiated question above a concrete version of the template question where the letters are replaced with process activity descriptions, when replacing these descriptions with their corresponding letters in the template, how would you rate the similarity between the revised instantiated question and the template question (where a 1 rate means they are identical and 0 they are completely different)? In your output, print only the rate value on a 0–1 scale. **OUTPUT:** 0.9 or 90% similarity.

3.3 Filtering Out Inadequate Questions

Our instantiated set of questions so far was relatively small. However, in full-scale development, covering the complete situation space and more domains, the number of questions will likely increase significantly. This process might generate domain-specific questions in a form that becomes incongruent with the original template questions used as its seed due to the non-deterministic nature of LLMs. Therefore, we also foresee the use of "LLM-as-a-judge" to curate the quality of the generated questions. That is, excluding the ones that do not maintain a faithful linkage to their corresponding template. To this end, prompt#2 in Table 5 was created to grade and remove questions scoring below an acceptable threshold.

4 Evaluation

An initial instantiation of the benchmark seed, the 'benchmark prototype' dataset, was populated for testing purposes, including one domain-specific question for each template question in situation #2. We used this prototype to assess its applicability and measure accuracy against state-of-the-art LLMs across three perspectives: process, causal, and their combination.

At first, we ran the three core sets of template questions from Table 3 with five different target LLMs: GPT3.5[4], GPT4[5], GPT4o[6], Mixtral-instruct-8x-7b (See Footnote 3), and Merlinite-7b[7]. We repeated each template question ten times with clean-slate prompts, preceding each question with the corresponding phrase describing the situation and instruction as illustrated in Table 6. We averaged the proportion of correctly answered questions for each reasoning perspective: process, causal, and combined. We then also added the extended set of questions (see Sect. 3.1) and revised the results accordingly.

As a second step, using the domain-specific questions, we repeated the benchmark testing for the two open-source LLMs (Mixtral-instruct-8x-7b and Merlinite-7b) ten times per prompt, measuring the proportion of correct answers for the three perspectives, both without and with the inclusion of the extension questions.

We acknowledge that our current evaluation caters strictly to the applicability of the benchmark to a handful of LLMs. Such an assessment lacks characterizing the benchmark quality. For this, we elaborate here on a set of relevant metrics that we intend to assess once the benchmark is developed in full scale.

Completeness refers to the range of realistic BP^C situations the benchmark covers. This is ensured by using causal "junctions" as per [17], which accommodate any causal structure, and by the richness of rules and their coverage by questions. Our design principle ensures each rule helps resolve at least one question and that the benchmark is extensible. A metric can capture the proportion of rules covered by questions. However, completeness is always limited by the core set of template questions and the finite domains involved. Therefore, it is crucial to disclose the list of questions and domains for any benchmark version.

Table 6. LLM Prompts used for benchmark testing

#	LLM prompts: process, causal, and both
1	**Phrase:** A, B, and C are activities in some process. ≪process related≫C occurs before A. A occurs before B. ≪causal related≫C causes the execution of A. C causes the execution of B. **Instruction:** Considering the above phrase about activities in a process, answer the following question. Your answer should be limited to either Yes or No and nothing else. **Question:** ≪process related (QP1)≫Does C occur before B? ≪causal related (QC1)≫Does A cause the execution of C? ≪causal & process related (QPC1)≫Does A cause the execution of B?

---

[4] https://beta.openai.com/docs/models/gpt-3.
[5] https://www.openai.com/research/gpt-4.
[6] https://openai.com/index/hello-gpt-4o/.
[7] https://huggingface.co/ibm/merlinite-7b.

Correctness refers to the degree each question's answer is adequate to the targeted process situation. For Yes/No questions, a correct answer indicates whether the condition holds (or not) in the corresponding situation. We use rules as a formal mechanism to determine answers and keep the same answers for the instantiated situations and corresponding questions.

Reliability reflects result consistency across multiple tests. This may also be influenced by the LLM's inherent consistency. It can be measured using conventional metrics like Cronbach's alpha. To ensure reliability, we reset the prompt to prevent prior context from affecting interactions.

Validity assesses how well each question captures the specific aspect of its domain. This can be measured through convergence metrics like factor loading, and discriminant validity when partitioning the instantiated questions by different perspectives and domains.

5   Results

We report the results for running our benchmark prototype in Table 7. This table shows the proportion of questions answered correctly by each LLM. The results are split between the template questions and domain-specific questions, and are also partitioned by the various perspectives, considering the causal perspective both without and with the addition of the extending questions. While the template questions provide exhaustive domain coverage, the domain-specific ones should be interpreted cautiously, as each template question has only one corresponding domain-specific question. As such, we expect that in a full-scale set of domain-specific questions, the performance is likely to get closer to the accuracy presented by the template questions. In addition, it is more likely the corpora employed for LLM training were domain-specific, hence incidentally implying an improved performance for domain-specific questions.

The benchmark can be used for testing LLM performance, as demonstrated here, and for training an LLM to improve its reasoning about BP^C. For testing, question answers serve as the "ground truth" to determine model accuracy, which can be analyzed by perspective and domain. For training, the questions and answers can be randomly split into training and testing subsets (e.g., 80/20). The testing subset is used before and after training to measure improvement.

Table 7. LLM accuracy results using the prototype benchmark for situation #2

	Num of questions	Template questions					Domain-specific questions	
		Merlinite 7b	Mixtral instruct 8x 7b	GPT 4o	GPT 4	GPT 3.5	Merlinite 7b	Mixtral instruct 8x 7b
Process	7	58%	71%	100%	85%	65%	100%	100%
Causal	6	66%	100%	100%	100%	56%	98%	100%
Process + Causal	2	100%	100%	100%	80%	15%	55%	95%
Extension	4	50%	50%	70%	65%	55%	73%	50%
Causal+Extension	10	60%	80%	88%	86%	56%	88%	80%
Total weighted avg		**63%**	**79%**	**94%**	**85%**	**55%**	**89%**	**89%**

6   Conclusion and Future Work

Setting aside the debate whether LLMs can generally reason, our goal here is more modest. We would like to equip (or measure) LLMs regarding their capacity to adequately infer sound conclusions when presented with knowledge about causal business processes. In this regard, our developed instrumentation can be employed in two (complementary) manners. It could be used to test such an ability with respect to a relatively wide variety of process domains. In addition, the tool can also be used as a model training dataset to augment an existing LLM with such ability and also to be adopted to additional problem domains for specific needs. For the former purpose, it could be used "as is" with only the questions component to facilitate model bench-marking, and with the ground truth answers for model training. For any newly embarked problem domains, the core seed of the template questions should be instantiated methodologically in the same process reported here to derive a corresponding set of domain-relevant items. This is also the rationale underlying our aim to release the model as open source, letting the community gradually contribute to the dataset to make its domain coverage broader. Our current version of the developed benchmark can be accessed here: https://huggingface.co/datasets/ibm/BPC.

The core contribution of this work is twofold. An open-source model that is developed and a methodology reporting how to construct a benchmark for a particular task to be facilitated by LLMs, in this case, the one of reasoning about BP^{C}s. At the time of submitting this paper, we embark on completing the specification of the template rules and questions corresponding to the other situations. As next steps, we plan to populate our benchmark with a more exhaustive set of domains and at a larger scale of instances per domain. Our vision is that LLM benchmarks may become standardized means to guide the choice of suitability of LLMs to specific tasks.

References

1. van der Aalst, W.: Process Mining. Springer, Heidelberg (2016)
2. Bellan, P., van der Aa, H., Dragoni, M., Ghidini, C., Ponzetto, S.P.: PET: an annotated dataset for process extraction from natural language text tasks. In: Lecture Notes in Business Information Processing. LNBIP, vol. 460 (2023)
3. Berti, A., Schuster, D., van der Aalst, W.M.P.: Abstractions, scenarios, and prompt definitions for process mining with LLMs: a case study. In: Business Process Management Workshops. BPM 2023. LNBIP, vol. 492 (2024)
4. Busch, K., Rochlitzer, A., Sola, D., Leopold, H.: Just tell me: prompt engineering in business process management. In: Lecture Notes in Business Information Processing. LNBIP, vol. 479 (2023)
5. Dumas, M., Fournier, F., Limonad, L., Marrella, A., et al.: AI-augmented business process management systems: a research manifesto. ACM Trans. Manag. Inf. Syst. **14**(1) (2023)
6. Fahland, D., Fournier, F., Limonad, L., Skarbovsky, I., Swevels, A.J.E.: How well can large language models explain business processes? arXiv (2024)
7. Fournier, F., Limonad, L., Skarbovsky, I., David, Y.: The WHY in business processes: discovery of causal execution dependencies (2023)
8. Gao, J., Ding, X., Qin, B., Liu, T.: Is ChatGPT a good causal reasoner? A comprehensive evaluation. In: Bouamor, H., Pino, J., Bali, K. (eds.) Findings of the Association for Computational Linguistics: EMNLP 2023, pp. 11111–11126. Association for Computational Linguistics, Singapore (2023)
9. Grohs, M., Abb, L., Elsayed, N., Rehse, J.R.: Large language models can accomplish business process management tasks. In: Business Process Management Workshops. BPM 2023. Lecture Notes in Business Information Processing, vol. 492, pp. 453–465. Springer, Cham (2024)
10. Huang, L., Yu, W., Ma, W., Zhong, W., et al.: A survey on hallucination in LLMs: principles, taxonomy, challenges, and open questions. arXiv (2023)
11. Kambhampati, S.: Can large language models reason and plan? Ann. N. Y. Acad. Sci. **1534**(1), 15–18 (2024)
12. Kampik, T., Warmuth, C., Rebmann, A., Agam, R., et al.: Large process models: business process management in the age of generative AI. arXiv (2023)
13. Klievtsova, N., Benzin, J.V., Kampik, T., Mangler, J., Rinderle-Ma, S.: Conversational process modelling: state of the art, applications, and implications in practice. In: Lecture Notes in Business Information Processing. LNBIP, vol. 490 (2023)
14. Narendra, T., Agarwal, P., Gupta, M., Dechu, S.: Counterfactual reasoning for process optimization using structural causal models. In: Lecture Notes in Business Information Processing, vol. 360 (2019)
15. Naveed, H., Khan, A.U., Qiu, S., Saqib, M., et al.: A comprehensive overview of LLMs. arXiv (2023)
16. Ning, Q., Wu, H., Han, R., Peng, N., Gardner, M., Roth, D.: TORQUE: a reading comprehension dataset of temporal ordering questions. In: Proceedings of the 2020 Conference on Empirical Methods in Natural Language Processing (EMNLP), pp. 1158–1172 (2020)
17. Pearl, J., Mackenzie, D.: The Book of Why: The New Science of Cause and Effect, 1st edn. Basic Books (2018)
18. Teubner, T., Flath, C.M., Weinhardt, C., van der Aalst, W., Hinz, O.: Welcome to the era of ChatGPT et al.: the prospects of large language models (2023)

19. Vidgof, M., Bachhofner, S., Mendling, J.: Large language models for business process management: opportunities and challenges. In: Lecture Notes in Business Information Processing. LNBIP, vol. 490, pp. 107–123 (2023)
20. Wei, J., Tay, Y., Bommasani, R., Raffel, C., et al.: Emergent abilities of large language models. Trans. Mach. Learn. Res. (2022)
21. Willig, M., Zečević, M., Dhami, D.S., Kersting, K.: Can foundation models talk causality? arXiv (2022)
22. Zhao, W.X., Zhou, K., Li, J., Tang, T., et al.: A survey of large language models. arXiv (2023)

Using Large Language Models to Generate Process Knowledge from Enterprise Content

Sandro Franzoi[1,2(✉)] [iD], Maxime Delwaulle[1] [iD], Julian Dyong[1] [iD], Jan Schaffner[1] [iD], Mara Burger[1,2] [iD], and Jan vom Brocke[1,2] [iD]

[1] University of Münster, 48149 Münster, Germany
`sandro.franzoi@uni-muenster.de`

[2] ERCIS – European Research Center of Information Systems, 48149 Münster, Germany

Abstract. Large language models (LLMs) have disrupted knowledge work in many application areas. Accordingly, the Business Process Management (BPM) community has started to explore how LLMs can be leveraged, resulting in a variety of promising research directions across the BPM lifecycle. Despite rapid adoption in practice and strong research interest, however, little is known about the actual design of BPM systems that leverage LLMs in organizational contexts. In this paper, we report on design science-based research in collaboration with a large multinational company to design a BPM system that leverages LLMs for process knowledge extraction from diverse enterprise content. Based on the development of our prototype, we observe that LLMs provide the means to organize and generate process knowledge independent of specific forms of representation. We present a conceptual framework that describes the role of LLMs in generating process knowledge from diverse input formats and, in turn, making it available in diverse output formats via prompting, resulting in representation-agnostic process knowledge. We also highlight implications of our study for BPM research and practice.

Keywords: Large Language Models · Process Knowledge · Generative Artificial Intelligence · Business Process Management

1 Introduction

The recent advent of transformer architectures has propelled generative artificial intelligence (AI), and particularly chatbots based on large language models (LLM), to the forefront of AI research and practice [32]. The capabilities of LLMs are also increasingly recognized and explored for Business Process Management (BPM) [6], resulting in many potential applications throughout the entire BPM lifecycle [34]. For example, LLMs can be used to discover and improve on issues in the process redesign phase, or to automatically identify process descriptions and models from heterogeneous sources in the process discovery phase [34]. Recent works have addressed the intersection of BPM and generative AI from various angles. For instance, prior research addresses challenges and opportunities of natural language processing (NLP)-based techniques for supporting

© The Author(s), under exclusive license to Springer Nature Switzerland AG 2025
K. Gdowska et al. (Eds.): BPM 2024 Workshops, LNBIP 534, pp. 247–258, 2025.
https://doi.org/10.1007/978-3-031-78666-2_19

process management practices [33], elaborates on the possibility of using generative pre-trained transformers (GPT) to create process models and enhance decision-making [1], creates virtual process assistants [4], or envisions AI-augmented process management systems in general [5]. From a practical perspective, companies often possess vast and diverse amounts of enterprise content, which can pose difficulties for identifying and using relevant process knowledge. Therefore, the capabilities of large language models offer promise in revolutionizing BPM, making process knowledge more accessible and useful by extracting important information from unstructured files with simple prompts. While scholars called for research on the use of LLMs in BPM and specific organizational applications [34], there is still a lack of understanding on how these models can support process knowledge representation and extraction. Therefore, this study aims to answer the question: *How can large language models be used to extract process knowledge from heterogeneous enterprise content?* We used a design science-based approach in collaboration with a multinational company to explore this question and develop an LLM-based prototype for process knowledge extraction. We contribute to BPM research by showcasing a novel approach to extract process knowledge from diverse enterprise content. Based on the insights gathered throughout this project, we present a conceptual framework of representation-agnostic process knowledge. This framework outlines how heterogeneous data sources can serve as input process knowledge, and prompts can be used to create equally heterogeneous output process knowledge (e.g., text, models, video, etc.).

The paper is structured as follows: First, in Sect. 2, we describe prior work that investigates the potentials and applications of generative AI in BPM and process knowledge representation as our research background. In Sect. 3, we outline the steps of our design science approach. In Sect. 4, we present the results of our work, including a description of the developed prototype and a conceptual framework for representation-agnostic process knowledge (RAPK). Finally, in Sect. 5, we discuss implications for BPM research and practice before concluding with the limitations of our study.

2   Research Background

2.1   Generative AI in Business Process Management

Generative AI refers to "computational techniques that are capable of generating seemingly new, meaningful content such as text, images, or audio from training data" [6]. Various techniques are used to analyze large amounts of data to identify patterns and relationships in an original dataset to generate novel output. In the field of business process management, these technologies can automate repetitive tasks, identify opportunities for process innovation, and increase employee or customer satisfaction [3, 6, 12]. Vidgof, Bachhofner and Mendling [34] describe various opportunities for leveraging LLMs throughout the entire BPM lifecycle. For instance, during the identification phase, employees often only have access to unstructured internal process knowledge, which necessitates manually searching for relevant information within these documents. LLMs can assist employees by quickly locating relevant information within the vast volumes of enterprise content. During the process discovery phase, Vidgof et al. [34] suggest that LLMs can be used to develop process models. This can be accomplished by analyzing documentation or communication logs, or by utilizing chatbots as interview partners for domain expert interviews. Similarly, other opportunities for applying LLMs in process discovery have been identified, such as producing process descriptions in various formats [6, 16]. In the process analysis phase, LLMs can summarize massive amounts of data to identify patterns and issues in the process. Problem identification and solution suggestion can also be conducted for process models [17, 34]. During the process redesign phase, LLMs can assist by redesigning process models and utilizing various techniques and methods to further develop potential solutions [34] for both explorative and exploitative BPM design strategies [6]. There are several ways of including LLMs in the process implementation phase [34]. For instance, users can interact with the LLM and ask for clarification on BPMN models, unstructured information, and document patterns. LLMs can also assist during the monitoring phase by extracting relevant information from a process dashboard [34]. Although LLMs offer promising opportunities for managing business processes, prior research indicates that there are few specific implementation scenarios of LLMs in BPM that would allow investigating the design of LLMs in empirical BPM contexts [6].

2.2 Process Knowledge Representation

Various forms can be used to represent information about business processes and capture the underlying knowledge [33]. These include textual descriptions [27], checklists [30], process models [23], spreadsheets [18], videos [17], and many more. Inherent to these different forms of representation are varying levels of structure, ranging from plain text descriptions to complex documents or video recordings. While elements on this spectrum of structuredness serve different purposes and requirements [33], the main objective is to convey information in a way that is understandable for the user [22, 33]. As a result, different representation formats with different levels of structuredness can be suitable for varying tasks. For example, checklists might offer a high-level description of all the necessary process steps, whereas textual descriptions might be more fitting to deliver information on complex problem-solving tasks [33].

Rapid advances in generative models enable the automatic creation of business process descriptions in various representation formats [6]. Furthermore, existing techniques for transitioning between different representation formats, such as generating BPMN models from textual descriptions [10] or vice versa [19] are enhanced by the increasing capabilities of generative models. These developments offer the potential to use the panoply of enterprise documents containing process information and knowledge, regardless of their format, as a foundation to train machine learning models, such as LLMs. With such systems in place, simple prompts can extract process information and knowledge in the most suitable representation format, offering prodigious opportunities for process improvement. In this regard, prior work focused primarily on text-related BPM tasks that use textual descriptions of business processes to mine imperative and declarative process models [11], extract business process entities and relations [2], or assess tasks regarding their suitability for automation [11].

3 Research Design

To study how LLMs can be used to extract process knowledge from heterogeneous enterprise content, we rely on a design science-based research approach in collaboration with a multinational company in Europe. Broadly speaking, design science research is used to generate design knowledge about innovative solutions to real-world problems [13]. In this study, we use the design science research grid [35] to structure our research into the dimensions of problem, input knowledge, research process, key concepts, solution description, and output knowledge.

Problem Description and Motivation: Despite the growing interest in the use of LLMs for BPM [34], little is known about actual BPM use cases that apply LLMs in organizational contexts. Organizations are finding it increasingly difficult to identify and understand process-related information and to navigate through the vast amounts of heterogeneous enterprise content. The question is whether LLMs can be leveraged as a solution for extracting process knowledge from diverse enterprise content.

Input Knowledge: We identified existing models, techniques, and methods that could be applicable to the specific use case. Our work was also informed by the large knowledge base of prior research in BPM.

Research Process: The research process can be divided into three main phases. During all phases, the demonstration and evaluation process included weekly alignments within the team and with the industry partner. The purpose of these weekly meetings with a company representative was to exchange relevant information about the requirements, share insights about the progress, receive quick feedback in the form of qualitative evaluations, and plan the development steps for the next week. These short iterations allowed for continuous close collaboration and informal evaluations of the prototype. The phases of development correspond to the iterations of our prototype development. In the following, they are described briefly from a methodological perspective and outlined in more detail in Sect. 4.1.

Phase 1: During the first four weeks of the project, we laid the foundation by validating and developing our *data preprocessing*. We tried different approaches and then used the most suitable for our prototype. Additionally, we investigated which specific LLM to use and how to incorporate domain-specific information. The phase was concluded with a joint workshop including company representatives and researchers.

Phase 2: In the following six weeks, we developed a *functional prototype*. After identifying a suitable platform, we implemented a chatbot that can be used by employees seeking to retrieve process knowledge from the entire enterprise content. We then evaluated the chatbot in the subsequent phase.

Phase 3: Following the implementation, we developed a framework to evaluate our chatbot. After this structured *evaluation*, we presented the prototype to the industry partner in a two-hour live demonstration. In this on-site workshop, we gained feedback from senior company representatives and discussed further use cases for the enterprise.

Key Concepts: The key concepts applied in this study are process knowledge extraction, generative AI, process knowledge generation, LLMs, and prompt engineering.

Solution Description: The solution proposed in this study includes an LLM-based chat-bot capable of recognizing process knowledge in (un-) structured and diverse enterprise content and answering questions about it.

Output Knowledge: The output knowledge includes implications for BPM research on the application of LLMs throughout the BPM lifecycle as well as a conceptual framework for the creation of representation-agnostic process knowledge.

4 Results

4.1 Prototype Development

The development and evaluation of the prototype was split into three phases, as described in Sect. 3. In the first phase, we preprocessed the data and determined which LLM to use. In the second phase, we used the preprocessed data as input for our functional chatbot, allowing users to retrieve process knowledge by querying the LLM. The third phase involves the evaluation of our chatbot using a self-developed evaluation framework.

(1) Data Preprocessing and LLM Selection: To enable the LLM to handle heterogeneous process information found in slide decks or BPMN models, we converted this data into uniform text. We developed an automation pipeline based on Microsoft Power Automate, for example by parsing PowerPoint slides with a Python script to extract the text from the slides. Using these uniform text documents, embeddings were created. The embeddings were stored in the vector database Pinecone [28], enabling Retrieval Augmented Generation (RAG) [20] to provide the LLM with long-term memory and domain-specific knowledge. This helps to avoid hallucinations when querying the LLM. Then, we decided which LLM to use for our prototype. At the point of development (December 2023), we considered GPT-4 by OpenAI [24] as the most suitable candidate.

(2) Functional Chatbot: In the second phase, we developed the prototype by using Flo-wise, a low-code tool that facilitates the development of LLM applications [9]. Figure 1 illustrates the architecture of our prototype.

Our prototype comprised several Flowise building blocks. First, we used a document loader to load the preprocessed data from the input folder. Second, the input data was broken down into smaller chunks using a recursive character splitter. Next, we used the OpenAI API to generate embeddings of the text chunks and store them in Pinecone. A "Conversational Retrieval QA Chain" [9] building block, which performs question-answering tasks with a retrieval component, combines all elements of the flow. The chain utilizes the ChatOpenAI building block to incorporate GPT-4 into the flow. This building block is a wrapper around OpenAI's LLMs and uses their chat endpoints [9]. The user can engage in conversational turns with the LLM, which provides answers based on the provided documents and the retrieval aspect of the chain. Chains are used to store and manage the conversational history and context for the LLM, enabling the model to comprehend the ongoing conversation and provide coherent and contextually relevant responses for the user. In addition, the chain as the endpoint allows to link the source documents used to build the response, providing more explainability [14, 29].

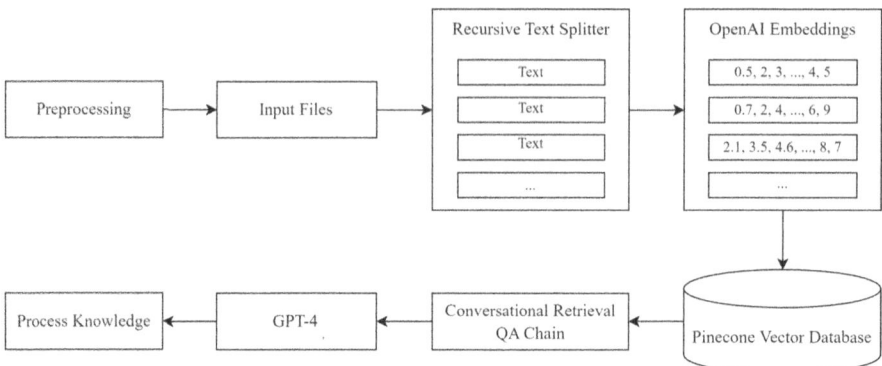

Fig. 1. Architecture of the Prototype

With a system message (*pre-prompt*), we directed the model's behavior by instructing how the LLM should respond to queries, resulting in better performance [15]. To prevent hallucinations, it ensures that the chatbot only relies on provided documents when answering a question. If the required information is not found in the documents, the chatbot points it out and offers to help with other questions – but politely refuses to answer the current question. As a last step, we integrated the entire flow of our prototype within a website using a pop-up window. The final prototype was evaluated with our industry partner. Figure 2 shows the user interface of the final prototype with an exemplary conversation.

(3) Evaluation: To evaluate the performance of the prototype, we used 29 self-created questions. The questions included single-choice and multiple-choice questions, open questions, as well as trick questions that cannot be answered with the given input data to check for potential hallucinations. Currently, objective evaluation strategies for the output of LLMs are lacking [8]. Thus, for each question, an exemplary answer was created, against which the output of the LLM was compared. To measure the performance, a subset of chatbot evaluation metrics [26] with varying points per metric is used: Factuality (e.g., "Is the answer correct?") is derived from the *Linguistic Perspective* and its category *Quality*; Completeness (e.g., "Does the answer provide all necessary information?") is derived from the *Information Retrieval Perspective* and its category *Accuracy*; Relevance (e.g., "Is the answer related to the question?") is derived from the *Linguistic Perspective* and its category *Relation*; Coherence (e.g., "Can the answer be understood?") is derived from the *User Experience Perspective* and its category *Performance*; Conciseness (e.g., "Is there unnecessary information in the answer?") is derived from the *Linguistic Perspective* and its category *Quantity*; Consistency (e.g., "Is the answer unambiguous?") is derived from the *Linguistic Perspective* and its category *Manner* and Grammatical Correctness (e.g., "Is the answer free of any grammatical errors?") is derived from the *Linguistic Perspective* and its category *Grammatical accuracy*. In this exact order, the points for each metric were: 5, 3, 4, 3, 1, 5, and 1. More points indicate that a metric is more important for a satisfying answer. If an answer fulfilled a metric, it received

its corresponding number of points. For an answer to be considered satisfactory, it had to receive a minimum of 11 points (out of 22). Our chatbot is intended to be used by employees seeking answers about processes. If the chatbot failed to fulfill Factuality or Completeness, the answer could have had adverse consequences in the intended business use case and we rejected the response regardless of points. Both criteria had to be fulfilled before we even started to rate all other metrics. Of the 29 answers, only one failed the evaluation.

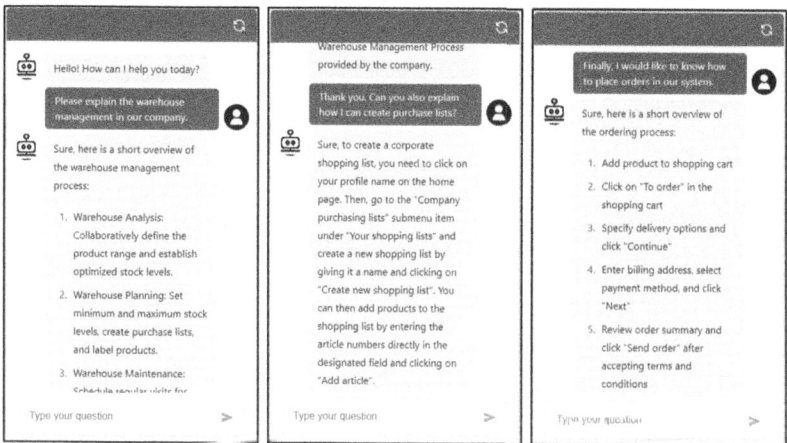

Fig. 2. Prototype with Exemplary Chat Conversation

Following the structured evaluation, the prototype was assessed in the last workshop with representatives of the industry partner. Here, we received favorable feedback on the usability and benefit of our prototype.

After the development of the final prototype, we observed a unique affordance that LLMs bring to BPM, which we refer to as *representation-agnostic process knowledge*. In the next section, we describe what we mean by representation-agnostic process knowledge and we conceptualize a more general framework.

4.2 Representation-Agnostic Process Knowledge

Information and knowledge about business processes is traditionally stored in various formats including process models, textual descriptions, checklists, and many more [33]. With recent technological advancements, such as LLMs, to support transformations between different representation formats (e.g., from BPMN to text or vice versa) [10, 19], there are unprecedented opportunities to utilize the vast and diverse forms of process knowledge stored in a variety of different enterprise documents [34].

Based on the development of our prototype, we conceptualize a more general depiction of process knowledge extraction. This approach introduces the concept of *representation-agnostic process knowledge (RAPK)*, which refers to the capability to represent process knowledge as data points irrespective of their original form of representation. Building on this concept, we present a framework that places an LLM at the center of an organization's BPM approach, creating a representation-agnostic process knowledge space. Within this space, process knowledge is gathered from diverse input forms and, in turn, made available in diverse output forms via prompting. This approach comprises three central elements: (1) process knowledge representations as an input, (2) a process knowledge space for transformation, and (3) process knowledge extraction to generate outputs. We conceptualize the approach by means of the RAPK framework in Fig. 3.

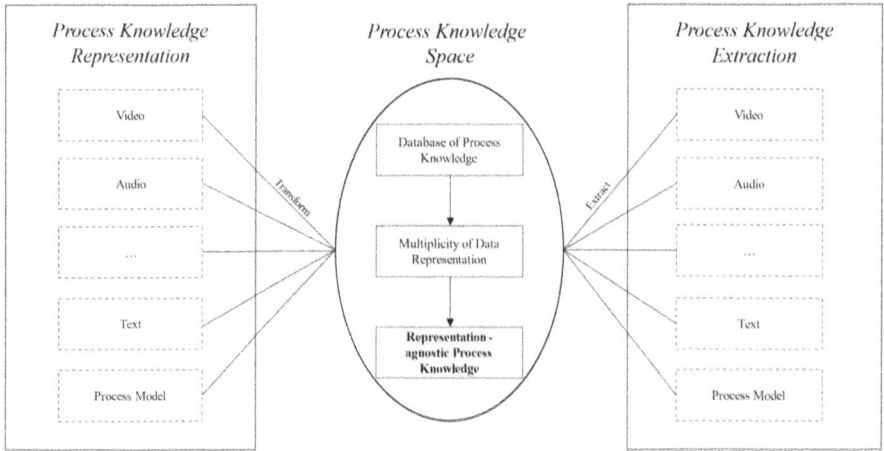

Fig. 3. Representation-agnostic Process Knowledge (RAPK) Framework

Process knowledge, typically stored in various enterprise documents, such as slides, drawings, texts, or recordings, serves as input. Our study shows that these representations are often heterogeneous and scattered throughout the organization, making it challenging for individuals to locate and access relevant process knowledge. The RAPK framework proposes using LLM-based approaches to extract process knowledge from such sources and store it agnostic to specific representation formats. This involves parsing diverse process knowledge representations into a machine-readable format to create a common data representation, such as in vector databases. This leads to a multiplicity of data representations that facilitates the storage of process knowledge irrespective of its specific form. Additionally, this enables the analysis of process knowledge across various sources. For example, process knowledge found in an internal slide show can be related to what can be found in process models, text documents, drawings, or other input sources. As a result, we see the emancipation of process knowledge beyond representation formats. Enabled by the representation-agnostic form of process knowledge, various forms of process knowledge can be generated through prompts. Increasingly capable generative models may be able to create process models [17] or even videos [25]. Furthermore, the system's output can be customized to meet the specific needs of the user, considering their role, expertise, or cognitive style [7]. For instance, the system would provide a different process description for a CIO compared to an intern requesting the same information. Therefore, process knowledge can be made available in a more adaptive and user-centric manner to all members of an organization. This can potentially increase overall process affinity within organizations, improve business processes, and lead to a move toward AI-augmented BPM [5].

In conclusion, adopting this view of LLM-based process knowledge representation has significant implications for BPM and provides promising avenues for future research, which we discuss in the following sections.

5 Discussion and Implications

5.1 Implications for BPM

In response to previous research calls [34], we developed a prototypical solution for extracting process knowledge from diverse enterprise content to demonstrate how it can be turned into actionable knowledge in an organizational setting. In doing so, we contribute to BPM research in two ways.

First, we showcase a BPM use case that highlights the specific benefits of LLMs in an organizational application scenario. While prior research has already emphasized the immense potential of LLMs throughout the entire BPM lifecycle [6, 34], our prototype demonstrates that these benefits can also be realized in an organizational setting. For example, Vidgof et al. [34] outline potentials such as process identification and process discovery from documentation, or the usage of LLM-based chatbots in process implementation. Our prototype has the greatest impact on process discovery. Specifically, our prototype can be used to discover processes from vast and heterogeneous enterprise content with simple prompts. Additionally, an LLM-based solution can provide an inquiring user with a process model or a process description [6, 18], thereby reducing the time and effort needed for process discovery. Lastly, our prototype can offer step-by-step instructions for specific business processes and concomitant tasks.

Second, our findings contribute to prior research on process information and knowledge representation. By demonstrating how LLMs can use enterprise content in various representation formats to generate equally diverse outputs, we suggest that LLMs can serve as the foundation for representation-agnostic process knowledge. We propose a representation-agnostic form of organizing process knowledge, which we conceptualize within the RAPK framework. The creation of RAPK using LLMs can have numerous positive impacts: Process knowledge that is naturally occurring and not called up by process management can be captured from large quantities of company content. Additionally, combining process knowledge from different and complementary sources can create novel connections that remain invisible when organizations limit themselves to certain forms of representation of process knowledge. Process knowledge can also be customized to meet various needs and situations, making it more widely used and valuable in organizations. Therefore, there is significant potential in further studying the transformation and creation of different representation formats (e.g., from text to BPMN [10]) or investigating novel transformations between representations (e.g., from BPMN to video).

Our research also has important practical implications. First, the accessibility of such systems, for example through prompting, can help spread process knowledge throughout organizations. This can strengthen organization-wide process thinking and foster a BPM-oriented culture. Second, LLM-based process knowledge extraction can simplify and streamline an organization's onboarding and training. New employees can use this tool to understand the organization and its business processes, learn about their new role, and continuously improve their knowledge of business processes. Third, implementing LLM-based process knowledge extraction in an organization captures diverse process knowledge and makes it reproducible. Using the entire process knowledge base, the risk of losing expert know-how is minimized. Fourth, we discovered that utilizing low-code platforms such as Flowise is well-suited for the development of LLM-based applications. They enable closer collaboration between business and IT professionals within organizations, provide an infrastructure that is accessible to experts with limited programming know-how, and accelerate development iterations [31].

5.2 Limitations and Outlook

As with any research endeavor, it is important to acknowledge the limitations of our study. One limitation lies in the input formats the prototype can process. Semi-structured data, such as XML files, are easier to process than unstructured data, such as slide shows. The prototype can extract text from slide shows, but visual information is lost, such as relationships and sequences of graphical elements. Future research could explore how to capture visual information more effectively, for example using multimodal LLMs. Besides input formats, the output formats of the prototype are another limitation. Although our prototype can handle different formats of process knowledge, future research could focus on processing and generating even more process representations. The recent emergence of video generation models, such as OpenAI's Sora, has enabled generative AI to create short videos [25], which could be utilized to create new process representations. Furthermore, our prototypical solution relies on the entire knowledge base to answer questions. This could be a problem when information is requested that goes beyond the authorization level of the inquiring employees. Therefore, future research could incorporate role and permission-based answering strategies. Additionally, we developed and evaluated our prototype within a specific organizational setting. Further research is needed to assess the generalizability of such solutions and their performance across various contexts and industries. Furthermore, handling context is a common challenge when working with large amounts of input data. When context is stored in a single vector database, the LLM's performance may decrease with increasing context size [21]. Therefore, a crucial challenge for future research is the implementation of a context architecture capable of effectively storing enterprise data without compromising the quality of LLM-generated responses. Additionally, we evaluated our prototype development primarily through qualitative feedback in weekly meetings and joint workshops. In further evaluations, we could use additional methods such as employee focus groups to enhance the existing solution. Moreover, cloud-based models such as the one we utilize for our prototype might not be possible for all organizational applications due to confidentiality. Hence, future research could adapt our modular solution by relying on local models and comparing their performance. Finally, the long-term effects of implementing such a system in an organizational environment are yet to be evaluated.

Disclosure of Interests. The authors have no competing interests to declare that are relevant to the content of this article.

References

1. Beheshti, A., et al.: ProcessGPT: transforming business process management with generative artificial intelligence. In: IEEE International Conference on Web Services, Chicago, USA, pp. 731–739 (2023)
2. Bellan, P., Dragoni, M., Ghidini, C.: Extracting business process entities and relations from text using pre-trained language models and in-context learning. In: Almeida, J.P.A., Karastoyanova, D., Guizzardi, G., et al. (eds.) Enterprise Design, Operations, and Computing. EDOC 2022. LNCS, vol. 13585, pp 182–199. Springer, Cham (2022)
3. Beverungen, D., Bujis, J.C.A.M., Becker, J., et al.: Seven paradoxes of business process management in a hyper-connected world. Bus. Inf. Syst. Eng. **63**, 145–156 (2021)
4. Bors, L., Samajdwer, A., van Oosterhout, M.: Oracle Digital Assistant: A Guide to Enterprise-Grade Chatbots, 1st edn. Apress, New York (2020)
5. Dumas, M., Fournier, F., Limonad, L., et al.: AI-augmented business process management systems: a research manifesto. ACM Trans. Manag. Inf. Syst. **14**(1), 1–19 (2023)
6. Feuerriegel, S., Hartmann, J., Janiesch, C., Zschech, P.: Generative AI. Bus. Inf. Syst. Eng. **66**(1), 111–126 (2024)
7. Figl, K., Recker, J.: Exploring cognitive style and task-specific preferences for process representations. Requirements Eng. **21**, 63–85 (2016)
8. Fill, H.-G., Fettke, P., Köpke, J.: Conceptual modeling and large language models: impressions from first experiments with ChatGPT. Enterp. Model. Inf. Syst. Archit. **18**, 1–15 (2023)
9. FlowiseAI. https://docs.flowiseai.com/. Accessed 07 June 2024
10. Friedrich, F., Mendling, J., Puhlmann, F.: Process model generation from natural language text. In: International Conference on Advanced Information Systems Engineering, pp. 482–496, Springer, London (2011)
11. Grohs, M., Abb, L., Elsayed, N., Rehse, J.-R.: Large language models can accomplish business process management tasks. In: De Weerdt, J., Pufahl, L. (eds.) Business Process Management Workshops. BPM 2023. LNBIP, vol. 492, pp 453–465. Springer, Cham (2024)
12. Haase, J., Hanel, P.H.P.: Artificial muses: generative artificial intelligence chatbots have risen to human-level creativity. J. Creativity **33**(3), 2–7 (2023)
13. Hevner, A.R., March, S.T., Park, J., Ram, S.: Design science in information systems research. MIS Q. **28**(1), 75–105 (2004)
14. Horvatic, D., Lipic, T.: Human-Centric AI: the symbiosis of human and artificial intelligence. Entropy **23**(3), 332–337 (2023)
15. Hu, T., Collier, N.: Quantifying the persona effect in LLM simulations (2024). arXiv:2402.10811
16. Kecht, C., Egger, A., Kratch, W., Rögliner, M.: Quantifying chatbots' ability to learn business processes. Inf. Syst. **113** (2023)
17. Kratsch, W., König, F., Röglinger, M.: Shedding light on blind spots: developing a reference architecture to leverage video data for process mining. Decis. Support. Syst. **158**, 1–34 (2022)
18. Krumnow, S., Decker, G.: A concept for spreadsheet-based process modeling. In: Mendling, J., Weidlich, M., Weske, M. (eds.) Business Process Modeling Notation. BPMN 2010. LNBIP, vol. 67, pp. 63–77. Springer, Heidelberg (2010)

19. Leopold, H., Mendling, J., Polyvyanyy, A.: Generating natural language texts from business process models. In: International Conference on Advanced Systems Engineering, pp. 64–79. Springer, Gdansk (2012)

20. Lewis, P., et al.: Retrieval-augmented generation for knowledge-intensive NLP tasks. In: Proceedings of the 34th International Conference on Neural Information Processing Systems, Vancouver, Canada, pp. 9459–9474 (2020)

21. Liu, N.F., et al.: Lost in the middle: how language models use long contexts. Trans. Assoc. Comput. Linguist. **12**, 157–173 (2024)

22. Mendling, J., Reijers, H., Cardoso, J.: What makes process models understandable? In: 5th International Conference on Business Process Management, pp. 48–63. Springer, Brisbane (2007)

23. Mendling, J., Strembeck, M., Recker, J.: Factors of process model comprehension – findings from a series of experiments. Decis. Support. Syst. **53**, 195–206 (2012)

24. OpenAI GPT-4. https://openai.com/index/gpt-4/. Accessed 07 June 2024

25. OpenAI Sora. https://openai.com/sora. Accessed 07 June 2024

26. Peras, D.: Chatbot evaluation metrics: review paper. In: Veselica, R., Dukic, G., Hammes, K. (eds.) Economic and Social Development (Book of Proceedings), 36th International Scientific Conference on Economic and Social Development, pp. 89–97 (2018)

27. Phalp, K.T., Vincent, J., Cox, K.: Improving the quality of use case descriptions: empirical assessment of writing guidelines. Softw. Qual. J. **15**, 383–399 (2007)

28. Pinecone. https://www.pinecone.io/. Accessed 07 June 2024

29. Ramchurn, S.D., Stein, S., Jennings, N.R.: Trustworthy human-AI partnerships. iScience **24**(8), 102891 (2021)

30. Reijers, H.A., Leopold, H., Recker, J.: Towards a science of checklists. In: Proceedings of the 50th Hawaii International Conference on System Sciences, Hawaii, USA, pp. 5773–5782 (2017)

31. Sundberg, L., Holmström, J.: Democratizing artificial intelligence: how no-code AI can leverage machine learning operations. Bus. Horiz. **66**(6), 777–788 (2023)

32. Teubner, T., Flath, C., Weinhardt, C., van der Aalst, W., Hinz, O.: Welcome to the era of ChatGPT et al. Bus. Inf. Syst. Eng. **65**, 95–101 (2023)

33. van der Aa, H., Carmona, J., Leopold, H., Mendling, J., Padró, L.: Challenges and opportunities of applying natural language processing in business process management. In: Proceedings of the 27th International Conference on Computational Linguistics, Santa Fe, USA, pp. 2791–2801 (2018)

34. Vidgof, M., Bachhofner, S., Mendling, J.: Large language models for business process management: opportunities and challenges. In: 21st International Conference on Business Process Management Forum, pp. 107–123. Springer, Utrecht (2023)

35. vom Brocke, J., Maedche, A.: The DSR grid: six core dimensions for effectively planning and communicating design science research projects. Electron. Mark. **29**, 379–385 (2019)

Efficient LLM-Based Conversational Process Modeling

Julius Köpke[(⊠)] and Aya Safan

Department of Informatics Systems, University of Klagenfurt, Klagenfurt, Austria
{julius.koepke,aya.safan}@aau.at
https://www.aau.at/en/isys/ics/

Abstract. Since the introduction of OpenAI's ChatGPT, the underlying technology of generative AI and large language models (LLMs) has gained tremendous interest in academia. Researchers began experimenting with LLMs' capabilities in various domains, including business process modeling. While these works indicate a promising potential of LLMs for this task, they do not consider the number of tokens of the prompting strategies and output formats. However, the token amount is the number one cost driver for LLM-based process modeling. In addition, an efficient representation of the conversation state has not been addressed so far. This paper addresses these concerns and introduces and evaluates an approach for efficient LLM-based conversational process modeling. We have implemented our approach as a publicly available online tool. In our experiments, we observed average input token reductions of 94% compared to an existing tool while maintaining even better levels of correctness. Furthermore, a user study at a public science fair indicates solid numbers for the tool's usefulness.

Keywords: Large Language Models · LLM · Conversational Process Modeling

1 Introduction

Large language models (LLMs) have recently gained tremendous attention from researchers in various fields. Numerous works such as [3,5–7] explore the potential of LLMs to transform textual descriptions to business process models. The paper [7] systematically analyzes LLMs for conversational process modeling and evaluates various prompting strategies. These works indicate promising capabilities of LLMs, particularly OpenAI's GPT-4 [12], for this task. With *ProMoAI* [9,10], there is already an online tool for conversational process modeling available. A user enters a textual description of a process; the tool then generates an initial model and presents it graphically to the user, who comments on the model in a feedback loop. While the tool's capabilities seem promising, it comes with significant costs. In initial experiments, with a 3 step process and two iterations of the feedback loop using GPT-4, we paid around 0.8 USD in OpenAI API usage fees.

© The Author(s), under exclusive license to Springer Nature Switzerland AG 2025
K. Gdowska et al. (Eds.): BPM 2024 Workshops, LNBIP 534, pp. 259–270, 2025.
https://doi.org/10.1007/978-3-031-78666-2_20

We argue that such high costs can significantly hinder the broader adoption of such tools. However, existing works only focus on the quality of the generated solutions, ignoring their costs. When using the OpenAI API, the amount of tokens directly determines the monetary costs of a request. While the cost models for self-hosted LLMs may differ, the number of tokens still significantly impacts the computational costs for generating a response [13,14].

To reduce costs, a context prompt should add minimal overhead to the user input, and the generated output format should be compact. However, the output format can also affect the quality of the generated models [7] and, thus, potentially the number of required context prompt tokens. Furthermore, even if monetary costs are not an issue, conversational modeling tools should be tuned for efficiency, considering scalability and the carbon footprint induced by energy consumption [14].

This paper introduces and evaluates a method for efficient LLM-based conversational process modeling. We present related works in Sect. 2. In Sect. 3, we identify the major cost drivers for conversational process modeling, propose an efficient approach, and introduce an optimized process meta-model for communicating processes with LLMs. Section 4 introduces the *BPMN-Chatbot* as an instantiation of our approach. In Sect. 5, we evaluate the chatbot against *ProMoAI* [9,10] and a prompting strategy from [7] regarding model quality and cost-efficiency. We suppose that a cost-efficient LLM-based modeling tool has the potential to make process modeling available to a broader audience, and we, therefore, additionally report on the acceptance of the tool by users of the general public at a public science fair. Finally, Sect. 6 concludes the paper.

2   Related Work

An analysis of potential applications of LLMs in the BPM domain was conducted in [16]. While there are approaches for generating process models from text using various techniques (see [15] for a survey), LLMs can potentially disrupt alternative approaches. Moreover, LLMs can handle scenarios without complete textual descriptions, allowing users to ask for a type of process, and the LLM answers based on its implicit background knowledge. Initial experiments in [3] show quite promising results on the modeling capabilities of ChatGPT (GPT-4) for conceptual modeling and business process modeling. Experiments in [5] use ChatGPT to generate imperative and declarative process models and RPA descriptions from text. In [6], the capabilities of LLMs for extracting relevant tasks from textual descriptions were evaluated. An extension of [6] is provided in [7], where the capabilities for extracting control flow from process descriptions were extensively evaluated. The work does not present a tool but assesses the outputs of various prompting strategies and output formats.

In [4], a framework for transforming textual descriptions into business processes is proposed. It uses Petri Nets as an intermediate format. However, in a second LLM step, the Petri Nets are transformed into JSON-Nets, including the graphical positioning of the nodes.

An online tool ProMoAI[1] for the LLM-based generation of process models was introduced in [9,10]. It supports various OpenAI LLMs and uses heavy-weight prompts to generate Python code, which, in turn, generates expressive intermediate POWL [11] models. Users can refine the models in a feedback loop.

3   Efficient LLM-Based Conversational Process Modeling

3.1   Conversational Process Modeling

We now discuss the general phases for conversational process modeling support-ing interactions like ProMoAI [10]: The user first provides an input text, and the system generates an initial model. Afterward, the user may repeatedly provide feedback, and the system adopts the generated model accordingly. This leads to two distinct phases:

Generation of an Initial Model: In this phase, the user enters a textual description d of a process. The tool then generates a prompt $p = (d, i, fi)$, where i is an instruction prompt for model generation and fi is the instruction for the intermediate output format f. The prompt is then sent to the LLM. It generates an answer a_0 in the intermediate format f. The system, in turn, translates a_0 to an output model m_0, which is presented graphically to the user. For this step, the larger p is, the larger the number of context tokens will be. Since we cannot control the size of d, we can only reduce the sizes of i and fi. Consequently, the intermediate format f is highly relevant in two aspects: Instances should be compact to reduce the number of output tokens. At the same time, f must be easily "understandable" by the LLM to efficiently achieve high-quality models.

Feedback Loop: In the n-th iteration of the feedback loop, the user comments on the model m_{n-1} with comment c_n. A new prompt $p' = (c_n, i', fi, st)$ is created where st is the conversation state. The prompt p' is sent to the LLM, and a new model a_n in format f is returned. It is then translated to the output model m_n and presented to the user.

Since existing LLM systems are stateless [18], the state of the conversation must be part of the prompt in the feedback loop. A conversation state that includes all prior message exchanges would dramatically increase the context token size. Moreover, since LLMs have a limited context window, the conversa-tion state cannot grow indefinitely.

3.2   Approach

For efficient LLM-based process modeling, we propose the following approach: In the initial step, we instruct the LLM to provide the answer a_0 in an optimized intermediate format. In the feedback loop, we construct the prompt as $p' = (i', fi, c, a_{n-1})$. Therefore, the state of the conversation is solely represented by the most recent model in the intermediate format.

---

[1] https://promoai.streamlit.app/. Last accessed 12.06.2024.

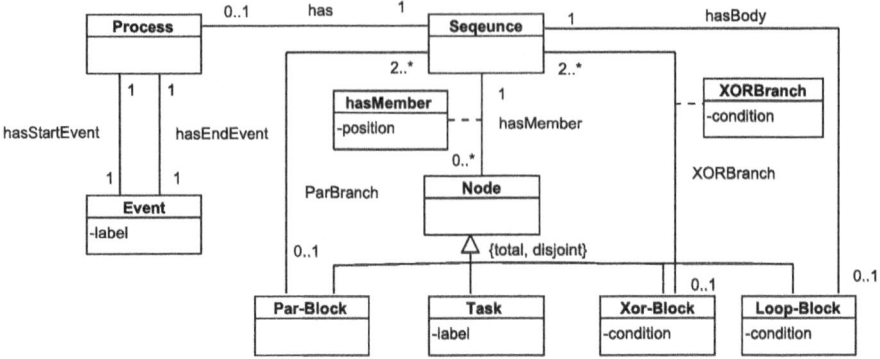

Fig. 1. Process Meta-Model for intermediate models

Process Meta-model: The intermediate format is a key factor in reducing the costs of conversational process modeling tools. It should be compact and "intuitive" for the LLM. Our preliminary experiments with BPMN-XML led to a costly, verbose, low-quality output. To achieve a small number of output tokens (and thus also context tokens for the feedback loop), we base our approach on the process meta-model shown in Fig. 1. It is a variant of full-blocked process models [8], where each split gateway has its corresponding join gateway. In addition, each process has exactly one Start and End event and one sequence of nodes. A node can be a task, a Par-Block, a Xor-Block, or a Loop-Block. Xor-Blocks have a condition and are connected to at least two sequences of nodes via Xor-Branches. Each Xor-Branch has its associated condition. Similarly, Par-Blocks are connected to at least two sequences via Par-Branches. Loop-Blocks have while semantics with a condition for repetition and are connected to a sequence via the loop body. This model covers the same BPMN core elements used in conversational process modeling in literature [3,7,9]. However, we force the output models to be full-blocked. Such processes have several benefits: They are correct by design by eliminating deadlocks and life locks. In addition, the model is structurally similar to block-structured programming languages. Based on the good programming performance of mainstream LLMs, we suggest that only small amounts of instructions are needed to create high-quality output models following this meta-model.

4 Introducing the BPMN-Chatbot

The *BPMN-Chatbot* is an instantiation of our proposed efficient conversational process modeling approach. It is freely available online[2]. However, upon the first start, the tool asks for an OpenAI API key.

---

[2] https://isys.uni-klu.ac.at/pubserv/BPMN-Chatbot/.

Usage Scenario: When a new session is started, the chatbot introduces itself and offers a tutorial on BPMN for novice users. The user can create a new process model by providing a description by typing or voice recording. The system then generates a response displayed as a BPMN diagram. The *BPMN-Chatbot* then asks the user if they would like to extend the model or fix errors. In these cases, the user provides a comment and an updated diagram is generated, and the feedback loop starts again. A screenshot of the tool is shown in Fig. 2. It shows the interface after one iteration of the feedback loop. Moreover, the tool allows navigation between model versions (using the arrow icons) to provide feedback on a particular version.

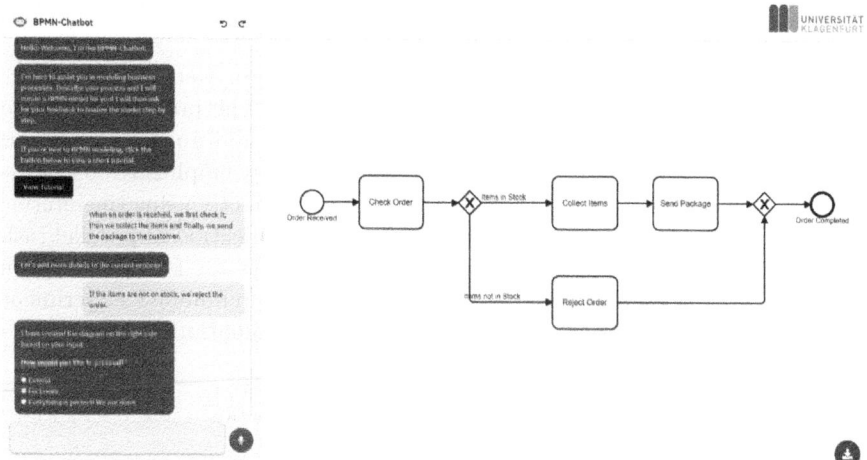

Fig. 2. Screenshot of the tool after one refinement.

Architecture: The prototype is implemented as a React single-page web application as shown in Fig. 3. The UI is built using React components, which manage rendering, state updates, and event handling. A Prompt Generator module takes user input and constructs a prompt for the LLM API. Once the prompt is ready, it sends a request to the LLM API, which processes the request and returns a JSON response containing the generated process model. The Model2Model Translator module then converts this JSON response into BPMN XML. Finally, the bpmn-js[3] library is used to render the diagram.

Prompt Generation: The implementation uses OpenAI's chat completions API[4], which allows the specification of the response format via JSON Schema. Therefore, we provide a schema derived from our process Meta-Model along with every request. Furthermore, we include the instruction prompt shown in Listing 1. The

---

[3] https://bpmn.io/toolkit/bpmn-js/.

[4] https://platform.openai.com/docs/api-reference/chat/create.

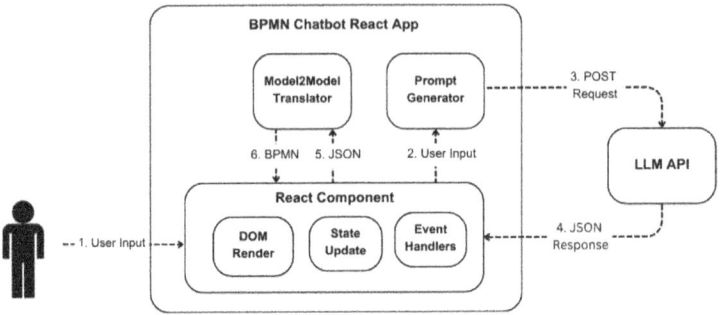

Fig. 3. BPMN Chatbot prototype architecture.

instruction prompt follows a combination of the Template, Meta Language Creation, and Persona prompt patterns cataloged in [17]. The prompt establishes the LLM's role as a business process modeling expert and guides it in identifying key elements within the process description. It also emphasizes on nested structures and process elements to guide the LLM in correctly using the intermediate format. The instructions also encourage the identification of parallel tasks for optimization and the use of clear and specific names for labels and conditions. The prompt was iteratively optimized based on preliminary experiments on process descriptions disjoint from the ones in the evaluation.

You are a business process modeling expert. I will provide you with a textual description of a business process. Generate a JSON model for the process.
Analyze and identify key elements:
1. Start and end events.
2. Tasks and their sequence.
3. Gateways (xor or parallel) and an array of "branches" containing tasks. For xor gateways, there is a condition for the decision point and each branch has a condition label.
4. Loops: involve repeating tasks until a specific condition is met.
Nested structure: The schema uses nested structures within gateways to represent branching paths.
Order matters: The order of elements in the "process" array defines the execution sequence.
When analyzing the process description, identify opportunities to model tasks as parallel whenever possible for optimization (if it does not contradict the user intended sequence).
Use clear names for labels and conditions.
Aim for granular detail (e.g., instead of "Task 1: Action 1 and Action 2", use "Task 1: Action 1" and "Task 2: Action 2").
Sometimes you will be given a previous JSON solution with user instructions to edit.

Listing 1. Instructions prompt for Meta-Model-based process modeling.

The OpenAI API accepts prompts in the form of message arrays. This could be used to include all the previous interactions with the chatbot as a state. However, following our approach, we use a much smaller prompt: For the initial prompt, we send an array starting with a system message containing the instructions prompt, followed by a user message containing the user input. In the feedback loop, we include the criticized model in the intermediate format as an assistant message between the instructions and the user message. In both cases, the specification of the intermediate process format is included in the prompt via a JSON Schema.

Model Transformation: Intermediate Model to BPMN: This step transforms the intermediate process model into a BPMN-XML-compliant model ready for rendering. By purpose, the input model does not contain any graphical representation information. This is included during model transformation. An important usability aspect is that small changes in the input model should also result in small changes in the graphical output model. We achieve this by making use of the block-structured input processes. In the first step, our algorithm deterministically adds the required positioning and size information by traversing the process tree. In particular, the dimensions of the elements are based on their type (event, task, or gateway), and the spacing between them is predefined. The starting coordinates assigned to the start event are initialized to fixed x and y values. The coordinates for each element are then calculated recursively. If an element has branches, the function iterates through them, updating the x and y coordinates accordingly. For each branch, it tracks the maximum height of nested elements. After processing all elements in the branch, the maximum y-coordinate is updated to ensure proper positioning of the following branches. Moreover, intermediate coordinates are calculated for sequence flows, adjusting for vertical and horizontal differences between connected elements.

5   Evaluation

Using our *BPMN-Chatbot* prototype, we assess our method's cost-efficiency and correctness for generating initial models in Sect. 5.1. The full system, including the feedback loop, is evaluated in a preliminary user study in Sect. 5.2.

5.1   Efficiency and Correctness Initial Model Generation

In this experiment, we use the same subset of the PET dataset [1] used in [7]. We have generated outputs for each process description with our *BPMN-Chatbot*, *ProMoAI* [10], and prompt pattern R with Mermaid JS output in [7]. We refer to the latter one as *patternR*. We have chosen *patternR* because this combination showed the best performance for direct model generation from input text without preprocessing in [7]. We have conducted all tests with GPT-4o. Each process model was requested three times by the different approaches. We ran the experiments for the *BPMN-Chatbot* on May 15, 2024. For *ProMoAI* [10], we used the OpenAI Dashboard to assess the number of tokens. We, therefore, executed each test case on a different day. These evaluations were conducted between May 15 and May 23. The experiments with *promptR* were executed on June 7. The dataset and all results are available online[5].

---

[5] https://github.com/BPMN-Chatbot/bpmn-chatbot-archive.

The resulting process models were evaluated anonymously by a commission of two BPMN experts (disjoint from the authors) based on the correctness definition of [7]: *A model is estimated to be correct if all logical aspects of the process description are captured. If some elements are missing, substituted, or new elements are added to the model, and these changes do not violate the information introduced in the process description, the model is considered correct.* Since the colleagues are experienced in assessing student submissions, they additionally graded the models on a 1 (best) to 5 (worst) school grade scale.

Results and Discussion: Regarding the number of tokens, our *BPMN-Chatbot* achieved an average reduction of the number of context tokens by 94% and of the output tokens by 75% compared to the baseline tool *ProMoAI*. Regarding *patternR*, we achieved an average reduction of the number of output tokens of 13% while the number of context tokens was 35% larger. Detailed results are shown in Fig. 4.

Our tool provides a substantial advantage over *ProMoAI*. However, the situation is more complex for *patternR*. According to our experiments, *patternR* is beneficial if the produced process models are small. However, in the case of larger models (see Case 1.3 in Fig. 4), the smaller output format leads to a substantial reduction of the output tokens. We require more context tokens since the schema definition of the output format also accounts for context tokens, while *patternR* makes use of the fact that the Mermaid format is known to GPT-4o. However, with the current pricing politics[6], this does not necessarily lead to a monetary advantage of *patternR* as output tokens cost three times more than context tokens. Moreover, in this experiment, we did not consider the costs of the feedback loop, where the output format is highly relevant, as a previous model is sent in each iteration of the loop.

Table 1 shows the correctness percentage and school grades. Overall, we achieved a correctness rate of 95% compared to 86% for *ProMoAI* and 81% for *patternR*. The high output quality is also reflected in the school grades, with an average grade of 2.09 compared to 2.9 for *ProMoAI* and 2.5 for *patternR*. Notably, in the school grade, *patternR* performs better than *ProMoAI*. One reason is the inability of *ProMoAI* to include expressions for conditional gateways and branches.

5.2   Technology Acceptance and Feedback Loop

We demonstrated the BPMN-Chabot at the public science fair *Lange Nacht der Forschung* in Klagenfurt on May 24th, 2024. This allowed us to gain feedback from visitors of the general public. The attendees were first shown an introductory video or an interactive introduction was provided. The introduction demonstrated the core BPMN control-flow elements and how modelers currently create process models. Afterward, the attendees designed processes using the tool. We chose an open task to avoid the issue of attendees simply entering the predefined

---

[6] https://openai.com/api/pricing/. Last accessed 09.06.2024.

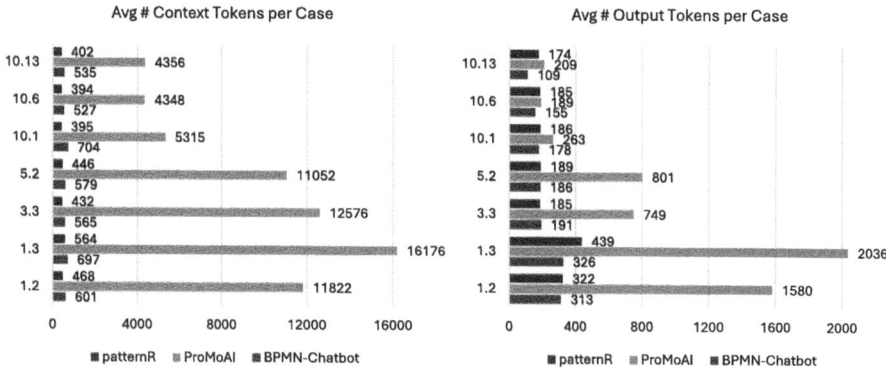

Fig. 4. Comparison of number of context tokens and output tokens

Table 1. Percentage of correctly generated models, avg school grade (1–5)

Case	#Words	#Tasks	BPMN-Chatbot	ProMoAI [10]	patternR [7]
10.13	39	3	(100%, 2.0)	(100%, 2.0)	(100%, 1.0)
10.6	30	4	(100%, 2.0)	(100%, 2.0)	(100%, 2.3)
10.1	29	4	(100%, 2.0)	(100%, 2.0)	(100%, 2.0)
5.2	83	7	(100%, 1.0)	(67%, 2.0)	(100%, 2.7)
3.3	71	7	(100%, 2.0)	(67%, 3.7)	(100%, 1.33)
1.3	162	11	(100%, 2.3)	(67%, 4.0)	(0%, 5.0)
1.2	100	10	(67%, 3.3)	(100%, 3.3)	(67%, 3.3)
Average			(95.24%, 2.09)	(86%, 2.9)	(81%, 2.5)

task description. After completing the task, we invited participants to complete a short questionnaire and donate their conversations for research. 40 Participants completed the questionnaire out of the total 76 conversations we collected.

Due to the public setting, we limited the questions on the demographics to their usage pattern of PC/Mac computers and their knowledge of business processes. 60% of the participants identified themselves as daily PC users, 22.5% as weekly users, and 17.5% as occasional users. 32.5% saw business processes for the first time in our experiment, 20% had seen business process models before, while 15% had previous knowledge from their work environment. Another 15% had already created business processes, with 12.5% creating them regularly. For 5%, no answer was available.

The questions, average results, and standard deviation (SD) are shown in Table 2. We used a 1 (strongly disagree) to 5 (strongly agree) Likert scale. Our questionnaire covered two aspects: On the one hand, we wanted to know if the prototype provided useful answers in general (Q2) and if, in particular, the answers in the feedback loop were useful (Q3). Question (Q1) asked if the participants completed their task with the tool successfully. On the other hand, we wanted to assess the general user acceptance. We have therefore opted for a technology acceptance test [2]. Since we surveyed people from the general public, we focused on perceived usefulness, attitude, and intention of use. We used three questions (Q5-Q7) to assess the perceived usefulness and report on the average values. The attitude and intention of use were assessed by one question (Q4) and (Q8), respectively. Since the original user acceptance questions target professional users, not general attendees, we have rephrased the questions for this scenario.

Table 2. Questionnaire results of the experiment at the science fair.

Category	Statement	Rating	SD
Usefulness of Response	(Q1) I successfully modeled my business process with the BPMN-Chatbot	4.15	0.88
	(Q2) The models created by the BPMN-Chatbot were helpful	4.18	0.86
	(Q3) The BPMN-Chatbot responded well to my feedback	4.25	0.8
	Usefulness of Response Average	*4.19*	*0.85*
Attitude	(Q4) It was fun using the BPMN-Chatbot	4.7	0.50
Perceived Usefulness	(Q5) I think the BPMN-Chatbot makes modeling business processes easier	4.45	0.67
	(Q6) I think the BPMN-Chatbot increases productivity in modeling business processes	4.25	0.80
	(Q7) I think the BPMN-Chatbot is useful for modeling business processes	4.5	0.67
	Perceived Usefulness Average	*4.4*	*0.71*
Intention of Use	(Q8) If I could use the BPMN-Chatbot, I would	4.28	1.02

Results and Discussion: Regarding the usefulness of response questions, we obtained 4.25 for the feedback loop (Q3) and 4.18 for the general answers provided by the system (Q2). For task completion, a score of 4.15 (Q1) was reached. Since 27.5% of the participants had modeled business processes before, we also report on this subgroup. Here, the scores were even better, with 4.55 (Q3) for the feedback loop, 4.64 for the answers (Q2), and 4.36 for task completion (Q1). We see this as a strong indication that the tool responds very well to user input in general, and the answers during the feedback loop were also considered very good. This clearly shows that our proposal for efficiently implementing the feedback loop using only a previous process model as the state still leads to highly valued responses.

Regarding the technology acceptance questions, the tool scored overall 4.4 for perceived usefulness, 4.7 for attitude, and 4.28 for the intention of use. In the subgroup of participants with at least some process modeling experience, the tool reached even better values with 4.7 for perceived usefulness, 4.82 for attitude, and 4.73 for intention of use. For novice users, the tool scored 4.28 for perceived usefulness, 4.63 for attitude, and 4.1 for intention of use. These results indicate substantial acceptance of the tool by participants with at least some process modeling experience. We still see the slightly lower numbers of novice users as a strong indication that LLM-based conversational process modeling has the potential to attract more people to process modeling.

Threats to Validity: The setting at the public science fair allowed us to collect feedback from a wide audience. Overall, it was very well received, and the participants enjoyed using it in a recreational setting. However, for a full technology acceptance test, a more defined setting is required, where people spend more time with the tool. Also, the fully open task may have led to higher acceptance scores as participants were not constrained to a fixed modeling goal. Therefore, we see the numbers only as a strong indication of the overall tool's usefulness. Regarding the experiments of the initial models, we have some randomness based on the nature of LLMs and uncontrolled factors. Therefore, we see the results as a strong indication of high output quality.

6 Conclusion and Future Works

LLM-based conversational process modeling systems should use LLMs resources efficiently. We have identified the core cost drivers for LLM-based conversational process modeling and introduced a meta-model for intermediate process models. We have instantiated the approach with our publicly available *BPMN-Chatbot*. The tool was evaluated with two experiments. In the first experiment, we compared the number of tokens and the quality of the solutions for initial models. The results show substantial reductions in the number of tokens (up to 94% compared to an existing tool) while achieving 95% correct models with an average school grade of 2.09 compared to 86% and 2.5 for the best competitors. An additional preliminary user study assessed the quality of the responses in the feedback loop and the overall user acceptance. The results indicate a strong acceptance and satisfying responses in the feedback loop.

Future works include an extension of the meta-model with more process elements, a technology acceptance test with modeling experts, experiments on the correctness with more input processes, and a larger jury. An evaluation with alternative and open-source LLMs is also considered relevant for future work.

References

1. Bellan, P., et al.: Process extraction from natural language text: the PET dataset and annotation guidelines. In: Proceedings of NL4AI'2022. CEUR Workshop Proceedings, vol. 3287, pp. 177–191. CEUR-WS.org (2022). https://ceur-ws.org/Vol-3287/paper18.pdf

2. Davis, F., Davis, F.: Perceived usefulness, perceived ease of use, and user acceptance of information technology. MIS Q. **13**, 319 (1989). https://doi.org/10.2307/249008

3. Fill, H., Fettke, P., Köpke, J.: Conceptual modeling and large language models: impressions from first experiments with ChatGPT. Enterp. Model. Inf. Syst. Archit. Int. J. Concept. Model. **18**, 3 (2023). https://doi.org/10.18417/EMISA.18.3

4. Forell, M., Schüler, S.: Modeling meets large language models. In: Modellierung 2024 Satellite Events (2024).https://doi.org/10.18420/modellierung2024-ws-003

5. Grohs, M., Abb, L., Elsayed, N., Rehse, J.R.: Large language models can accomplish business process management tasks. In: Business Process Management Workshops, pp. 453–465. Springer, Cham (2024)

6. Klievtsova, N., Benzin, J.V., Kampik, T., Mangler, J., Rinderle-Ma, S.: Conversational process modelling: state of the art, applications, and implications in practice. In: International Conference on Business Process Management, pp. 319–336. Springer (2023)

7. Klievtsova, N., Benzin, J.V., Kampik, T., Mangler, J., Rinderle-Ma, S.: Conversational process modeling: can generative ai empower domain experts in creating and redesigning process models? (2024). https://arxiv.org/abs/2304.11065v2

8. Kopp, O., Martin, D., Wutke, D., Leymann, F.: The difference between graph-based and block-structured business process modelling languages. Enterp. Model. Inf. Syst. Archit. Int. J. Concept. Model. **4**(1), 3–13 (2009).https://doi.org/10.18417/EMISA.4.1.1

9. Kourani, H., Berti, A., Schuster, D., van der Aalst, W.M.P.: Process modeling-withÂ large language models. In: Enterprise, Business-Process and Information Systems Modeling, pp. 229–244. Springer, Cham (2024)

10. Kourani, H., Berti, A., Schuster, D., van der Aalst, W.M.: ProMoAI: process modeling with generative AI. In: Proceedings of IJCAI-24, pp. 8708–8712. IJCAI Organization (2024). https://doi.org/10.24963/ijcai.2024/1014. Demo Track

11. Kourani, H., van Zelst, S.J.: POWL: partially ordered workflow language. In: Di Francescomarino, C., Burattin, A., Janiesch, C., Sadiq, S. (eds.) Business Process Management, pp. 92–108. Springer, Cham (2023)

12. OpenAI: GPT-4 technical report (2024). https://arxiv.org/abs/2303.08774

13. Patel, P., et al.: Splitwise· efficient generative LLM inference using phase splitting. Power **400**(700W), 1–75 (2023)

14. Samsi, S., et al.: From words to watts: benchmarking the energy costs of large language model inference (2023). https://arxiv.org/pdf/2310.03003

15. Schüler, S., Alpers, S.: State of the art: automatic generation of business process models. In: Business Process Management Workshops. BPM 2023, pp. 161–173. Springer (2024)

16. Vidgof, M., Bachhofner, S., Mendling, J.: Large language models for business process management: opportunities and challenges. In: Business Process Management Forum, pp. 107–123. Springer, Cham (2023)

17. White, J., et al.: A prompt pattern catalog to enhance prompt engineering with ChatGPT (2023)

18. Yu, L., Li, J.: Stateful large language model serving with pensieve (2024). https://arxiv.org/pdf/2312.05516

Leveraging Generative Vision Models for Extracting Process Models from Documents

Marvin Voelter[1,2(✉)], Raheleh Hadian[1], Timotheus Kampik[1],
Marius Breitmayer[2], and Manfred Reichert[2]

[1] SAP, Berlin, Germany
{marvin.voelter,raheleh.hadian,timotheus.kampik}@sap.com
[2] Ulm University,Ulm, Germany
{marius.breitmayer,manfred.reichert}@uni-ulm.de

Abstract. This paper investigates the vision capabilities of multimodal Generative Pre-trained Transformers (GPTs) to auto-generate structured process models from diagram- and text-based documents. We introduce a dataset of 123 process models and corresponding documentation, emphasizing real-world element distributions. Using evaluation metrics for process model similarity, this enables ground truth-based assessment of process model generation. We evaluate commercial GPT capabilities with zero-, one-, and few-shot prompting strategies. Our results indicate that generative vision models can be useful tools for semi-automated process modeling based on multimodal documents. More importantly, the dataset and evaluation metrics as well as the open-source evaluation code provide a structured framework for continued systematic evaluations moving forward.

Keywords: Generative Vision Models · Multimodal Large Language Models · Document Analysis · Business Process Management

1 Introduction

With the advent of generative artificial intelligence, the auto-generation of structured process models based on unstructured input data has re-surged as a line of research [3,4,10,14,17,25]. Here, the main input modality is text. However, unstructured process models are in practice often persisted in image-based or image-like formats (e.g., PNG or PDF files) containing flow charts as well as corresponding textual information (e.g., a process description). Hence, leveraging the vision capabilities of multimodal generative models to handle image-like formats is potentially promising. In contrast to existing traditional approaches, where the image modality is handled using a Convolutional Neural Network (CNN) (cf. [1,12,23,24]), generative multimodal models offer several advantages: First, generative AI approaches, and Large Language Models (LLMs) in particular, are known for their transfer learning capabilities. The latter allow handling

© The Author(s), under exclusive license to Springer Nature Switzerland AG 2025
K. Gdowska et al. (Eds.): BPM 2024 Workshops, LNBIP 534, pp. 271–282, 2025.
https://doi.org/10.1007/978-3-031-78666-2_21

different tasks, for example, various types of diagrams, better compared to traditional CNNs trained for one specific purpose [6,11]. Further, it allows an "off-the-shelf" usage of foundational models without the need of effortful training. In addition, only one model is required for multiple modalities, reducing both maintenance and development efforts.

Therefore, this paper presents an investigation of using the vision capabilities of multimodal GPTs to auto-generate structured process models from graphical figures with surrounding texts. We provide a systematic evaluation approach to ensure a scientific comparison and consider future model developments and improvements. We introduce a dataset with 123 models reflecting real-world element distributions, each matched to a PDF-based process documentation artifact. We then evaluate the commercial GPT capabilities using ground truth-based metrics by passing the documents as images using zero, one-shot and few-shot prompting strategies, and discuss their implications.

2   Background and Related Work

Applying LLMs in the BPM domain is considered a promising research direction. Vidgof et al. suggest using LLMs to produce process models from documentations [28]. Van der Aa et al. discuss opportunities of NLP in BPM [15]. Beheshti et al. and Rizk et al. recommend training LLMs on business process data [2,22]. These position papers lack detailed feasibility assessments.

A nascent body of research exists regarding the generation of process models from text with NLP in general and GPTs/LLMs in particular. Klievtsova et al. evaluated application scenarios for modeling business processes with LLMs based on textual conversations on a real-world data set [17]. Bilal et al. found 11 NLP and 8 BPMN tools for generating models from textual requirements utilizing NLP [19]. Friedrich et al. and Sholiq et al. evaluated NLP techniques to generate process models from natural language [10,25]. Bellan et al. use LLMs for extracting business process entities and relations from texts [3,4]. Grohs et al. apply LLMs to the generation of both imperative and declarative process models from textual descriptions [14]. While these papers process textual descriptions they are not able to handle image-like inputs.

Regarding the use of images to generate process models, Schaefer et al. introduce a CNN model to create BPMN models from hand-drawn sketches [23,24]. Kang et al. transform BPMN business process images into Petri nets with traditional deep learning techniques [16]. Antinori et al. recreate business process models from BPMN images using traditional object recognition and Optical Character Recognition (OCR) methods [1]. Gantayat et al. create BPMN models from images using traditional CNNs [12]. In contrast to these papers, our approach has two main differences: First, we use a generative approach instead of the discussed traditional CNN approaches. This promises greater versatility since an LLM can perform multiple tasks using fundamental reasoning skills without being exclusively trained on a specific type of image or text format. It allows using foundation models that do not require effortful special-purpose training. Second, we explore a more realistic use case where both graphical figures and textual process descriptions are combined within documents. In the papers discussed above, only the figures were considered and not embedded into documents.

To the best of our knowledge, no previous work has investigated the vision capabilities of multimodal generative models to auto-generate structured process models from intersected image- and text-like documents in detail. The paper aims to contribute towards closing this gap.

3   Method

We present both a dataset and corresponding evaluation metrics for ground truth-based assessment of the process model generation. Subsequently, we employ commercial GPT capabilities for the generation process. The complete source code, including the dataset, evaluation metrics, and generation part, is publicly available[1].

3.1   Dataset

Motivation. Although several datasets of process models [13,26], model-text pairs [5,17,18,21] and model-image pairs [1,23] exist, none of them provide documents combining graphical figures and textual process descriptions with corresponding ground truths. To ensure systematic and more realistic evaluations, we therefore introduce a new data set.

Creation. A subset of the SAP-SAM dataset, which contains several hundred thousands of process models created by researchers, teachers, and students [26], is selected by filtering for high-quality and representative models. These process models are then parsed into a simplified JSON format and, consequently, serve as ground truth process models. Afterwards, the cleaned models are used to generate multimodal process documentation in a PDF format with the SAP Signavio Process Manager, a commercial process management tool[2]. The resulting PDFs are then converted to corresponding PNG replicas.

Characteristics. Overall, the dataset comprises 123 models (2,56 times larger than the current gold standard for process description [5]), including original SAP-SAM data, multimodal process documentation in both PDF and PNG formats as well as the ground truth in JSON format. In a nutshell, the documentation comprises a cover page, a table of contents, the meta-information, a BPMN diagram, and short descriptions for each element. Depending on the process model, the documentation spans between 4–18 pages. Typically, the textual

---

[1] https://github.com/SAP-samples/multimodal-generative-ai-for-bpm.
[2] Cf. https://www.signavio.com/products/process-manager/; accessed at 23-03-2023.

descriptions match the visual models, while occasionally additional information or a brief overview is provided.

The JSON ground truth schema consists of different objects, each referenced by corresponding IDs. Objects may represent a task, an event, a gateway, a pool, a lane, a message flow, or a sequence flow. Tasks, events, and gateways have a name and a type, while the name is optional for gateways. A pool has a name and can have multiple lanes. A lane has a name and can contain multiple referenceable objects. A message flow has an optional label and references exactly one source and one target object. Sequence flows are defined equivalently to message flows, although the label attribute is called a *condition*. Overall, the schema covers the most used BPMN elements. It includes all elements used in more than 25% of publicly available repositories examined by Compagnucci et al. [7, pp. 87–88]. Covering all elements used in more than 3% of models, the schema can be extended by adding new lists for associations and object elements with the respective types: data object, text annotation, message, group, and a flow type for default flow.

The number of ground truth elements in the dataset varies between 13 and 88, with an average of 39. This illustrates the adherence of the model to the guideline of limiting elements to 50 per model, suggesting high quality [20]. Predominantly, tasks and sequence flows constitute the elements, featuring 10 task types with abstract and manual tasks being most common. Moreover, the dataset includes 17 event types, notably *end none* and *start none* events, and identifies four gateway types, with exclusive and parallel gateways being favored. These findings closely match the distributions reported by Compagnucci et al. [7, pp. 87–88], demonstrating real-world representativeness.

3.2   Evaluation Framework

Element-Based Breakdown. First, the ground truth and to-be-evaluated models (generated models) are broken down into individual elements. We define the following multisets: TN as task names, TT as task types, EN as event names, ET as event types, GN as gateway names and GT as gateway types. Further, we define LN as a multiset of lane names. A lane name is defined as a tuple (p, l) where p is the pool label and l is the label of the lane. And LR is a multiset of lanes with reference. A lane with reference is a triple (p, l, r) where p is the pool label, l is the label of the lane, and $r \in TN \cup EN \cup GT$. GT is chosen over GN because the names of the gateways are optional, while the type is mandatory. In addition, let SF be a multiset of sequence flows. A sequence flow is defined as a triple (s, c, t) where $s, t \in TN \cup EN \cup GT \cup LN$ and c provides a name for the sequence flow. Equivalent, let MF be a multiset of message flows. A message flow is defined as a triple (s, l, t) where $s, t \in TN \cup EN \cup GT \cup LN$ and l provides a name for the message flow. The model is then described as a joint set of elements:

$$M = \{TN, TT, EN, ET, GN, GT, LN, LR, SF, MF\} \tag{1}$$

As a short-hand, the elements are referred to as E_i:

$$M = \{E_i : 1 \leq i \leq 10\} \tag{2}$$

Score Calculation. Next, the similarity scores of the two models are calculated. Let $M_1 = \{E_{i,1} : 1 \leq i \leq 10\}$ be the ground truth model and $M_2 = \{E_{i,2} : 1 \leq i \leq 10\}$ be the generated model. The overall similarity score is calculated as follows:

$$\text{sim}(M_1, M_2) = \frac{\sum_{i=1}^{10} w_i \cdot dice_{SFA}(E_{i,1}, E_{i,2})}{\sum_{i=1}^{10} w_i} \quad \text{with } w_i = |E_{i,1}| + |E_{i,2}| \tag{3}$$

where $dice_{SFA}(E_{i,1}, E_{i,2})$ is an element-wise adjusted Sørensen–Dice coefficient and the weights are element cardinalities. The original Sørensen–Dice coefficient is defined as [8,27]:

$$dice(E_{i,1}, E_{i,2}) = \frac{2|E_{i,1} \cap E_{i,2}|}{|E_{i,1}| + |E_{i,2}|} \tag{4}$$

Further aggregated scores for tasks (TN and TT), events (EN and ET), gateways (GN and GT), flows (SF and MF) and lanes (LN and LR) are calculated analogously to the overall score.

$dice_{SFA}$. $dice_{SFA}$ is a semantic- and frequency-aware adjustment of the Sørensen–Dice coefficient to calculate the similarity of two lists. First, semantically similar items from the two lists are matched, even when syntactically dissimilar. To achieve that, a sentence transformer model (stsb-mpnet-base-v2) for item vectorization is employed and items are considered similar if their cosine similarity exceeds a 0.7 threshold. Note that this threshold has been determined experimentally and may be adjusted for different contexts. Items are only matched once, prioritizing matches with the highest similarity first. Next, indexing items enables the Sørensen–Dice coefficient to process multisets by converting them into sets with preserved entries. Finally, the original Sørensen–Dice coefficient is calculated. Note that the corresponding source code of the evaluation framework is provided in GitHub[3].

3.3   Generation Process

We leverage the vision capability of a multimodal LLM to generate process models from the process documentation in the dataset. We decided on GPT-4V (gpt-4-0125-preview, default temperature) because it performed best in benchmarks of the following categories: existence recognition, element counting, position understanding, OCR, common sense reasoning, and code reasoning [11].

Listing 3.3 describes the meta prompt used for the LLM and follows different prompting strategies. First, it applies priming: "You are a BPMN expert."

---

[3] https://github.com/SAP-samples/multimodal-generative-ai-for-bpm.

Then, it clearly and precisely describes the to-be-performed task: extracting the process information from the passed document. Next, the prompt constrains the output to only generate the JSON according to the passed schema. The schema corresponds to the defined ground truth schema. The meta prompt has been refined based on iterative experiments. For example, explicitly mentioning to add the sequence and message flow resulted from them often being empty. We generated the models in three different scenarios [6]:

1. Zero-Shot Prompting: Giving the LLM only the meta prompt and no example at all to perform the task.
2. One-Shot Prompting: Providing the LLM the meta prompt and one example to perform the task.
3. Few-Shot Prompting: Providing the LLM the meta prompt and three examples to perform the task.

The first three models of the dataset are used as examples of the one- and few-shot approach. Each approach is then applied to the remaining 120 models.

```
meta_prompt = """### Instruction ###
    You are a BPMN expert. Your task is to extract process
        information out of
    the passed documents which are parsed as a list of
        images where each image
    represents one page of the document. Make sure that you
        include the
    sequence and message flow. Use numbers for the ids
        starting from zero.
    Generate json according to the following schema for
        extracting the process
    information. Only output the generated json.

    ### Schema ###
    """ + str(bpmn_schema)
```

Listing 1.1. Meta prompt

4   Results

Figure 1 visualizes the score distributions with medians as the central tendency along with shaded areas between 25% and 75% quantiles which represent score variability. We have chosen to show the medians and quantiles because the score distributions are skewed. Additionally, Table 1 lists the score averages. The overall score median for the one-shot scenario is 89%, for the few-shot scenario 87%, and 83% for the zero-shot scenario. For all approaches, the score for task names, task types, and event types is higher than the overall score, while the event names are slightly below it with larger variability. The largest variability is seen in gateway names which also has the lowest similarity score. Message flow has higher score variability than sequence flow. The overall score for flows lies slightly below the overall score for all elements.

The performance of the one-shot and few-shot scenario is very similar, and both significantly outperform the zero-shot scenario and reduce variability in scores. On average, the difference between one-shot and few-shot is below 1 percent point, while the difference to few-shot is around 7 - 8 percent points.

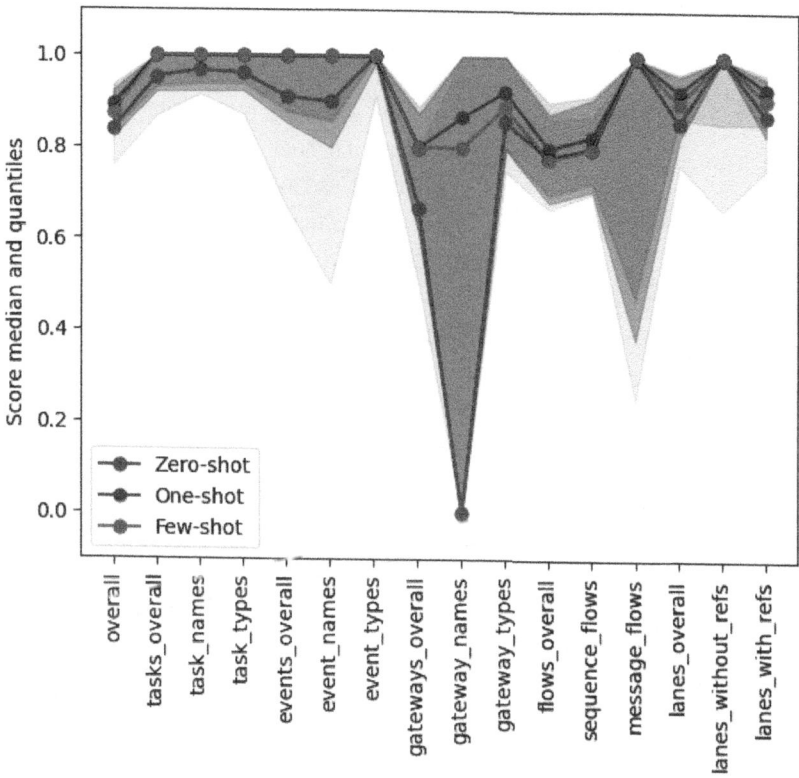

Fig. 1. Median scores and quantiles for zero-, one-, and few-shots.

5 Discussion

5.1 Implications

The results, with an average overall similarity score of 87 percent for the one-shot approach, indicate that creating process models from documents using generative vision models is feasible. However, human feedback might still be required to improve the results. Our experiments provide first evidence that using the vision capability of a multimodal GPT can be an alternative to processing

Table 1. Comparison of average zero-, one-, and few-shot scores.

Score Name	Zero-Shot	One-Shot	Few-Shot
Overall	0.812654	0.871079	0.861744
Tasks Overall	0.889146	0.939833	0.944912
Task Names	0.919780	0.958201	0.946046
Task Types	0.858431	0.922100	0.944000
Events Overall	0.814431	0.908772	0.893474
Event Names	0.697021	0.865807	0.848409
Event Types	0.924640	0.945658	0.932955
Gateways Overall	0.609449	0.696482	0.686032
Gateway Names	0.284444	0.575058	0.601764
Gateway Types	0.856959	0.883706	0.865670
Flows Overall	0.736907	0.776942	0.751045
Sequence Flows	0.750651	0.792892	0.770219
Message Flows	0.682600	0.723168	0.696521
Lanes Overall	0.813442	0.881780	0.868453
Lane Names	0.825298	0.872698	0.878962
Lane Refs	0.807339	0.877842	0.859720

text and images separately for process modeling. Processing the information together is a realistic use case and allows the GPT to access more information to improve the performance. Further, the proposed GPT approach requires only one machine learning model that may then also be used for alternative tasks, thus reducing both development and maintenance efforts. In addition, the GPT does not require effortful special-purpose training and can be used "off-the-shelf", unlike traditional CNN approaches that have been used for image processing in the past. We expect that future enhancements, including the introduction of function-calling capabilities and advancements in models like Gemini 1.5, along with optimized prompting, can further solidify evidence regarding the feasibility of multimodal GPTs for automated process model generation based on unstructured documents.

For the task at hand, GPTs can effectively handle JSON schemas and one-shot prompting often suffices without additional examples to significantly improving outcomes compared to zero-shot prompting. While GPTs excel at identifying tasks and events, they struggle with flows, possibly due to relational complexity. We further recognize challenges related to identifying gateway labels. A possible reason might be that the presence of gateway labels is optional and label placement varies substantially.

5.2   Limitations

Several factors limit the results of the presented study. First, the relative contribution of the diagram images and textual information to the overall score has not been analyzed. Second, the study exclusively utilized one GPT, limiting insights into potential variations in performance or challenges with other models. Rapid advancements in technology, evidenced by the release of GPT-4V and Gemini 1.5 during the paper-writing period, highlight the dynamic nature of the field. A stable benchmarking environment was established to address this, facilitating future technological comparisons. Third, resource constraints prevented fine-tuning strategies, and prompt optimization could further be enhanced with additional effort. Finally, while empirically somewhat robust, the evaluation methodology is imperfect. Two specific issues arise: First, employing a sentence transformer model for similarity assessments with a fixed threshold may lead to potential deviations from linearity. Second, set overlap results in unequal weighting of elements, presenting an additional challenge due to the ambiguity in determining the appropriate priority or weight for each element. Despite these intricacies, the empirical data consistently affirm the methodology's intuitive precision and evaluation uniformity. However, future work can potentially further enhance the evaluation approach, for example by considering existing research on process model similarity assessment [9].

6   Conclusion

The paper presented an investigation of the vision capabilities of multimodal GPTs to auto-generate structured process models from graphical figures with surrounding text. We first created a dataset of 123 models (2.56 times larger than the current gold standard for process description [5]) containing both process documentations and ground truths. It indicates high real-world element-distributions and is. Then, we introduced a ground truth-based evaluation framework incorporating an element-wise breakdown, semantic and frequency awareness as well as the Sørensen–Dice coefficient. We evaluated GPT-4V capabilities using zero-, one-, and few-shot prompting strategies. Our results, with an average evaluation score of 87%, indicated that GPTs can be useful tools for semi-automated process modeling based on visual documents. In addition, the dataset, the evaluation metrics, and the open-source evaluation code provide a structured framework for continued systematic evaluations moving forward.

Future research should benchmark text-only and CNN based image-only models against the dataset to evaluate their performance relative to the current method. Additionally, analyzing the relative contribution of text and image information to the generated process model could be valuable. Extending the ground truth to include a wider array of BPMN elements could further improve generated models. Expanding the dataset to encompass a broader variety of diagrams, images, and textual descriptions could also enrich the benchmarking possibilities. Moreover, new models and approaches can be benchmarked using the introduced dataset. Exploring chain-of-thought prompting, function calling, and fine-tuning may also be promising, though resource-intensive.

References

1. Antinori, A., Coltrinari, R., Corradini, F., Fornari, F., Re, B., Scarpetta, M.: BPMN-redrawer: From images to BPMN models. In: Janiesch, C., et al. (eds.) Proceedings of the Best Dissertation Award, Doctoral Consortium, and Demonstration & Resources Track at BPM 2022 co-located with 20th International Conference on Business Process Management (BPM 2022), Münster, Germany, September 11th to 16th, 2022. CEUR Workshop Proceedings, vol. 3216, pp. 107–111. CEUR-WS.org (2022). https://ceur-ws.org/Vol-3216/paper_246.pdf
2. Beheshti, A., et al.: ProcessGPT: transforming business process management with generative artificial intelligence. In: Ardagna, C.A., et al. (eds.) IEEE International Conference on Web Services, ICWS 2023, Chicago, IL, USA, July 2-8, 2023, pp. 731–739. IEEE (2023). https://doi.org/10.1109/ICWS60048.2023.00099
3. Bellan, P., Dragoni, M., Ghidini, C.: Extracting business process entities and relations from text using pre-trained language models and in-context learning. In: Almeida, J.P.A., Karastoyanova, D., Guizzardi, G., Montali, M., Maggi, F.M., Fonseca, C.M. (eds.) Enterprise Design, Operations, and Computing: 26th International Conference, EDOC 2022, Bozen-Bolzano, Italy, October 3-7, 2022, Proceedings, Lecture Notes in Computer Science, vol. 13585, pp. 182–199. Springer, Cham (2022).https://doi.org/10.1007/978-3-031-17604-3_11
4. Bellan, P., Dragoni, M., Ghidini, C.: Leveraging pre-trained language models for conversational information seeking from text. CoRR **abs/2204.03542** (2022).https://doi.org/10.48550/ARXIV.2204.03542
5. Bellan, P., van der Aa, H., Dragoni, M., Ghidini, C., Ponzetto, S.P.: PET: an annotated dataset for process extraction from natural language text tasks. In: BUSINESS PROCESS MANAGEMENT WORKSHOPS, pp. 315–321. SPRINGER INTERNATIONAL PU, [S.l.] (2023).https://doi.org/10.1007/978-3-031-25383-6_23
6. Brown, T.B., et al.: Language models are few-shot learners. In: Larochelle, H., Ranzato, M., Hadsell, R., Balcan, M., Lin, H. (eds.) Advances in Neural Information Processing Systems 33: Annual Conference on Neural Information Processing Systems 2020, NeurIPS 2020, December 6-12, 2020, virtual (2020). https://proceedings.neurips.cc/paper/2020/hash/1457c0d6bfcb4967418bfb8ac142f64a-Abstract.html
7. Compagnucci, I., Corradini, F., Fornari, F., Re, B.: Trends on the usage of BPMN 2.0 from publicly available repositories. In: Buchmann, R.A., Polini, A., Johansson, B., Karagiannis, D. (eds.) BIR 2021. LNBIP, vol. 430, pp. 84–99. Springer, Cham (2021). https://doi.org/10.1007/978-3-030-87205-2_6
8. Dice, L.R.: Measures of the amount of ecologic association between species. Ecology **26**(3), 297–302 (1945). https://doi.org/10.2307/1932409, http://www.jstor.org/table/1932409
9. Dijkman, R., Dumas, M., van Dongen, B., Käärik, R., Mendling, J.: Similarity of business process models: Metrics and evaluation. Inf. Syst. **36**(2), 498–516 (2011).https://doi.org/10.1016/j.is.2010.09.006, https://www.sciencedirect.com/science/article/pii/S0306437910001006, special Issue: Semantic Integration of Data, Multimedia, and Services

10. Friedrich, F., Mendling, J., Puhlmann, F.: Process model generation from natural language text. In: Mouratidis, H., Rolland, C. (eds.) CAiSE 2011. LNCS, vol. 6741, pp. 482–496. Springer, Heidelberg (2011). https://doi.org/10.1007/978-3-642-21640-4_36

11. Fu, C., et al.: MME: a comprehensive evaluation benchmark for multimodal large language models. https://arxiv.org/pdf/2306.13394.pdf

12. Gantayat, N., Sridhara, G., Sankaran, A., Dechu, S., Mani, S., Dasgupta, G.B.: Towards creating business process models from images. In: Pahl, C., Vukovic, M., Yin, J., Yu, Q. (eds.) ICSOC 2018. LNCS, vol. 11236, pp. 100–108. Springer, Cham (2018). https://doi.org/10.1007/978-3-030-03596-9_7

13. GitHub: CAMUNDA/BPMN-for-research: a collection of BPMN diagrams that can be used for research (2024-02-20). https://github.com/camunda/bpmn-for-research

14. Grohs, M., Abb, L., Elsayed, N., Rehse, J.R.: Large language models can accomplish business process management tasks. In: de Weerdt, J., Pufahl, L. (eds.) Business Process Management Workshops, Lecture Notes in Business Information Processing, vol. 492, pp. 453–465. Springer, [S.l.] (2024).https://doi.org/10.1007/978-3-031-50974-2_34

15. HVan der Aa, H., Carmona Vargas, J., Leopold, H., Mendling, J., Padro, L.: Challenges and opportunities of applying natural language processing in business process management. In: Proceedings of the 27th International Conference on Computational Linguistics, pp. 2791–2801 (2018), https://aclanthology.org/C18-1236/

16. Kang, Q., Yu, Y., Dai, F.: Automatic recognition of business process images. In: 2019 IEEE International Conference on Computer Science and Educational Informatization (CSEI), pp. 104–113. IEEE (2019). https://doi.org/10.1109/CSEI47661.2019.8938949

17. Klievtsova, N., Benzin, J.V., Kampik, T., Mangler, J., Rinderle-Ma, S.: Conversational process modelling: state of the art, applications, and implications in practice. In: Di Francescomarino, C. (ed.) Business Process Management Forum : BPM 2023 Forum, Utrecht, the Netherlands, September 11-15, 2023, Proceedings, pp. 319–336. Springer (2023).https://doi.org/10.1007/978-3-031-41623-1-19

18. Li, X., Ni, L., Li, R., Liu, J., Zhang, M.: Mad: A dataset for interview-based bpm in business process management. In: 2023 International Joint Conference on Neural Networks (IJCNN), pp. 1–8. IEEE (2023)

19. Maqbool, B., et al.: A comprehensive investigation of BPMN models generation from textual requirements—techniques, tools and trends. In: Kim, K.J., Baek, N. (eds.) ICISA 2018. LNEE, vol. 514, pp. 543–557. Springer, Singapore (2019). https://doi.org/10.1007/978-981-13-1056-0_54

20. Mendling, J., Reijers, H., van der Aalst, W.M.P.: Seven process modeling guidelines (7PMG). Inf. Softw. Technol. 52(2), 127–136 (2010).https://doi.org/10.1016/j.infsof.2009.08.004, https://www.sciencedirect.com/science/article/pii/S0950584909001268

21. Nataliia Klievtsova: Nataliia / conversational_modeling · gitlab (2024-02-20). https://gitlab.com/Klievtsova/conversational_modeling

22. Rizk, Y., Venkateswaran, P., Isahagian, V., Narcomey, A., Muthusamy, V.: A case for business process-specific foundation models. In: de Weerdt, J., Pufahl, L. (eds.) BUSINESS PROCESS MANAGEMENT WORKSHOPS, Lecture Notes in Business Information Processing, vol. 492, pp. 44–56. Springer, [S.l.] (2024).https://doi.org/10.1007/978-3-031-50974-2_4

23. Schäfer, B., van der Aa, H., Leopold, H., Stuckenschmidt, H.: Sketch2process: End-to-end BPMN sketch recognition based on neural networks. IEEE Trans. Software Eng. **49**(4), 2621–2641 (2023). https://doi.org/10.1109/TSE.2022.3228308

24. Schäfer, B., van der Aa, H., Leopold, H., Stuckenschmidt, H.: Sketch2BPMN: automatic recognition of hand-drawn BPMN models. In: Advanced Information Systems Engineering : 33rd International Conference, CAiSE 2021, Melbourne, VIC, Australia, June 28 - July 2, 2021 : Proceedings, pp. 344–360. Springer (2021).https://doi.org/10.1007/978-3-030-79382-1_21

25. Sholiq, S., Sarno, R., Astuti, E.S.: Generating BPMN diagram from textual requirements. J. King Saud Univ. - Comput. Inf. Sci. **34**(10), 10079–10093 (2022). https://doi.org/10.1016/j.jksuci.2022.10.007, https://www.sciencedirect.com/science/article/pii/S1319157822003585

26. Sola, D., Warmuth, C., Schäfer, B., Badakhshan, P., Rehse, J., Kampik, T.: SAP signavio academic models: a large process model dataset. In: Montali, M., Senderovich, A., Weidlich, M. (eds.) Process Mining Workshops - ICPM 2022 International Workshops, Bozen-Bolzano, Italy, October 23-28, 2022, Revised Selected Papers. Lecture Notes in Business Information Processing, vol. 468, pp. 453–465. Springer (2022).https://doi.org/10.1007/978-3-031-27815-0_33

27. Sørensen, T.: A method of establishing groups of equal amplitude in plant sociology based on similarity of species content and its application to analyses of the vegetation on Danish commons. Biologiske skrifter **5**, 1–34 (1948)

28. Vidgof, M., Bachhofner, S., Mendling, J.: Large language models for business process management: opportunities and challenges. In: Francescomarino, C.D., Burattin, A., Janiesch, ., Sadiq, S.W. (eds.) Business Process Management Forum - BPM 2023 Forum, Utrecht, The Netherlands, September 11-15, 2023, Proceedings. Lecture Notes in Business Information Processing, vol. 490, pp. 107–123. Springer (2023).https://doi.org/10.1007/978-3-031-41623-1_7

2nd International Workshop on Object-centric Processes from A to Z (OBJECTS 2024)

2nd International Workshop on Object-centric Processes from A to Z (OBJECTS 2024)

The topic of object-centric processes has been gaining momentum in the last few years, with many works addressing foundational and practical problems on the interplay of processes and objects, where behaviour arises from the complex interplay among multiple business objects and their one-to-many/many-to-many relationships. Despite the surging number of results on the topic, many related problems have not yet been addressed. One such problem relates to correct modelling and analysis of such processes, where a suitable trade-off between expressiveness and feasibility of analytic techniques must be found. Another problem concerns how object-centric processes can be handled in Process Mining (PM), including novel, efficient PM techniques and suitable event data/log formats to operate over complex event data for such processes and fully unleash the insights hidden therein. The OBJECTS workshop provides a platform for researchers from the fields of Business Process Management and Process Mining who work on object-centric processes where they can share their ideas and current research addressing the aforementioned problems as well as discuss challenges and future directions of the field.

OBJECTS 2024 was the second edition of the workshop, and attracted 6 high-quality international submissions. Each paper was reviewed by at least three members of the Program Committee. All of the submitted manuscripts were accepted for presentation, two of them as regular contributions. These two papers are published in these proceedings.

Arnold and Reichert presented an approach to support creation of correct object-centric coordination processes by performing on-the-fly syntax checking and background validation against a set of predefined properties. The approach was implemented and tested within the PHILharmonicFlows framework.

Khayatbashi, Hartig and Jalali proposed a formally defined temporal extension of Event Knowledge Graphs (EKGs) to capture dynamic attributes in object-centric processes. Beyond that, the work also proposed a conversion algorithm from the Object-Centric Event Log (OCEL) 2.0 format to temporal EKGs.

The workshop also featured an invited talk and two invited tutorials. The talk, titled "Declarative Object-Centric Processes as DCR Graphs with Data: Modelling, Execution and Mining", was given by Thomas Troels Hildebrandt. Thomas presented his view on how declarative behavioural constraints should be combined with data and objects, and presented how this vision manifests in an extension of the notation and (object-aware) execution semantics of DCR (Dynamic Condition Response) Graphs, and showed how this extension can be used in the context of object-centric Process Mining. He also discussed future challenges of declarative object-centric modelling, execution and discovery for DCR Graphs.

The first invited tutorial, presented by Karolin Winter, covered the discovery of instance-spanning constraints and exceptions, while the second tutorial, delivered by

Marco Montali, provided a concise overview of key modelling concerns and language constructs for object-centric processes.

We thank the authors for their interesting contributions and the members of the Program Committee for their invaluable help in the reviewing and discussion phase of the manuscripts. We hope that, by reading these papers, the reader will know more about the latest advances in research on object-centric processes.

Organization

Organizing Committee

Marco Montali	Free University of Bozen-Bolzano, Italy
Andrey Rivkin	Technical University of Denmark, Denmark
Jan Martijn van der Werf	Utrecht University, The Netherlands

Program Committee

Han van der Aa	University of Mannheim, Germany
Wil van der Aalst	RWTH Aachen University, Germany
Johannes De Smedt	KU Leuven, Belgium
Rik Eshuis	Eindhoven University of Technology, The Netherlands
Dirk Fahland	Eindhoven University of Technology, The Netherlands
Ekkart Kindler	Technical University of Denmark, Denmark
Irina Lomazova	HSE University, Russia
Jan Mendling	Humboldt-Universität zu Berlin, Germany
Giovanni Meroni	Technical University of Denmark, Denmark
Jorge Munoz-Gama	Pontificia Universidad Católica de Chile, Chile
Artem Polyvyanyy	University of Melbourne, Australia
Natalia Sidorova	Eindhoven University of Technology, The Netherlands
Monique Snoeck	KU Leuven, Belgium
Pnina Soffer	University of Haifa, Israel
Dominique Sommers	Eindhoven University of Technology, The Netherlands
Francesca Zerbato	Eindhoven University of Technology, The Netherlands

Coordination Process Verification for Object-Centric Business Processes

Lisa Arnold[✉][iD] and Manfred Reichert[iD]

Institute of Databases and Information Systems, Ulm University,
James-Franck-Ring 1, 89081 Ulm, Germany
{lisa.arnold,manfred.reichert}@uni-ulm.de

Abstract. The accuracy and efficacy of an object-centric business process are of paramount importance during its execution. In the context of the PHILharmonicFlows framework, user interaction forms are automatically generated from the structure of the business process. Consequently, errors (e.g. deadlocks) in the business process result in malfunctioning during execution. It is therefore of the utmost importance to identify and rectify any errors in the business process at the earliest possible stage, namely at the point of specification. However, the concept of object-centric process management is sophisticated and requires a high level of expertise to implement effectively. In particular, modelling the coordination processes that control the business process in order to represent the interactions between multiple business objects represents a significant challenge. In light of this, a verification algorithm has been developed, comprising two mechanisms (prevention and alerting), to assist process modellers in creating coordination processes. This verification algorithm was subjected to testing during the emulation of three existing business processes. This revealed a number of flaws, including the presence of cycles that comprise several coordination processes. These deficiencies had previously evaded the detection of numerous modelling experts.

Keywords: Verification · Error detection · Modelling Tool · Web-based

1 Introduction

The testing and execution of a potentially buggy or incomplete business process to detect and avoid execution problems such as deadlocks is a costly undertaking for organisations. Pilot studies can be time-consuming and resource-intensive if business processes are required to be repeated several times due to avoidable errors in previous versions that could have been detected with sophisticated verification. The utilisation of a verification algorithm is of paramount importance for the effective and efficient modelling of object-centric business processes. The provision of support to a modeller throughout the creation of complex business processes can result in a reduction in costs, resources and time. This is achieved

© The Author(s), under exclusive license to Springer Nature Switzerland AG 2025
K. Gdowska et al. (Eds.): BPM 2024 Workshops, LNBIP 534, pp. 287–299, 2025.
https://doi.org/10.1007/978-3-031-78666-2_22

by avoiding the occurrence of errors. For instance, a modeller can be assisted by being alerted to incorrect process models or prevented from creating incorrect process models directly [1].

The concept of object-centric process management is challenging and requires a shift in perspective and thinking. The objective of developing the object-centric process modelling environment was to reduce the complexity for the modeller by providing a sophisticated verification algorithm. The algorithm is designed to alert the modeller to potential issues during the creation of a business process and, if possible, to prevent them from occurring.

In the modelling of large and complex business processes, coordination processes can rapidly become confusing and error-prone. For this reason, two mechanisms have been developed. Firstly, an active validation mechanism is employed to prevent the incorrect modelling of the coordination process. This involves the blocking of movement and the highlighting of invalid targets in red, while valid targets are highlighted in green. Secondly, passive validation is employed, which returns an error message when a problem is identified that is calculated with a validation algorithm in the background. The verification algorithm was subjected to analysis and evaluation through the emulation of three existing coordination processes. The evaluation assesses whether the prevention logic impedes the process modelling by blocking the occurrence of correct interactions in the coordination process or allows the introduction of incorrect modifications. Furthermore, it determines whether the error messages that arise during the modelling process are displayed correctly.

The remainder of this paper is structured as follows. Section 2 provides a concise overview of the key principles and characteristics of object-centric business processes. Section 3 provides an insight into our web-based, object-centric business process framework, PHILharmonicFlows, as well as rules for modelling a coordination process. Section 4 presents two mechanisms for verifying a coordination process. Section 5 evaluates the developed verification algorithm. Section 6 concludes the paper.

2   Fundamentals of Object-Centric Business Process

In our object-centric process management paradigm, a business process is described in terms of interacting business objects that correspond to real-world entities. A running example (cf. Ex. 1) is employed to elucidate the concepts and artefacts of object-centric processes.

Ex. 1: (Recruitment Business Process) In order to fill a vacancy, management will typically advertise and publish the vacancy in a number of different ways, including on the company's website or on a job portal. Subsequently, candidates are afforded the opportunity to prepare their application and submit it to the company. All applications received are then subjected to at least three reviews. Consequently, the applicant's suitability for the position is assessed. Should the majority of reviewers conclude that the candidate is suitable for the position, he will be invited for an interview. Should the candidate fail to receive the requisite majority of votes, two additional reviewers may also be possible. The interview is prepared in advance, which involves scheduling an appointment and formulating questions. Subsequently, the candidate is interviewed, after which a decision is made as to whether or not to offer an employment contract. Once the position has been filled, the vacancy is closed.

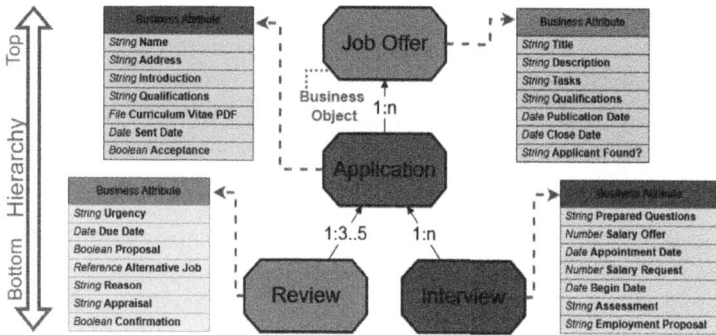

Fig. 1. RPS of the recruitment business process with a number of the objects' business attributes.

The interactions between the objects, as well as their relations, including their cardinalities, hierarchical structure, and semantic relations, are manifested in the **Relational Process Structure** (RPS) [2]. Furthermore, business attributes may be defined for each business object, specifying the business process. The RPS corresponding to the recruitment business process of Ex. 1, along with a number of its business attributes, is depicted in Fig. 1. During execution, any number of object instances may be created by the business objects, provided that the constraints imposed by their cardinalities are observed.

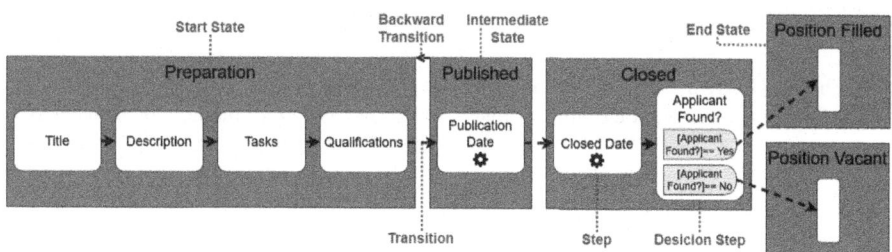

Fig. 2. Lifecycle of the object *Job Offer*.

The runtime behaviour of these business objects is defined in terms of **object lifecycles** (lifecycle for short) [3]. The lifecycle of the object *Job Offer* is depicted in Fig. 2. In general, a lifecycle comprises of *states*, with one start state (*Preparation*) and at least one end state (*Position Filled* and *Position Vacant*), as well as any arbitrary number of intermediate states (*Published* and *Closed*). In order to facilitate user interaction at runtime for each state, an automatically generated *form sheet* is built from the lifecycle structure. The runtime behaviour of each object instance is defined by its own lifecycle instance. In particular, the states of a lifecycle define the form sheets and their steps, which in turn design the input fields. These are built from the object attributes. The result is a data-driven business process. It is possible to define backward transitions to previous states within the lifecycle with the objective of reading or verifying previously entered data. However, by default, there are no backward transitions, as they must be explicitly set by a modeller. It should be noted that backward transitions are not intended to create loops, whereby a new instance is created. Instead, a previous form is displayed again with the input variables already entered.

A **coordination process** controls the interactions between the lifecycles of multiple objects and defines the sequence of states between multiple lifecycle states. Consequently, every object-centric business process must contain at least one coordination process. A coordination process is defined from the perspective of one business object. In other words, the lifecycle of one object is extended with the lifecycle states of other objects to represent their correlations and interactions. In general, a coordination process can be considered as a graph, where the vertices represent the *coordination steps* and the edges represent the *coordination transitions* (cf. Fig. 3). The coordination process graph (cf. Fig. 4) is a directed, acyclic and connected graph that precludes backward transitions or loops to preceding coordination steps. Otherwise, cyclic dependencies and, consequently, deadlocks are possible. Therefore, the acyclicity of coordination processes is not a limitation of expressivity but a necessity for correctness [4]. A business object that has a coordination process is called a *Coordinating Business Object*.

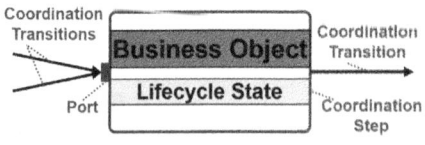

(a) The **AND-Join** is a realization with a single port and two incoming transitions.

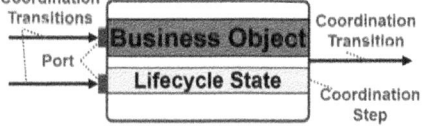

(b) The **OR-Join** realization is achieved with the use of multiple ports, each of which has one incoming transition.

Fig. 3. Coordination steps with different joins.

In particular, a coordination process must contain each lifecycle state of its coordinating business object in the same order as defined in the lifecycle itself. Furthermore, the start coordination step must correspond to the start state of the lifecycle, and the end coordination step(s) must correspond to the end state(s) of the lifecycle. Moreover, the coordination process may encompass any

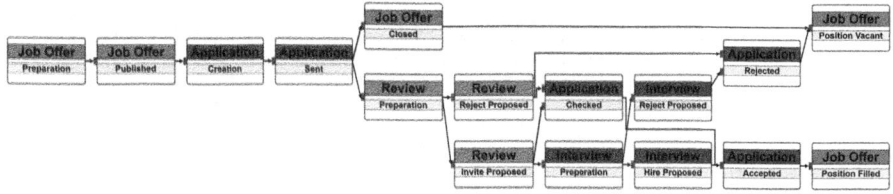

Fig. 4. The coordination process of the coordinating business object *Job Offer*.

other lifecycle states of other business objects. In addition, the coordination steps in the coordination process can be arranged in a sequential, parallel, or concurrent manner. However, it is important to note that coordination steps based on the same object instance cannot be executed in parallel. Moreover, the ports of a coordination process can be utilised to create logical AND (cf. Fig. 3a) or OR (cf. Fig. 3b) semantics. In order for a coordination step to be activated, it is necessary for all incoming transitions (i.e. AND semantics) of a port to be activated. Conversely, multiple ports represent the OR semantics.

The coordination process of the coordinating business object, Job Offer, which is derived from the recruitment business process of Ex. 1, is depicted in Fig. 4. In the coordination process, the sequence of the lifecycle states of the business objects is coordinated. For instance, prior to the submission of an application to a company, the job offer must be prepared and made public.

An implementation of this paradigm is provided by the PHILharmonicFlows framework [5,6]. This framework facilitates the dynamic evolution of object-centric business processes, allowing for both built-in flexibility and ad-hoc process changes at runtime [7,8].

3   Modelling a Coordination Process

In order to model a coordination process of a coordination object ① (cf. Fig. 5), a new coordination step is selected from the coordination step template ② and then dragged to the modelling interface ③. For the newly created coordination step ④, no references to a business object ⑤ with its corresponding lifecycle ⑥ are initially set. Subsequently, the user may specify them via the drop-down menus, which automatically provide all possible business objects and the life-cycle states ⑦ of the selected business object. Two coordination steps may be linked by means of a coordination transition by dragging a connection between them. This can be achieved by directly dragging a coordination transition onto a coordination step or onto an existing port. The act of dragging a coordination transition onto a coordination step will result in the creation of a new port that connects the two aforementioned coordination steps. This mechanism allows the creation of multiple ports, which may be considered logical OR-joins. When a coordination transition is dragged directly to an existing port of a coordination step, a logical AND-join is created. In order to ensure the accurate representation of a coordination process, it is essential to adhere to the following rules:

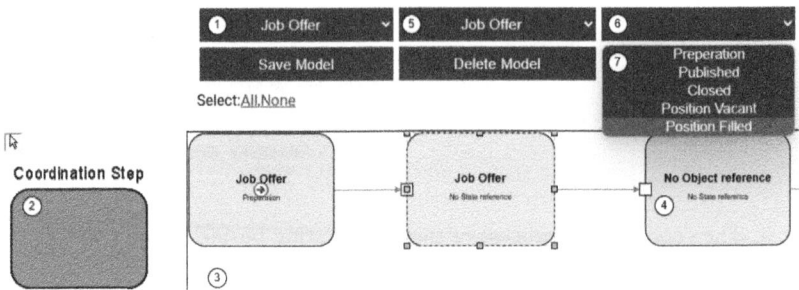

Fig. 5. A screenshot of the coordination process modelling interface from the framework `PHILharmonicFlows`

1. Each object-centric business process must include at least one coordination process.
2. Each coordination process must be a directed, acyclic, and connected graph.
3. Each lifecycle state of the coordinating business object must occur in the order defined by its lifecycle within the coordination process.
4. The start coordination step corresponds to the start state in the lifecycle of the coordinating business object.
5. The end coordination step(s) correspond(s) to the end state(s) in the lifecycle of the coordinating business object.
6. No duplication of coordination steps is permitted in a coordination process.
7. No cycle spanning multiple coordination processes is permitted.

In essence, an object-centric business process must include at least one coordination process that defines the interactions between the various business objects (1.). A coordination process is represented by a directed, acyclic, and connected graph (2.), which connects coordination steps via coordination transitions. In addition, a coordination process may be regarded as an extension of the lifecycle, from the perspective of a coordinating business object. Consequently, all lifecycle states of the coordinating business object must be included in the coordination process in the defined sequence from its lifecycle (3.). This necessitates that the start coordination step aligns with the start state of the coordinating business object's lifecycle (4.). In addition, the end coordination step(s) correspond(s) to the end state(s) in the lifecycle of the coordinating business object (5.).

As stated in Sect. 2, the object-centric approach is data-driven, with forms generated from the lifecycle structure. This is achieved by generating the form sheet from states and input fields derived from steps. It is not the objective of this approach to include repetitions or loops in the form sheet sequence. In essence, the emergence of cycles can be attributed to three types. Firstly, cycles may arise when the same coordination steps occur on multiple occasions within a given coordination process. Such cycles may take the form of self-loops, whereby the same coordination steps occur immediately after each other, or cycle comprising several coordination steps (6.). Secondly, a cycle can be created by two or more coordination processes (7.). To illustrate this, *Objects A* and *B* both exhibit the *Lifecycle States 1, 2,* and *3.* The coordination processes of the coordination *Objects A* and *B* are illustrated in Fig. 6a and 6b. Both coordination processes are valid when considered in isolation. However, by supplementing the lifecycles of both objects by their corresponding coordination transitions (in red), a cycle (consisting of *Object A State 2* and *Object B State 2*) is created, as illustrated in Fig. 6c. Thirdly, a similar cycle can be created with a lifecycle process and a coordination process (3.) if the modelled sequence of coordination steps differs from that of the lifecycle. As these forms of loops are not permitted, they must be identified by the verification algorithm in order to alert the modeller.

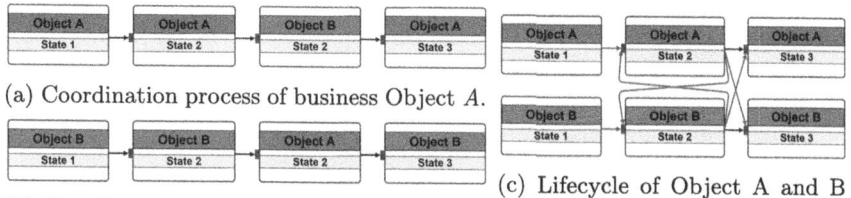

(a) Coordination process of business Object *A*.

(b) Coordination process of business Object *B*.

(c) Lifecycle of Object A and B with all transitions (in red).

Fig. 6. Example of a cycle that extends over several coordination procedures.

4   Verification Procedure

In the context of large business processes comprising numerous interactions between multiple business processes, the coordination process can rapidly become opaque and intricate. Furthermore, the complexity of coordination processes can be significantly augmented when there are more than one such process. It is inevitable that multiple coordination processes will be required for business processes that represent complete organisational structures. It is of the utmost importance that they do not result in deadlocks due to interdependencies. For this reason, the framework PHILharmonicFlows is extended by a sophisticated validation of coordination processes, which is based on two mechanisms:

1. **Mechanisms - Active validation:** Avoid incorrect modelling of the coordination process: Block the modification and highlight the invalid target(s) in red.
2. **Mechanisms - Passive validation:** Checks the coordination process for errors. Returns error messages to the modeller in case of errors.

The **active validation** serves to prevent the user from modelling incorrect processes. This mechanism employs highlighting and prevention logic and is based on the modeller's user interactions. This is utilised to generate the coordination process graph. To illustrate, if a coordination is dragged from one coordination step to a port of another coordination step, all potential coordination steps are highlighted in green, while those that generate cycles are highlighted in red. However, this cycle detection is limited to the one coordination process currently being modelled. Consequently, this mechanism is unable to detect and prevent cycles that may be generated across multiple coordination processes, which are addressed in the second mechanism.

In general, a coordination transition may only exist between a coordination step and a port of a subsequent coordination step. Furthermore, no coordination transition may commence from a port or a coordination transition. In addition, the following three coordination transitions starting from a coordination step are also not permitted and are prevented by the first mechanism:

   I A transition to a coordination transition
  II A transition to a previous coordination step or port.
 III A transition to an already connected port or coordination step.

It is not permitted to drag a coordination transition that commences at a coordination step and terminates at a coordination transition (I.). Consequently, when a coordination transition is dragged, all existing coordination transitions are highlighted in red and are rendered inaccessible. The coordination transition from a coordination step to a previous coordination step or port is also highlighted in red and is blocked (II.) as it generates a cycle at the coordination process. A coordination step may have several outgoing transitions. Nevertheless, a second coordination transition to the same port or coordination step as the first one is also prohibited (III.). Consequently, all already connected ports or coordination steps are also highlighted in red and are blocked. This mechanism is designed to operate on the graph (i.e. the vertices and edges) of a coordination process. It does not verify the accuracy of the contents.

The **passive validation** generates and displays error messages for content errors (e.g. invalid setting of business objects or their lifecycle state references) or for complex cycles (i.e. spanning multiple coordination processes). The coordination process is continuously monitored, and error messages are returned to the modeller in the event that the business process rules are violated. A total of nine distinct cases are validated directly (e.g. the duplication of identical coordination steps). The validation algorithm generates a validation graph in the background, which is designed to detect all kinds of cycles. These may include, for example, a sequence of lifecycle states that is incorrect or spans several coordination processes. The validation graphs contain all lifecycle processes and their respective lifecycle states. In order to identify the cycles, all existing coordination transitions from all created coordination processes are included in the validation graph by linking the corresponding states (cf. Fig. 6c red coloured transitions). Subsequently, the validation graph is analysed and cycles are identified through a depth-first search.

The following nine statements are subjected to continual investigation during the modelling of a coordination process. In the event that the coordination process is invalid, the modeller will be informed of this via an error message.

Case 1. *The coordination process encompasses all lifecycle states of the coordinating object.*

Upon initiating the creation of a coordination process, a list of error messages is automatically generated and displayed, indicating all missing lifecycle states of the coordinating object that are not currently referenced in a coordination step. This is achieved by comparing the coordination process steps with all lifecycle states of the coordinating object, and generating an error message when a lifecycle state is not present and has not been referred to as a coordination step.

Error message: 'Lifecycle state *¡NameOfState¿* of Object *¡NameOfObject¿* is not referenced.'

Case 2. *All coordinated steps are referenced to an assigned object.*

When a coordination step is dragged onto the modelling interface, it is not yet associated with an object or lifecycle state reference. Consequently, an error message is generated as long as the coordination step lacks a referenced object. This error message indicates that at least one coordination step lacks an object reference.

Error message: 'Object reference is missing.' **Clicking on the error message:** Highlighting the coordination steps without an object reference in red.

Case 3. *All coordination steps are referenced to a specific lifecycle state of the associated object.*

In addition to the object references, all coordination steps are referred to the lifecycle state, which is derived from the assigned object. The procedure is identical to that described in Case 2.

Error message: 'State reference is missing.' **Clicking on the error message:** Highlighting the coordination steps without a state reference in red.

Case 4. *A coordination process has exactly one start coordination step (i.e. without incoming transition).*

A coordination process is initiated by a single start coordination step, devoid of incoming transitions. All other coordination steps have at least one incoming coordination transition. In order to ascertain the coherence of the coordination process, the verification algorithm is employed to ascertain the presence of multiple start coordination steps. This is achieved by the algorithm continuously monitoring the absence of incoming coordination transitions (i.e. ports) at each coordination step, and issuing an error message if more than one is identified.

Error message: 'Coordination process can only have one start coordination step.' **Clicking on the error message:** Highlighting all coordination steps without incoming transitions in red.

Case 5. *The start coordination step is equivalent to the start lifecycle state of the coordination object.*

From the perspective of the coordinating business object, the process must commence with the start state of the corresponding lifecycle. In the event that this condition is not met, an error message will be generated and presented to the modeller. This is accomplished by maintaining a continuous check to ensure that the start coordination step aligns with the start state of the coordinating object's lifecycle process.

Error message: 'Lifecycle state *¡NameOfState¿* of *¡NameOfObject¿* must be the start coordination step .'

Case 6. *The end coordination step(s) is/are equivalent to the end lifecycle state(s) of the coordination object.*

Analogous to Case 5, the end coordination step(s) must correspond(s) to the end lifecycle state(s) of the coordinating object.

Error message: 'Lifecycle state *¡NameOfState¿* of *¡NameOfObject¿* must be the end coordination step.'

Case 7. *No duplication of coordination steps occurs.*

Two coordination steps will indirectly generate cycles if they refer to the same lifecycle state. In the data-driven paradigm of PHILharmonicFlows, it is not permitted for coordination steps to be embedded multiple times in the coordination process, resulting in the generation of multiple forms. **Error message:** 'Coordination step *¡NameOfObject - NameOfLifecyclState¿* occurs several times in the coordination process.' **Clicking on the error message:** Highlighting all duplicates in red.

Case 8. *No cycles across different coordination processes exit.*

The validation graph is capable of detecting cycles across different coordination processes (cf. Example of Fig. 6). In the event of a cycle being identified, an error message will be displayed.

Error message: 'Detecting cycles across different coordination processes.'

Case 9. *All coordination steps (i.e. states of all referenced objects) are present in the manifest lifecycle sequence.*

Analogous to Case 8, a coordination process in conjunction with the lifecycle process can result in a cycle if the defined sequence of the lifecycle states does not correspond to the sequence of the coordination steps.

Error message: 'Coordination step *¡NameOfObject - NameOfLifecyclState¿* occurs in the wrong sequence as defined in the lifecycle process.'

Clicking on the error message: Highlighting the coordination steps that occurred in the wrong sequence in red.

5   Testing the Verification Algorithm

In order to evaluate the accuracy and efficacy of the coordination process of the defined verification procedure, three object-centric business processes are subjected to analysis. The aforementioned processes were developed within the local PHILharmonicFlows framework [5,6] and subsequently transferred to the web-based PHILharmonicFlows framework, which provides coordination verification. The following section presents the three business processes and their background, including the process dimensions (e.g. number of objects). Furthermore, the manner in which the verification algorithm behaved during the modelling process is elucidated and subjected to critical analysis.

Business Process 1: Recruitment.

Objects: 4 Coordination Processes: 1 Coordination steps: 16 The process description is provided in Ex. 1. The process components are illustrated in Fig. 1

(i.e. RPS), Fig. 2 (i.e. lifecycle of the object *Job Offer*), and Fig. 4 (i.e. coordination process of the coordination object *Job Offer*).

Business Process 2: PHoodle.

Objects: 9 Coordination Processes: 1 Coordination steps: 6 An e-learning platform designed to facilitate the management of lectures at Ulm University.

Business Process 3: Car Manufacturing.

Objects: 27 Coordination Processes: 8 Coordination steps: 25, 28, 30, 31, 20, 25, 30, and 20

A car manufacturing plant that produces and assembles all of its own parts. Such components include, for example, the chassis, engine compartment and interior, as well as tyres with rims and rubber, crankshafts, pistons or spark plugs.

In order to assess the accuracy and efficacy of the three business processes, they are imported (i.e. emulated) into the web-based version of PHILharmonicFlows. The initial stage of analysis is to evaluate the active mechanism, which encompasses the prevention and highlighting logic. This entails the meticulous modelling of each of the three business processes in a step-by-step manner. Furthermore, any potential interference with the process modelling by blocking the correct interactions during the creation of the coordination process is noted (if they occur). Nevertheless, the modelling of coordination processes presents no difficulties in terms of prevention logic for any of the three business processes. The second step of the evaluation process concerns the passive mechanism, which encompasses the content and behaviour of error messages.

The data model of the *PHoodle* business process is larger than the RPS of the *Recruitment* business process (cf. Fig. 1). Nevertheless, this business process is characterised by a single coordination process comprising six distinct steps. In conclusion, this business process is relatively straightforward to manage. Upon modelling the coordination process, it was observed that only the error messages pertaining to Cases 1 to 4 were generated. This is due to the fact that not all coordination steps of the corresponding objects are included at the beginning, and no object and state references are set when a new coordination step is added.

The coordination process of the *Recruitment* business process is more extensive than that of the *PHoodle*, comprising 16 distinct coordination steps derived from all four business objects. The verification algorithm yields the same observations (i.e., output error messages from Case 1 to Case 4) as those observed in the *PHoodle* business process. Nevertheless, Case 9 indicates that the coordination step *Job Offer - Closed* does not occur in the sequence preceding the coordination step *Job Offer - Position Filled*. The case remained unresolved, as it had never been concluded. This issue arises exclusively in the context of a single application for a vacancy that is immediately filled. In this instance, the job offer can be closed by generating a livelock at the port of the coordination step *Job Offer - Position Vacant* or the business process instance can be terminated without closing its job offer by generating a livelock at the port of the coordination step *Job Offer - Closed*. In order to ensure the integrity of the model, it is necessary to draw a coordination transition from the coordination step *Job Offer - Closed* to the coordination step *Job Offer - Position Filled* (cf. Fig. 4).

The coordination processes of the *Car Manufacturing* business process are considerably more intricate. The process comprises 27 business objects and a total of 8 coordination processes, each with at least 20 coordination steps. During the modelling of the coordination process, the same error messages (i.e. Case 1 to Case 4) were observed as in the other business processes (cf. Business Processes 1 and 2). Nevertheless, upon completion of the modelling of all coordination processes, the error message from Case 1 persists in each coordination process, indicating that not all lifecycle states of the coordinating object are referenced. For each coordination process, between two and seven lifecycle state references are absent. Furthermore, the verification algorithm has identified a previously undetected cycle (i.e. Case 8) involving two coordination processes.

6   Summary and Outlook

The objective of this paper is to examine the verification of coordination processes during their development. The verification process is based on two distinct mechanisms: an active and passive one. The objective of the verification procedure described above is to ensure that the coordination process is error-free and correct once modelling is complete. The developed verification algorithm is analysed and evaluated through three business processes and their 13 associated coordination processes. Firstly, the evaluation assesses whether the prevention logic impedes the process modelling by obstructing the correct interactions in the coordination process. Secondly, it determines whether the error messages that arise during modelling are correctly displayed. Among others, the evaluation identified a cycle in a business process that is spanned across multiple coordination processes, which had previously been overlooked by various modellers. In future work, the verification procedure will be evaluated through an eye-tracking study. This will facilitate an understanding of how modellers respond to error messages. Furthermore, the modelled coordination processes will be incorporated into the runtime engine with the objective of elucidating their actual behaviour in greater detail.

Acknowledgment. This work is part of the ProcMape project, funded by the KMU Innovativ Program of the Federal Ministry of Education and Research, Germany (F.No. 01IS23045B).

References

1. Dumas, M., Rosa, L., Mendling, J., Reijers, A.: Fundamentals of Business Process Management. Springer, Cham (2018)
2. Steinau, S., Andrews, K., Reichert, M.: The relational process structure. In: CAiSE 2018, pp. 53–67. Springer (2018)
3. Steinau, S., Andrews, K., Reichert, M.: Executing lifecycle processes in object-aware process management. In: SIMPDA 2017, pp. 25–44. Springer (2017)

4. Steinau, S., Andrews, K., Reichert, M.: Coordinating large distributed relational process structures. Softw. Syst. Model. **20**(5), 1403–1435 (2021)
5. Steinau, S., Andrews, K., Reichert, M.: A modeling tool for PHILharmonicFlows objects and lifecycle processes. In: BPMD 2017, CEUR Workshop Proceedings, CEUR-WS.org (2017)
6. Andrews, K., Steinau, S., Reichert, M.: A tool for supporting ad-hoc changes to object-aware processes. In: EDOCW 2018, pp. 220–223. IEEE (2018)
7. Andrews, K., Steinau, S., Reichert, M.: Enabling runtime flexibility in data-centric and data-driven process execution engines. Inf. Syst. (2021)
8. Andrews, K., Steinau, S., Reichert, M.: Enabling ad-hoc changes to object-aware processes. In: EDOC 2018, pp. 85–94. IEEE (2018)

Transforming Object-Centric Event Logs to Temporal Event Knowledge Graphs

Shahrzad Khayatbashi[1(✉)], Olaf Hartig[1], and Amin Jalali[2]

[1] Linköping University, Linköping, Sweden
{shahrzad.khayatbashi,olaf.hartig}@liu.se
[2] Stockholm University, Stockholm, Sweden
aj@dsv.su.se

Abstract. Event logs play a fundamental role in enabling data-driven business process analysis. Traditionally, these logs track events related to a single object, known as the case, limiting the scope of analysis. Recent advancements, such as Object-Centric Event Log (OCEL) and Event Knowledge Graph (EKG), capture better how events relate to multiple objects. However, attributes of objects can change over time, which was not initially considered in OCEL or EKG. While OCEL 2.0 has addressed some of these limitations, there remains a research gap concerning how attribute changes should be accommodated in EKG and how OCEL 2.0 logs can be transformed into EKG. This paper fills this gap by introducing Temporal Event Knowledge Graph (tEKG) and defining an algorithm to convert an OCEL 2.0 log to a tEKG.

Keywords: Event Knowledge Graphs · Object-Centric Event Data · Object-Centric Process Mining

1 Introduction

Business processes involving participants, resources, and systems can be analyzed from different perspectives [9]. These perspectives include different objects based on which a process can be analyzed for further improvement. Traditional analysis focuses on a single object (a.k.a. case), making it challenging to answer questions considering multiple objects and perspectives simultaneously. Object-Centric Process Mining (OCPM) addresses this limitation, aiming to uncover insights by capturing interrelations between objects and events in event logs. Data that includes the relation between events to multiple objects is known as Object-Centric Event Data (OCED) [15], promising the discovery of more insights and addressing the limitations of traditional analysis methods.

Around 2020, two data models were introduced to record OCED: Object-Centric Event Log (OCEL) [6] and Event Knowledge Graph (EKG) [5]. OCEL 1.0 [6] proposed a conceptual model for event logs, enabling the recording of events related to multiple objects and facilitating the development of OCPM

K. Gdowska et al. (Eds.): BPM 2024 Workshops, LNBIP 534, pp. 300–313, 2025.
https://doi.org/10.1007/978-3-031-78666-2_23

algorithms, e.g. [7, 8, 16]. EKG presented an alternative technique to record event logs in a Knowledge Graph [3, 4].

Transforming logs between these formats not only enables the application of techniques developed for each format but also facilitates comparing limitations, which can be used to extend these models for further analysis. For instance, a recent study on transforming EKG to OCEL 1.0 highlights the lack of support in capturing relations between objects in OCEL [10], a concern now addressed by OCEL 2.0 [2].

OCEL 2.0 extends its predecessor with support to record information on object relationships, to qualify relationships, and to capture the change of values for attributes of objects over time [2]. This extension allows capturing the temporal value of objects in practice. As an example, the price of an item in an online webshop can change over time. If these price changes aren't accurately tracked, it becomes difficult to analyze why an item suddenly becomes popular. This is because we lack the correct price data for when customers viewed the item at different times.

While EKG has also undergone improvements, it lacks support for such temporal aspects. Additionally, there exists a gap in transforming OCEL 2.0 to EKG, hindering a direct comparison and tool reuse between these two formats. To fill these gaps, this paper focuses on the following research questions:

RQ1: How can temporal aspects of object attributes be captured in an EKG?
RQ2: How can an OCEL 2.0 log be transformed into a temporal EKG?

To address RQ1 we extend the EKG model into a model of Temporal Event Knowledge Graph (tEKG). To address RQ2 we define an algorithm for transforming an OCEL 2.0 file into our proposed tEKG representation. We have implemented this algorithm and provide the source code of this implementation publicly.[1] Structure of the paper: Section 2 gives a background using a running example, and Sect. 3 introduces tEKG informally. Section 4 defines tEKG formally, Sect. 5 defines the transformation algorithm, and Sect. 6 concludes the paper.

2   Background

This section introduces the relevant background on EKGs based on a running example.

The example revolves around a fictional education-related process where a student failed to pass a course and must retake it the following year. In the first year, the student reads instructions for an assignment and submitted it accordingly. Subsequently, the teacher decided to increase the points allocated for the assignment from 2 to 3, as it was discovered that the assignment was considerably more challenging than anticipated. Here, we provide a high-level overview of this process to convey the essential concepts necessary for understanding EKG

---

[1] https://github.com/shahrzadkhayatbashi/BPM2024.

and tEKG. To ensure simplicity, we do not include representations of students, teachers, or other entities typically involved in such a process. A key aspect, however, is that the number of points of the assignment can change over time and must be correctly recorded for the different years.

Figure 1 illustrates a part of an EKG using nodes and edges to record data of our running example. In EKGs, nodes can be labeled as Log, Entity, Class, or Event. These nodes capture information about log files, objects, event types, and events, respectively. The label Entity is used for nodes representing objects in EKGs. In this paper, we use the terms "object" and "entity" interchangeably when referring to an object in OCEL and EKG, respectively. In the figure, only nodes labeled Entity and Event are displayed. For instance, c1 and a1, depicted as ovals, represent the course and the assignment, respectively, in our example. An event, denoted by e1 and shown as a diamond, captures the event of the student reading the instruction at time t2. Each node can be annotated with key-value pairs called properties. For example, an assignment may have a specific number of points that students can receive upon submitting the assignment.

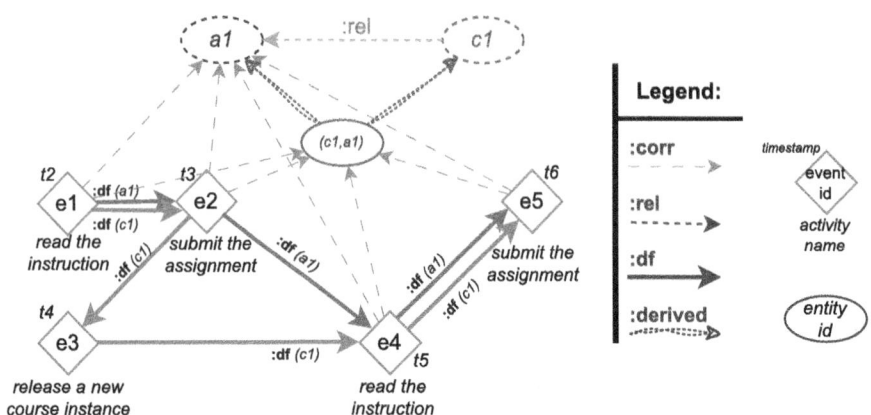

Fig. 1. A part of an Event Knowledge Graph

Edges in EKGs establish relationships among nodes, and these edges can be labeled to indicate the type of relationship between nodes. For example, relationships between entities are represented using edges labeled as rel. In our example, c1 is connected to a1 via such an edge, indicating that the course has an assignment. Edges between nodes labeled Event and nodes labeled Entity are labeled corr, and the label df is used for edges representing directly-follows relationships among nodes representing events.

Such directly-follows relationships between events are fundamental in process mining. In EKGs, two events, say e1 and e2, can be connected by a df edge if i)there is a shared entity to which both events have a corr edge, ii)e1 occurred before e2, and iii) there are no other events that fulfill the first condition and occurred between e1 and e2. Such a df edge is associated with two properties called EntityID and EntityType, representing the identifier and type of the corresponding entity to which the two connected events are related via a corr edge. As an example, in the figure, e1 is linked to e2 by an edge labeled df, with a1 as the value of EntityType.

In this graph, all events have a corr edge to c1 since they are events that occurred within this course. However, we have not depicted these edges in this figure to avoid overwhelming complexity. Instead, we have visualized the resulting df edges. It is apparent that the event flow related to the assignment differs from that of the course, primarily because releasing a new course instance is unrelated to the assignment.

There are scenarios where it becomes necessary to link an object to the relationship between objects, which cannot be achieved directly in EKGs because an edge cannot connect a node to another edge. In EKGs this limitation is addressed by adding helper nodes known as reified entities. An example of a reified entity in Fig. 1 is (c1, a1) which reifies the rel edge from c1 to a1. These two entities are connected to the reified entity by edges with the label derived. All events connected to the aforementioned entities will also be connected to the reified entity. For more detailed information on these concepts, and on EKGs in general, we refer readers to Fahland's work [5]. We emphasize that the lacks the capability to capture changes in the values of attributes of entities. In our example, the value of the Points property of a1 was modified to 3 in the second year when the student retook the course. Consequently, the recorded information for the first year, where the assignment had 2 Points, would be lost due to the overwrite. This discrepancy can lead to erroneous analysis results. The tEKG model that we propose in this paper addresses this limitation.

3   Temporal Event Knowledge Graphs

This section introduces our proposal informally and discusses our design choices.

Our initial design choice is to ensure backward compatibility with EKG. This choice aims to facilitate the reuse of existing solutions, such as inferring missing entity identifiers [14] or aggregating event knowledge graphs for task analysis [13]. Therefore, we define tEKG as an extension of the EKG model that supports all EKG features as well as additional features for handling temporal entities.

The values of attributes of entities can change over time, and there are various methods to track these changes in information systems. One approach involves recording transactions for attribute modifications, while another entails capturing snapshots of the state of an object at different points in time. The latter method is commonly employed in data warehousing to store facts in periodic snapshot fact tables [12], which prioritizes query performance for data analysis over transactional performance. Therefore, we have chosen a similar design choice to enhance query performance, which involves generating snapshots of entities when the values of their attributes change.

tEKGs contain multiple nodes per entity; one such node represents the entity in general, independent of the temporal dimension (i.e., exactly as in an EKG), and the other nodes represent snapshots of the entity at specific times. The identifier of each snapshot node is the combination of a timestamp and the identifier of the corresponding entity. Figure 2(a) illustrates such snapshot nodes; in particular, the course c1 and assignment a1 each have two snapshots at times t1 and t4, respectively. The relationship between each entity and its snapshots is established through edges with label snapshot.

Edges with the label rel can be used to capture relationships between snapshots, as illustrated in Fig. 2(b). Such an edge has an attribute named qual with a value of update if the connected snapshots are for the same entity. Essentially, such edges document the lifecycle of an entity within a tEKG. For instance,

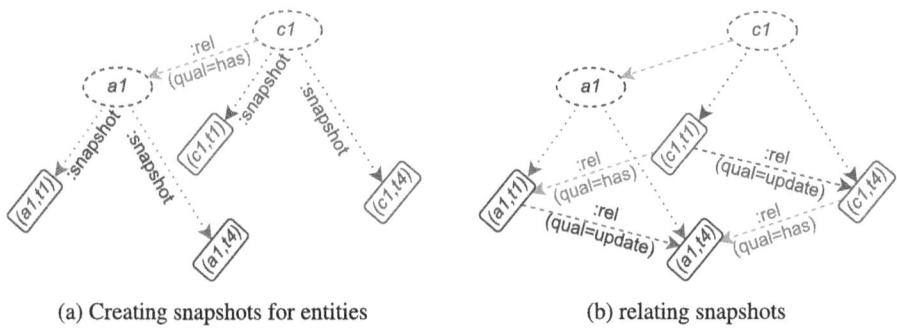

(a) Creating snapshots for entities   (b) relating snapshots

Fig. 2. Creating snapshots to capture changes in object's attributes over time

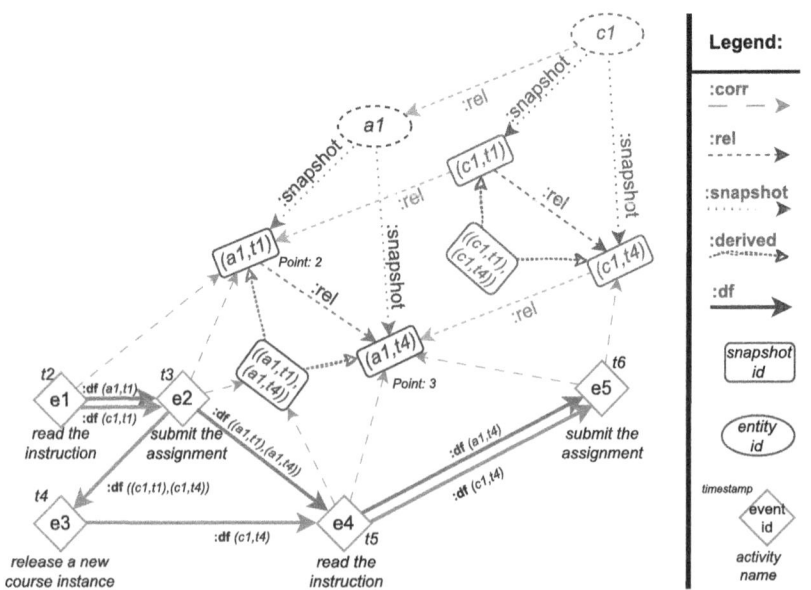

Fig. 3. An example of a Temporal Event Knowledge Graph

assignment a1 is initially created at time t1 and subsequently updated at time t4, as captured by the snapshot nodes (a1, t1) and (a1, t4), respectively. The edges labeled rel between entities will be copied to their snapshots, with the condition that each snapshot is connected only to snapshots that have existed in its lifetime.

We adapt the design choices made for EKGs to reify entities for snapshots, to connect events to snapshots, and to create directly-follows edges between events corresponding to the same snapshots. This results in additional edges in our graph compared to Fig. 1, a portion of which is illustrated in Fig. 3. For instance, we have included df edges for snapshots. To avoid over-complicating the illustration, we have omitted drawing previous edges. In particular, we depict corr edges only for the snapshots of a1, which caused the creation of df edges associated with the snapshots of a1.

The tEKG in Fig. 3 is more feature rich than a pure EKG, allowing analysts to monitor temporal aspects of entities. For instance, event e2 is connected to the (a1, t1) snapshot and not (a1, t4), which highlights that the student reads the assignment when it had 2 points. Notice also that such connections between events and snapshots of entities at specific points in time are only implicitly present in OCEL 2.0. Making them explicit in a tEKG enables direct access to them for temporal analysis of event logs.

4   Formalization

This section defines our notion of a tEKG. We begin with a recap of definitions of OCEL and EKG that we build on and that are adapted from the related publications [1, 2, 5].

4.1   Preliminaries

Definition 1. $\mathbb{U}_\Sigma$ is a **universe** consisting of the following, pairwise disjoint sets [2]:
$\mathbb{U}_{eid}$ is the universe of event identifiers, $\mathbb{U}_{val}$ is the universe of attribute values,
$\mathbb{U}_{oid}$ is the universe of object identifiers, $\mathbb{U}_{time}$ is the universe of timestamps,
$\mathbb{U}_{etype}$ is the universe of event types, $\mathbb{U}_{qual}$ is the universe of qualifiers, and
$\mathbb{U}_{otype}$ is the universe of object types, $\mathbb{U}_{lbl}$ is the universe of labels.
$\mathbb{U}_{att}$ is the universe of attribute names,

Definition 2. An **Object-Centric Event Log (OCEL)** L is a tuple $(E, O, EA, OA, evtype, evid, time, objtype, objid, eatype, oatype, eaval, oaval, E2O, O2O)$ where [2]:

- E and O are disjoint sets of events and of objects, respectively,
- $EA \subseteq \mathbb{U}_{att}$ and $OA \subseteq \mathbb{U}_{att}$ are sets of attributes for events and objects, respectively,
- $evtype : E \to \mathbb{U}_{etype}$ is a function that assigns event types to events,

- $evid : E \rightarrow \mathbb{U}_{eid}$ is a function that assigns event id to events,
- $time : E \rightarrow \mathbb{U}_{time}$ is a function that assigns timestamps to events,
- $objtype : O \rightarrow \mathbb{U}_{otype}$ is a function that assigns object types to objects,
- $objid : O \rightarrow \mathbb{U}_{oid}$ is a function that assigns object id to objects,
- $eatype : EA \rightarrow \mathbb{U}_{etype}$ is a function that assigns event types to event attributes,
- $oatype : OA \rightarrow \mathbb{U}_{otype}$ is a function that assigns object types to object attributes,
- $eaval : (E \times EA) \nrightarrow \mathbb{U}_{val}$ is a partial function that assigns values to (some) event attributes such that $evtype(e) = eatype(ea)$ for all $(e, ea) \in dom(eaval)$,
- $oaval : (O \times OA \times \mathbb{U}_{time}) \nrightarrow \mathbb{U}_{val}$ assigns values to object attributes such that $objtype(o) = oatype(oa)$ for all $(o, oa, t) \in dom(oaval)$,
- $E2O \subseteq E \times \mathbb{U}_{qual} \times O$ are the qualified event-to-object relations, and
- $O2O \subseteq O \times \mathbb{U}_{qual} \times O$ are the qualified object-to-object relations.

While the *oaval* function of OCEL assigns values to object attributes at particular points in time, the idea is that such a value remains valid until the next time point at which *oaval* assigns a new value to the attribute. Yet, for all time points in between, the *oaval* function is undefined for the corresponding attribute. To denote the *current value* that an object $o \in O$ has for an attribute $oa \in \mathbb{U}_{att}$ at some arbitrary point in time $t \in \mathbb{U}_{time}$, the OCEL specification writes $oaval^t_{oa}(o)$ which we formalize as follows.

- If there exists a timestamp $t' \in \mathbb{U}_{time}$ such that i)$t' \leq t$, ii)$(o, oa, t') \in dom(oaval)$, and iii)there is no $t'' \in \mathbb{U}_{time}$ such that $t' < t'' \leq t$ and $(o, oa, t'') \in dom(oaval)$, then $oaval^t_{oa}(o)$ is $oaval(o, oa, t')$.
- If no such t' exists, then $oaval^t_{oa}(o)$ is $\bot$, where $\bot$ is a special value not in $\mathbb{U}_{val}$.

Definition 3. A **Labeled Property Graphs (LPG)** G is a tuple $(N, R, \gamma, \lambda, \rho)$ (adopted from [1,5]), where:

- N and R are finite sets of nodes and of edges (relationships), respectively,
- $\gamma : R \rightarrow N \times N$ is a function assigning pairs of source and target nodes to edges,
- $\lambda : (N \cup R) \rightarrow \mathbb{U}_{lbl}$ is a function assigning a label to every node or every edge,
- $\rho : (N \cup R) \times \mathbb{U}_{att} \nrightarrow \mathbb{U}_\Sigma \cup (\mathbb{U}_\Sigma \times \mathbb{U}_\Sigma) \cup ((\mathbb{U}_\Sigma \times \mathbb{U}_\Sigma) \times (\mathbb{U}_\Sigma \times \mathbb{U}_\Sigma))$ is a partial function assigning (potentially composite) values to attributes of nodes and edges.

Given an LPG $G = (N, R, \gamma, \lambda, \rho)$ and a label $\ell \in \mathbb{U}_{lbl}$, we write N^ℓ to denote the subset of N consisting of all the nodes with label l; i.e., $N^\ell = \{n \in N \mid \lambda(n) = \ell\}$. Similarly, for edges: $R^\ell = \{r \in R \mid \lambda(r) = \ell\}$. We now introduce Event Knowledge Graphs (EKGs) as a special kind of LPGs that use a specific schema $\mathcal{S}$, which we capture as a set of 3-tuples:

$$\mathcal{S} = \{(\mathsf{Log}, \mathsf{has}, \mathsf{Event}), (\mathsf{Event}, \mathsf{observed}, \mathsf{Class}), (\mathsf{Class}, \mathsf{dfc}, \mathsf{Class}), (\mathsf{Event}, \mathsf{df}, \mathsf{Event}),$$
$$(\mathsf{Event}, \mathsf{corr}, \mathsf{Entity}), (\mathsf{Entity}, \mathsf{rel}, \mathsf{Entity}), (\mathsf{Entity}, \mathsf{derived}, \mathsf{Entity})\}.$$

Each such 3-tuple represents one of the types of edges in EKGs, where the second element of the 3-tuple provides the label of these edges, and the first and the third element captures the labels of corresponding source and target nodes, respectively. Formally, we say that an LPG $G = (N, R, \gamma, \lambda, \rho)$ *conforms to* S if, for every edge $r \in R$ with $\gamma(r) = (n, n')$, there exists $(s, l, t) \in S$ such that $n \in N^s$, $n' \in N^t$, and $r \in R^l$.

The ρ function assigns values to an attribute of nodes and edges, so its range is defined to cover different scenarios that are informally explained in Fig. 3 such as a singular value (e.g. a_1), a tuple (e.g. (a_1, t_1)), a tuple of tuples (e.g. $((a_1, t_1), (a_1, t_4)))$.

Definition 4. An **Event Knowledge Graph (EKG)** is an LPG $(N, R, \gamma, \lambda, \rho)$ that conforms to the schema S and every node $n \in N$ has the following properties (as per [4,5]):

- If $n \in N^{\mathsf{Event}}$, then $\rho(n, \mathsf{id}) \in \mathbb{U}_{eid}$, $\rho(n, \mathsf{act}) \in \mathbb{U}_{etype}$, and $\rho(n, \mathsf{time}) \in \mathbb{U}_{time}$.
- If $n \in N^{\mathsf{Entity}}$, then $\rho(n, \mathsf{id}) \in \mathbb{U}_{oid} \cup (\mathbb{U}_{oid} \times \mathbb{U}_{oid})$ and $\rho(n, \mathsf{type}) \in \mathbb{U}_{otype}$.

By Definition 4, nodes with the label Event in an EKG have attributes id, act, and time, with the value of an event identifier, an event type and a timestamp, respectively. Similarly, nodes with the label Entity have attributes id and type, with the value of an object identifier and an object type, respectively. The id value can be a single identifier or a tuple thereof. Entities with an id value of the latter type are called *reified entities*. In contrast to the original definition of EKGs [5], Definition 4 is more relaxed, as it does not enforce the existence of specific properties and edges. This flexibility allows our transformation algorithm to construct and add nodes, edges, and properties incrementally. The same approach is followed in the next definition.

4.2 Temporal Event Knowledge Graphs

To define tEKGs that capture temporal objects, we extend the aforementioned schema S by adding four more 3-tuples as follows:

$$S' = S \cup \{(\mathsf{Event}, \mathsf{corr}, \mathsf{Snapshot}), (\mathsf{Snapshot}, \mathsf{rel}, \mathsf{Snapshot}),$$
$$(\mathsf{Entity}, \mathsf{snapshot}, \mathsf{Snapshot}), (\mathsf{Snapshot}, \mathsf{derived}, \mathsf{Snapshot})\}.$$

Definition 5. A **temporal Event Knowledge Graph (tEKG)** is an LPG $(N, R, \gamma, \lambda, \rho)$ that conforms to the schema S' and every node $n \in N$ has the properties as in an EKG (see Definition 4) as well as the following property:

- If $n \in N^{\mathsf{Snapshot}}$, then $\rho(n, \mathsf{id}) \in (\mathbb{U}_{oid} \times \mathbb{U}_{time}) \cup ((\mathbb{U}_{oid} \times \mathbb{U}_{time}) \times (\mathbb{U}_{oid} \times \mathbb{U}_{time}))$ and $\rho(n, \mathsf{type}) \in \mathbb{U}_{otype}$.

By Definition 5, every node with the label Snapshot in a tEKG has attributes id and type, with the value of a snapshot identifier and an object type, respectively. The snapshot identifier is a tuple of an object identifier and a time, or a tuple of such tuples. In the latter case, the corresponding snapshot is called a *reified snapshot*.

5   Transformation

Given the notion of a tEKG, we now specify the transformation algorithm that converts logs from the OCEL 2.0 format into a corresponding tEKG representation. Algorithm 1 defines the main part of the transformation, which is complemented by a procedure for creating the directly-follows edges (Algorithm 2). Additionally, the algorithm uses a procedure called AddNode for adding a node to the tEKG, and a procedure called AddEdge for adding an edge; due to space constraints, the detailed pseudo code for the latter two procedures is available only in the extended version of this paper [11].

After initializing the tEKG to be populated (line 1 in Algorithm 1), the algorithm initializes a data structure—captured by function $\wp$—for tracking the nodes in the tEKG that have been created based on specific elements of the log (line 2). Next, a node for the log itself is added to the tEKG (lines 3–4). The next step involves adding a node with label Class to the tEKG for each event type in the log (lines 5–7). After that, for each event in the log, a node with label Event is added and connected to the previously-created nodes for both the log and the corresponding event type (lines 8–12).

The algorithm then iterates over all objects in the log (lines 13–23). For each object, it adds a node with label Entity (line 14). Next, it identifies all timestamps at which the value of an attribute of the object has changed (line 16). The algorithm then iterates over these timestamps, adding a node with label Snapshot to the tEKG and linking it to the corresponding entity (lines 17–20). Snapshots are added if the object has a value over time. If an object has no attribute and value in OCEL (i.e., it is not initiated or related to an event), the algorithm creates no snapshot for it. Finally, the algorithm creates edges between such snapshot nodes to represent the updates occurring over time (lines 21–23).

The next step involves adding a rel-labeled edge for every object-to-object relationship between any two objects o_1 and o_2 (line 25). After that, the algorithm iterates over all snapshot edges created for the node corresponding to o_1 in tEKG. It then collects all snapshots of the node corresponding to o_2 in tEKG that occurred before the snapshot of o_1, and puts them into a set named $\mathcal{R}_{st}$ (line 27). This set is used to filter the last valid snapshot for o_2 at the time of o_1, to which we can link the snapshot (line 29). In our running example (see Fig. 3), the snapshot $(c1, t1)$ could be linked to $(a1, t1)$, which is a snapshot of the related object with a timestamp that is less than or equal to $(c1, t1)$.

The next step focuses on reified entities and reified snapshots. Here, the algorithm iterates over all rel-labeled edges in the tEKG and adds a node for each of them, as well as an edge from this node to the start and end node of each of these edges (lines 32–36).

Next, the algorithm iterates over all event-to-object relationships in the log (lines 37–46), performing the following operations for each of them: *First*, it adds an edge between the corresponding event and entity (line 38). *Second*, it iterates over all reified entities derived from the corresponding entity and adds an edge from the corresponding event to each of them to tEKG aligned with

---

Algorithm 1: Converting an OCEL 2.0 log L into a tEKG G.

Input: $L = (E, O, EA, OA, evtype, evid, time, objtype, objid, eatype, oatype,$
$\qquad\qquad eaval, oaval, E2O, O2O)$

Output: $G = (N, R, \gamma, \lambda, \rho)$

1 Create G as an initially empty tEKG;

2 Let $\wp : E \cup O \cup \mathbb{U}_{etype} \cup (O \times \mathbb{U}_{time}) \cup (O \times O) \to N$ be an initially empty
$\quad$ helper function that maps elements of L to nodes created for them;

$\quad$ // add a node for the log

3 $N \leftarrow N \cup \{log\}$, where log is a new node that is not in N (i.e., $log \notin N$);

4 Extend λ such that $\lambda(log) = \mathsf{log}$;

$\quad$ // add a node for each event type

5 **foreach** $c \in \mathbb{U}_{etype}$ **do**

6 $\quad\quad$ $n \leftarrow$ AddNode$(c, \mathsf{Class}, G, L)$;

7 $\quad\quad$ Extend $\wp$ such that $\wp(c) = n$;

$\quad$ // add a node for each event and connect it to both ...

8 **foreach** $e \in E$ **do**

9 $\quad\quad$ $n \leftarrow$ AddNode$(e, \mathsf{Event}, G, L)$;

10 $\quad\quad$ Extend $\wp$ such that $\wp(e) = n$;

11 $\quad\quad$ AddEdge$(G, log, n, \mathsf{has}, \emptyset)$; $\qquad\qquad$ // ... the log node and the

12 $\quad\quad$ AddEdge$(G, n, \wp(evtype(e)), \mathsf{observed}, \emptyset)$; $\qquad$ // node of its class

$\quad$ // add a node for each object

13 **foreach** $o \in O$ **do**

14 $\quad\quad$ $n \leftarrow$ AddNode$(o, \mathsf{Entity}, G, L)$;

15 $\quad\quad$ Extend $\wp$ such that $\wp(o) = n$;

16 $\quad\quad$ $\mathcal{O}_{st} \leftarrow \{t \in \mathbb{U}_{time} | (o, oa, t) \in dom(oaval)$ for some $oa \in OA\}$;

$\quad\quad$ // add a node for each object snapshot and connect ...

17 $\quad\quad$ **foreach** $t \in \mathcal{O}_{st}$ **do**

18 $\quad\quad\quad\quad$ $n' \leftarrow$ AddNode$((o, t), \mathsf{Snapshot}, G, L)$;

19 $\quad\quad\quad\quad$ Extend $\wp$ such that $\wp((o, t)) = n'$;

20 $\quad\quad\quad\quad$ AddEdge$(G, n, n', \mathsf{snapshot}, \emptyset)$; $\qquad$ // ... it to the object node

$\quad\quad$ // connect the object snapshots in their temporal order

21 $\quad\quad$ **foreach** $t_1, t_2 \in \mathcal{O}_{st}$ **do**

22 $\quad\quad\quad\quad$ **if** $t_1 < t_2$ and there is no $t_3 \in \mathcal{O}_{st}$ such that $t_1 < t_3 < t_2$ **then**

23 $\quad\quad\quad\quad\quad\quad$ AddEdge$(G, \wp((o, t_1)), \wp((o, t_2)), \mathsf{rel}, \mathsf{update})$;

$\quad$ // connect objects and their snapshots using qualifiers

24 **foreach** $(o_1, q, o_2) \in O2O$ **do**

25 $\quad\quad$ AddEdge$(G, \wp(o_1), \wp(o_2), \mathsf{rel}, q)$;

26 $\quad\quad$ **foreach** $r \in R^{snapshot}$ with $\gamma(r) = (\wp(o_1), os_1)$ **do**

27 $\quad\quad\quad\quad$ $\mathcal{R}_{st} \leftarrow \{r' \in R^{snapshot} \mid \rho(os_2, \mathsf{time}) \leq \rho(os_1, \mathsf{time})$ with
$\quad\quad\quad\quad\quad\quad \gamma(r') = (\wp(o_2), os_2)\}$;

28 $\quad\quad\quad\quad$ **foreach** $r' \in \mathcal{R}_{st}$ for which there is no $r'' \in \mathcal{R}_{st}$ with $\gamma(r'') = (\wp(o_2), os'_2)$
$\quad\quad\quad\quad$ such that $\rho(os_2, \mathsf{time}) < \rho(os'_2, \mathsf{time})$ **do**

$\quad\quad\quad\quad\quad\quad$ // connect existing snapshots at a time ...

29 $\quad\quad\quad\quad\quad\quad$ AddEdge$(G, os_1, os_2, \mathsf{rel}, q)$;

```
   // add nodes for reified entities and reified snapshots
32 foreach r ∈ Rʳᵉˡ with γ(r) = (℘(o₁), ℘(o₂)) do
33     n ← AddNode((o₁, o₂), label, G, L), where label = λ(℘(o₁));
34     Extend ℘ such that ℘((o₁, o₂)) = n;
35     AddEdge(G, n, ℘(o₁), derived, ∅);
36     AddEdge(G, n, ℘(o₂), derived, ∅) ;
37 foreach (e, q, o) ∈ E2O do
       // connect event nodes to corresponding entity nodes
38     AddEdge(G, ℘(e), ℘(o), corr, q);
39     foreach r ∈ Rᵈᵉʳⁱᵛᵉᵈ with γ(r) = (o', o) do
40         AddEdge(G, ℘(e), o', corr, q);
       // connect event nodes to corresponding snapshot nodes
41     R_st ← {r ∈ Rˢⁿᵃᵖˢʰᵒᵗ | ρ(os₁, time) ≤ ρ(℘(e), time) with
             γ(r) = (℘(o), os₁)};
42     foreach r ∈ R_st with γ(r) = (℘(o), os₁) do
43         if there is no r' ∈ R_st with γ(r') = (℘(o), os₂) such that
              ρ(os₁, time) < ρ(os₂, time) then
44             AddEdge(G, ℘(e), os₁, corr, q);
45             foreach r'' ∈ Rᵈᵉʳⁱᵛᵉᵈ with γ(r'') = (os₃, os₁) do
46                 AddEdge(G, ℘(e), os₃, corr, q);

47 G ← AddDFs(G) ;                                    // add directly-follows edges
48 return G;
```

design choice made in [4] (line 40). *Third*, it retrieves a set of snapshots for the corresponding object that existed at the time of the event (line 41). *Fourth*, it connects the corresponding event to the last valid snapshot (line 44), as well as connecting the event to all derived snapshots of the given snapshot aligning with the same design choice for reified entities made in [4] (line 46).

In the end, Algorithm 2 is called (line 47). This algorithm receives the current tEKG as input, adds relevant directly-follows edges to it, and returns the updated graph as output. The algorithm consists of three *phases*: adding all directly-follows edges, identifying edges that add new information, and removing the ones that do not.

More specifically, in the *first phase*, the algorithm iterates over any two corr edges that are targeting the same entity from two events. If there are no other events occurring in between that have a corr edge to the same entity, the algorithm adds an edge with label df between those two events (line 4). It also sets the value of the ent and type attributes of the added edge to the value of id and type of the entity (line 5).

The *second phase* identifies all df-labeled edges that provide new information (lines 6–10). To this end, the algorithm applies the same rule as defined by Fahland [5], stating that not all df edges created for derived entities provide additional information. Specifically, if there is a derived node o related to o' for which there exist df edges between two events, the df edge created for the derived entity o does not add new information.

In the *last phase*, the algorithm removes df edges that do not add new information with the condition that there shall not be any similar df edges both before and after them that are among the added information df.

Algorithm 2: Extending a given tEKG with directly-follows edges.

---

1 **Function** AddDFs:

   Input: $G = (N, R, \gamma, \lambda, \rho)$

   Output: G, extended with directly-follows edges

   // add directly-follows edges between event nodes

2   **foreach** $r_1, r_2 \in R^{corr}$ *with* $\gamma(r_1) = (e_1, o)$ *and* $\gamma(r_2) = (e_2, o)$ *such that* $e_1 \neq e_2$ **do**

3   **if** *there is no* $r_3 \in R^{corr}$ *with* $\gamma(r_3) = (e_3, o)$ *such that* $e_1 \neq e_2 \neq e_3$ *and* $\rho(e_1, \text{time}) < \rho(e_3, \text{time}) < \rho(e_2, \text{time})$ **then**

4          $r \leftarrow$ AddEdge$(G, e_1, e_2, \text{df}, \emptyset)$;

5   Extend ρ such that $\rho(r, \text{type}) = \rho(o, \text{type})$ and $\rho(r, \text{ent}) = \rho(o, \text{id})$;

   // identify directly-follows edges providing new information

6    $I \leftarrow \emptyset$;

7   **foreach** *label* $\in$ {Entity, Snapshot} **do**

8   **foreach** $r \in R^{df}$ and $o \in N^{label}$ such that $\rho(r, \text{ent}) = \rho(o, \text{id})$ **do**

9   **if** there is no $r' \in R^{df}$ and $o' \in N^{label}$ and $r'' \in R^{derived}$ such that $\rho(r', \text{ent}) = \rho(o', \text{id})$ and $\gamma(r) = \gamma(r')$ and $\gamma(r'') = (o, o')$ **then**

10             $I \leftarrow I \cup \{r\}$;

   // remove directly-follows edges not providing new information

11   **foreach** $r_1, r_2 \in R^{df}$ *such that* $r_1 \neq r_2$ *and* $\gamma(r_1) = \gamma(r_2)$, *with* $\gamma(r_1) = (e_1, e_2)$ **do**

12   **if** $r_1 \notin I$ **then**

13   **if** there are no $r_3, r_4 \in R^{df}$ with $\gamma(r_3) = (e', e_1)$ and $\gamma(r_4) = (e_2, e'')$ such that $\rho(r_1, \text{ent}) = \rho(r_2, \text{ent}) = \rho(r_3, \text{ent})$ and $r_3 \in I$ and $r_4 \in I$ **then**

14             $R \leftarrow R \backslash \{r_1\}$;

15             $dom(\gamma) \leftarrow dom(\gamma) \backslash \{r_1\}$;

16             $dom(\lambda) \leftarrow dom(\lambda) \backslash \{r_1\}$;

17   **foreach** $att \in \mathbb{U}_{att}$ **do**

18   **if** $(r_1, att) \in dom(\rho)$ **then**

19                   $dom(\rho) \leftarrow dom(\rho) \backslash \{(r_1, att)\}$;

20   **return** G;

6 Concluding Remarks

This paper introduces and formalizes temporal Event Knowledge Graphs (tEKGs), which are designed to record object-centric event data and to facilitate the tracking of changes in entity attributes over time - like OCEL 2.0 specification but for logs to be stored and processed as graphs. For instance, consider the price of item, which can fluctuate; tEKGs allow for the analysis of events with respect to the accurate price at any given time, before or after any changes. This capability is crucial for conducting effective data-driven analyses in real-world scenarios. Moreover, the paper presents and implements an algorithm to transform Object-Centric Event Logs (OCEL) 2.0 into tEKGs.

As a future direction, we aim to provide a complete formal definition of temporal event knowledge graphs by eliciting requirements for object-centric event data based on different case studies. Investigating the practical applications of tEKGs could provide deeper insights into business processes and decision making.

Acknowledgements. Khayatbashi's and Hartig's contributions to this work were funded by Vetenskapsrådet (the Swedish Research Council, project reg. no. 2019-05655).

References

1. Angles, R., Arenas, M., Barceló, P., Hogan, A., Reutter, J., Vrgoč, D.: Foundations of modern query languages for graph databases. ACM Comput. Surv. **50**(5) (2017)
2. Berti, A., et al.: OCEL (Object-Centric Event Log) 2.0 specification. https://www.ocel-standard.org/2.0/ocel20_specification.pdf (2023)
3. Esser, S., Fahland, D.: Storing and querying multi-dimensional process event logs using graph databases. In: Di Francescomarino, C., Dijkman, R., Zdun, U. (eds.) BPM 2019. LNBIP, vol. 362, pp. 632–644. Springer, Cham (2019). https://doi.org/10.1007/978-3-030-37453-2_51
4. Esser, S., Fahland, D.: Multi-dimensional event data in graph databases. J. Data Semant. **10**(1–2), 109–141 (2021)
5. Fahland, D.: Process mining over multiple behavioral dimensions with event knowledge graphs. In: Process Mining Handbook, pp. 274–319. Springer (2022)
6. Bellatreche, L., et al. (eds.): ADBIS 2021. CCIS, vol. 1450. Springer, Cham (2021). https://doi.org/10.1007/978-3-030-85082-1
7. Gherissi, W., El Haddad, J., Grigori, D.: Object-centric predictive process monitoring. In: International Conference on Service-Oriented Computing, pp. 27–39. Springer (2022)
8. Jalali, A.: Object type clustering using Markov directly-follow multigraph in object-centric process mining. IEEE Access **10**, 126569–126579 (2022)
9. Jalali, A., Johannesson, P.: Multi-perspective business process monitoring. Int. Workshop Bus. Process Model., Dev. Support (2013)
10. Khayatbashi, S., Hartig, O., Jalali, A.: Transforming event knowledge graph to object-centric event logs: a comparative study for multi-dimensional process analysis. In: International Conference on Conceptual Modeling, pp. 220–238. Springer (2023)

11. Khayatbashi, S., Hartig, O., Jalali, A.: Transforming object-centric event logs to temporal event knowledge graphs (extended version) (2024). https://arxiv.org/abs/2406.07596v1
12. Kimball, R., Ross, M.: The Data Warehouse Toolkit: The Complete Guide to Dimensional Modeling. John Wiley & Sons, Hoboken (2011)
13. Klijn, E.L., Mannhardt, F., Fahland, D.: Aggregating event knowledge graphs for task analysis. In: International Conference on Process Mining, pp. 493–505. Springer (2022)
14. Swevels, A., Dijkman, R., Fahland, D.: Inferring missing entity identifiers from context using event knowledge graphs. In: International Conference on Business Process Management (2023)
15. van der Aalst, W.: Object-centric process mining: unraveling the fabric of real processes. Mathematics 11(12), 2691 (2023)
16. van der Aalst, W., Berti, A.: Discovering object-centric petri nets. Fund. Inform. 175(1–4), 1–40 (2020)

2nd International Workshop on Change, Drift, and Dynamics of Organizational Processes (ProDy 2024)

2nd International Workshop on Change, Drift, and Dynamics of Organizational Processes (ProDy 2024)

The Workshop on Change, Drift, and Dynamics of Organizational Processes (ProDy) brought together researchers who are interested in how socio-technical processes are performed and change over time. There are at least three perspectives on socio-technical processes that the workshop aimed to encompass. First, the workshop aimed to attract managerial contributions that explain how organizations can recognize and capitalize on change opportunities in processes. Second, we encouraged algorithmic contributions that help to detect and make drift in processes visible. Third, we also welcomed contributions that further develop our understanding of why and how processes change over time. We organized and held the 2nd Workshop on Change, Drift, and Dynamics of Organizational Processes in conjunction with the 22nd International Conference on Business Process Management in Krakow, Poland. We received a total of six submissions. Each submission underwent regular peer-review by three members of the program committee. We accepted three papers for presentation at the workshop and invited an additional speaker to present an ongoing research project.

The workshop comprised two sessions. Both sessions consisted of paper presentations followed by a discussant presentation given by one of the workshop chairs, and discussion among the general audience. First, Vidgof [1] presented his paper on how different types of concept drift impact process complexity. Second, Buder gave his invited presentation on the change of organizational practices and how they can lead to changes in the mission of an organization. Third, Löhr et al. [2] introduced the LongSWORD, an enhanced framework for workaround detection. Fourth, Zhang et al. [3] suggested how research on process change can use topological data analysis to visualize dynamics of outpatient medical clinics.

We would like to thank everyone who contributed to making the workshop possible in one way or another. We thank all authors, committee members, the workshop chairs, and all attendees for their contributions to the workshop. For the next iterations of the workshop, we encourage all research that contributes to our knowledge of change, drifts, and dynamics of organizational processes. We encourage papers from different research streams, following different methods, and from different epistemological positions.

References

1. Vidgof, M.: First insights into the impact of concept drift on process complexity. In: BPM 2024 Workshop Proceedings. LNBIP (2024)
2. Löhr, B., Bartelheimer, C., Köhne, F., Nordlohne, S., Alile, D., Latten, A.: Forging the LongSWORD: exaptation and enhancement of the SWORD framework for workaround detection. In: BPM 2024 Workshop Proceedings. LNBIP (2024)

3. Zhang, L., Wolf, J.R., Pentland, A.P., Pentland, B.T.: Visualizing routine dynamics in outpatient Medical clinics with topological data analysis. In: BPM 2024 Workshop Proceedings. LNBIP (2024)

Organization

Workshop Chairs

Bastian Wurm	LMU Munich School of Management, Germany
Waldemar Kremser	Johannes Kepler Universität Linz, Austria
Jan Mendling	Humboldt-Universität zu Berlin, Germany

Program Committee

Han van der Aa	University of Vienna, Austria
Markus Becker	University of Southern Denmark, Denmark
Iris Beerepoot	Utrecht University, the Netherlands
Phil Hennel	University of Cologne, Germany
Marta Indulska	University of Queensland, Australia
Kalle Lyytinen	Case Western Reserve University, USA
Brian Pentland	Michigan State University, USA
Michael Rosemann	Queensland University of Technology, Australia
Christoph Rosenkranz	University of Cologne, Germany
Maxim Vidgof	Vienna University of Economics and Business, Austria

Forging the LongSWORD: Exaptation and Enhancement of the SWORD Framework for Workaround Detection

Bernd Löhr[1]([✉])(iD), Christian Bartelheimer[1](iD), Frank Köhne[2],
Sina Nordlohne[2], Daniel Alile[2], and Andrees Latten[2]

[1] Faculty of Business Administration and Economics, Paderborn University,
Paderborn, Germany
bernd.loehr@uni-paderborn.de
[2] viadee Consulting AG, Münster, Germany
http://www.viadee.de

Abstract. Full Paper - Regular Research Paper: Workarounds are goal-driven adaptations of business processes that employees implement to overcome perceived constraints at work. While process deviations can be easily detected with data-driven methods like process mining, it is hard to distinguish workarounds from related, yet distinct, concepts. The SWORD framework constitutes a state-of-the-art method for the data-driven detection of workarounds in business process event logs based on pre-defined patterns extracted from support processes in the healthcare domain. However, currently, SWORD has solely been applied to highly standardized processes with low variation and knowledge intensity, while it can be assumed that workarounds more often appear in the latter and also bear bigger potential for innovating processes. Furthermore, SWORD neither comprises data preparation steps nor enables the analysis of workarounds and their implications for the organization. In this paper, we develop an exaptation of the existing SWORD framework, coined LongSWORD, together with two industrial case organizations. Our contribution to theory and practice is threefold. First, we present a framework that enables the preparation of a meaningful event log in alignment with according process model or routine. Second, the detection and analysis of workarounds in core industrial processes is enabled by adding two new cross-case patterns. Third, the LongSWORD method enables others to assess the implications of workarounds beyond its individual implementers.

Keywords: Workarounds · Workaround Detection · Process Science · Causal Inference

1 Introduction

Process Mining provides methods and tools for the data-driven analysis and improvement of business processes. Corresponding techniques are classified into six subgroups: process discovery, conformance checking, performance

© The Author(s), under exclusive license to Springer Nature Switzerland AG 2025
K. Gdowska et al. (Eds.): BPM 2024 Workshops, LNBIP 534, pp. 319–331, 2025.
https://doi.org/10.1007/978-3-031-78666-2_24

analysis, comparative process mining, predictive process mining, and action-oriented process mining [1]. Organizations, however, regularly only implement two techniques—discovery and conformance checking. While process discovery enables the data-driven development of process models, conformance checking aims to detect deviations of process behavior in the data traces compared to-be process models.

Traditional views of Business Process Management (BPM) assume that deviations from to-be process models refer to inefficiencies in process execution and need to be prevented [7]. Recent research has identified that, among others, such deviations can also refer to so-called workarounds [9,28]. Workarounds are goal-driven adaptations that participants implement individually to overcome obstacles induced by misfits [3]. They can trigger bottom-up process innovation by either adapting the process or resolving the underlying misfit [3]. However, manual identification and analysis is a complex and time-consuming task.

Alongside qualitative approaches such as interviews or observation [4], several data-driven approaches exist, which, e.g., leverage deep learning techniques [28]. These, however, require a considerable time investment or pre-existing knowledge about the regarded workarounds. One mature approach for the data-driven workaround detection in business processes is the so-called SWORD method (Semi-automated WORkaround Detection) method [25–27]. It leverages different patterns that manifest in the control flow, data, resource, or time perspective of a business process. However, pattern identification, the development of the detection method, and its demonstration were conducted in a healthcare [27] and public administration context [26]. Therefore, it still needs to be investigated whether the approach can be extended to other domains without adaptation.

In this paper, we follow the strategy of exaptation outlined by Gregor and Hevner (2013) [12] to revise and enhance the SWORD method to be applicable in an industry context. We instantiated the Design Science Research (DSR) methodology [21] and adapted, demonstrated, and evaluated our IT artifact coined *LongSWORD* in the production and procure-to-pay processes of two German organizations. From a theoretical perspective, we contribute by extending the SWORD method to allow for the application in an manufacturing context. Further, LongSWORD provides guidance on data preparation and the interpretation of process drift. From a managerial perspective, we provide a holistic method for identifying workarounds, which can be applied to diverse event logs.

The remainder of the paper is structured as follows. In Sect. 2, we outline related work on workarounds and detection methods. In Sect. 3, we present our DSR approach. In Sect. 4, we develop and demonstrate the IT artifact. Section 5 discusses the theoretical and managerial implications of our results and concludes the paper in Sect. 6.

2   Theoretical Background

The potential of workarounds as a trigger for bottom-up process improvements has been recognized only recently [3,5]. From a conceptual view, many related, yet distinct, concepts exist that describe individuals' deviations from prescribed activities in business processes [22]. However, compared to other concepts, workarounds highlight the creative problem-solving capabilities of actors to overcome misfits [3,13,15]. We take up the definition presented by Alter (2014) [2]: A workaround is a goal-driven adaptation, improvisation, or other change to one or more aspects of an existing work system in order to overcome, bypass, or minimize the impact of obstacles, exceptions, anomalies, mishaps, established practices, management expectations, or structural constraints that are perceived as preventing that work system or its participants from achieving a desired level of efficiency, effectiveness, or other organizational or personal goals [2].

Workarounds can be detected with qualitative [4] and quantitative approaches [28]. Qualitative approaches mostly refer to interviews and observations. However, while they can be used to detect workarounds and other types of process deviations [17], such techniques do not allow continuous monitoring of identified deviations due to resource intensity. This property limits their applicability in organizations and reinforces the need for efficient (semi-)automated methods.

Quantitative approaches employ data analytics techniques to identify workarounds based on data traces of business processes and the analysis of additional external information [27]. Beyond the SWORD method, Weinzierl et al. (2022) [28] leveraged a deep learning approach, evaluated on a synthetically enhanced dataset containing manually labeled workarounds in the training set. Wijnhoven et al. (2023) [29] present a process mining approach to identify workaround candidates by comparing the de jure model and the de facto process models. However, human input is still required to categorize the identified workaround candidates and assess their potential impact. Earlier approaches indicate similar challenges and discuss the importance of domain knowledge [18].

The SWORD method [19,25–27] is the most recent data-driven method for detecting workarounds in event logs. The method applies 22 pre-defined workaround patterns identified in a healthcare and retail context. These patterns are sorted into the four categories Control Flow (eight patterns), Data (three patterns), Resource (five patterns), and Time (six patterns), where each checks for common deviations from the intended process model, such as unexpected activities, data values outside boundaries, etc. A notable limitation is, that each pattern can only indicate a workaround, the validation is still done by a domain expert. The updated version of the method [27] is streamlined to optimize the use of domain experts' time and provides more orientation concerning the strength of deviations and the (ex-post) evaluation of the usefulness of patterns. However, the method has predominantly been applied in healthcare [27] and public administration processes [26].

3   Applying the Standard SWORD Patterns

The SWORD method's patterns, initially developed in healthcare [25], and applied to public administration procurement processes [19]. Consequently, current limitations of the approach stem from healthcare-specific factors—such as regulations, financial constraints depending on the patient, and ethical considerations—that potentially constrain the innovation potential of workarounds. Thus, adapting the method for other contexts enables considering a broader range of workaround types [8]. In this paper, we develop a data-driven method for discovering and evaluating workarounds in business processes. We enhance the SWORD method for workaround discovery in healthcare [5, 27] by iteratively adapting it to manufacturing processes based on two case companies.

Safety Solutions Inc. is a German SME with international clients, 240 employees, and a revenue of 43 million Euros in 2022. They produce pressure relief technology using a knowledge-intensive Industry 4.0 process, prioritizing product quality due to the critical nature of their products. The standard process is linear, starting with an unrecorded preparation and scheduling task. The first logged event is batch number assignment, followed by production, quality control, and optional packaging. Quality control may also be performed during production. Re-producing faulty products, including partially completed or completed orders, is part of the production step. The event log (2019–2023) contains 252,850 events across 41,526 orders. 21,097 orders (50.8%) follow the standard process.

Fuel Logistics Inc. specializes in industrial gases, hydrogen, energy solutions, and petrol stations. This family-owned company operates over 20 sites in Germany, the Netherlands, Belgium, France, Switzerland, and Austria, with 2,000 employees and a 2022 revenue of 2.3 billion Euros. The Purchase-to-Pay (P2P) process, supported by an enterprise system, includes Order Requests, Purchase Orders, Order Confirmations, Goods Receipts, Invoice Receipts, and Billing Document activities. The event log (2022–2023) includes 21 ERP system tables comprising 336,734 observations in 103,625 cases.

We used different BI-tooling methods like PowerBI to identify the different SWORD patterns. We further used R and respective packages for the analysis, including a process map generated with bupaR [14] in both cases. In the following, we will explore the applicability of the provided patterns and their findings. However, several patterns were not applicable due to the properties of the process or the event log. These are marked in Table 1. We generally categorized the outcome as follows: True Positives denote found and verified workarounds, and True Negatives/False Negatives denote no findings or workarounds, which remained hidden. False positives denote candidates that were no workaround according to domain experts.

Control Flow Patterns: Patterns (2), (3), and (7) were frequently observed in the Safety Solutions Inc. data set. Despite the linear to-be process, domain experts had reasonable explanations for the workarounds. Therefore, we classified them as false positives. Pattern (4) yielded a true positive, identifying a pre-packaging step used in cases with limited time or storage space. At Fuel Logistics Inc., this pattern revealed cases where invoices are received before the booking of the goods delivery. Exploring pattern (6) at Safety Solutions Inc., we identified a true positive, where a batch number assignment was followed by a quality control task, which was a workaround to deal with large orders. At Fuel Logistics Inc., in some cases, Goods Receipt was followed by Order Confirmation, deviating from the to-be process model, this however was a false positive, as it was a rare, but intended edge case. Pattern (7) Occurence of directly repeating activities and (8) A specific activity is missing in the trace were not identified at Fuel Logistics Inc. A relevant number of Safety Solutions Inc. production cases are *priority orders* with severe time pressure. For such cases, the documentation of events is a subsequent activity. If performed at the end of the process, it invalidates all timestamps for these cases, therefore yielding a false positive for the pattern (7). A similar workaround of delayed documentation was found in the original SWORD paper [25] and labeled as 'batching.'

Data Patterns: The data pattern (9) Objects with value outside boundary was not applicable in both cases since the data sets did not contain appropriate fields. The data pattern (10) Change in between events was also not applicable at Safety Solutions Inc. since no intermediate data are stored, with only the final state of the process data being available. At Fuel Logistics Inc., the enterprise system logged changes automatically. Nonetheless, anomaly detection only revealed an audit-related data change, which cannot be classified as a workaround. The data pattern (11) Information is logged in free-text fields instead of dedicated fields was applicable and led to identifying multiple potential workarounds at Safety Solutions Inc. Test-runs were marked with a prefix in a free text field as well as the production of half-products. In the Fuel Logistics Inc. dataset, only a few text fields exist with slight variance, thus leading to no findings.

Resource Related Patterns: Due to system constraints, resource patterns were either not applicable or did not yield any interesting results for Fuel Logistics Inc. At Safety Solutions Inc., pattern (13) Activities executed by multiple resources led to the identification of a pre-packaging step. Pattern (14) is only a false positive, as we found some intentionally automated process instances.

Time Related Patterns: The patterns for the time component are also only applicable to a limited extent. For example, there are no unusual time windows at Fuel Logistics Inc. (17) due to, e.g., emergency assignments taking place on weekends, on public holidays, or outside standard working hours. Timestamps solely exist if an activity was completed. However, the duration from the start of a trace to an activity (18), the duration between activities (19), and the duration of a complete trace (21) can be analyzed, which yielded false positives, where the duration of trace was exceptionally short or long, but covered the handling of rare edge cases. At Safety Solutions Inc., patterns (17) to (21) were applicable, with pattern (18) revealing a workaround, which covers the early creation of quality-related documents. The other patterns led to false positives. Interestingly, we also identified test data within the productive system by applying the patterns, which did not qualify as workarounds.

4   LongSWORD: An Exaptation of the SWORD Method

We added activities around the core pattern application based on our findings and experience applying the SWORD method. We created two additional patterns, introducing a cross-case perspective. Thus, LongSWORD extends the original method in both temporal directions. We first explain our extension to the core patterns before elaborating on the added steps.

Pattern 23: Cross Case—Workaround Chains: Upon closer inspection of the Safety Solutions Inc. data set and analysis of the interviews, it became clear that workaround chains might exist. For example, the existing misfit of time pressure in *priority orders* leads to the workaround of using unbooked material. This uncertainty created in the inventory leads to backup ordering of larger quantities of raw materials to make sure that *priority orders* can be performed quickly from the stock of raw materials at all times.

Misfit:	→	**Workaround:**	→	**Workaround:**
Priority Order		Unbooked Material		Backup Ordering

Therefore, it seems reasonable to investigate workarounds' causes and effects in a forward/backward analysis. Thereby, analysts can find related workarounds as well as their underlying misfits and generate an understanding of the overall effects of a workaround. Potential tools are root cause analyses as discussed in the process quality literature, six-sigma or the Theory of Constraints [11].

Pattern 24: Cross Case—Drift of Process KPI: A workaround can be dominant and effectively replace the former To-Be process. Therefore, it becomes invisible for all methods that rely on variance detection in a given time frame, even if the analysis is not limited to individual cases. Hence, we propose a drift analysis [23] as an additional SWORD pattern on the global scope. Unlike the existing SWORD patterns, this pattern does not analyze specific process instances but merely identifies hints of process evolution on the process level. Using changes in KPIs as sensors enables estimating the consequences a potential workaround might have on a process or organizational level.

Before—Choose Relevant Patterns: Besides the additional patterns, we extend the core method by several activities. Applying the SWORD method is resource-intensive because it requires both (data) analytics skills and domain knowledge. We propose that only patterns relevant to the case should be applied. Employing causal-directed acyclic graphs (DAG) can help identify the relevant dimensions ex-ante. We will explain this procedure with the following example:

Before & After—Process Instance Interference and Causal Effects: We build upon the example in pattern 23: The *priority order* process at Safety Solutions Inc. will regularly brush aside planned schedules for a day, including the production of semi-finished products. This aligns with the organization's goals and is expected for *priority orders*. Hence, it cannot be identified by investigating variance within *priority order* process data. When investigating regular processes, *priority order* effects are indistinguishable from other delay or quality issues. However, a correlation of *priority order* frequency and regular order performance could indicate the existence of the semi-finished products workaround.

To evaluate the impact of this workaround, it is necessary to investigate the duration of regular orders as a target variable in the same time frame. With workarounds and target variables in the same graph, it is evident that we need to consider the effect of backup orders on the duration of regular orders as well. If the workaround is effective, the duration effect of repurposed material will be reduced at the cost of additional inventory.

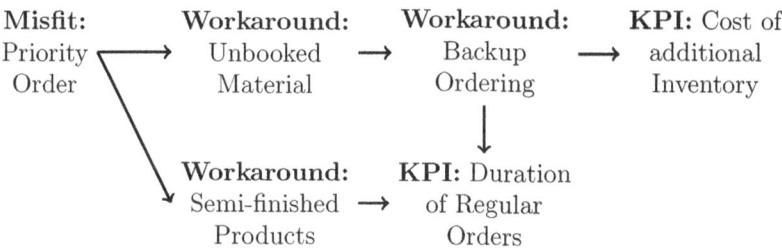

In summary, we show a causal-directed acyclic graph (DAG) [20], where workarounds can be considered as treatment variables. Because after we discovered a workaround, it could be encouraged or discouraged. This decision can be made by investigating the average treatment effect (ATE). Martini et al. (2015) introduce the technical debt metaphor, where short-term and long-term prioritization needs to be balanced and rely on the awareness of the interest that needs to be paid in the future [16]). These DAGs evolve iteratively during the application of the LongSWORD method. Primarily, they help to identify relevant patterns, while secondary, helping to estimate effects and derive actions based on these.

During—Iterative Guidance: While the SWORD method [27] is presented as a linear process, which expects a fixed to-be process and event log, LongSWORD stresses the necessity to focus on both IT artifacts and contextual aspects before and after conducting the activities of the SWORD method iteratively. This is based on the following observations:

1. To-be processes can be a starting point of reference. However, the event data might be on a lower level, where informal rules and organizational routines have to be gathered and codified into the new reference.
2. The event log quality and the corresponding understanding evolve over time. Its first version is often incomplete. Hence, new data sources must be added and integrated with existing data to make workarounds better visible.

The data set and the to-be process evolve iteratively during the analysis and in cooperation with the domain experts. Besides these methodical changes and additions, we also added two new patterns, which take a cross-case perspective.

The whole LongSWORD method is depicted in Fig. 1. It represents an iterative method that acknowledges that the structure of the data set and the

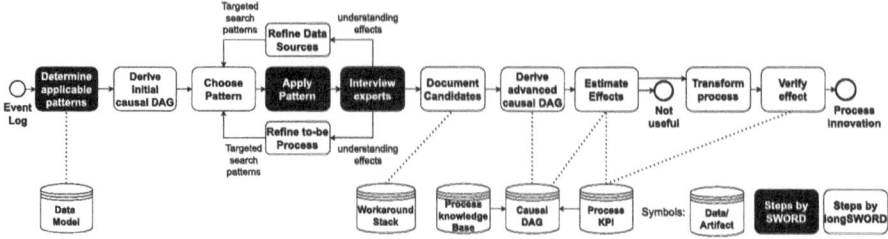

Fig. 1. The LongSWORD method and key artifacts

understanding of the To-Be process will likely (or strategically) evolve during its application. First, it starts with the existing step to determine the applicable patterns based on the provided dataset. From there, an initial causal DAG is derived based on the targeted processes. The estimated effects will then give a hint on which patterns are relevant. Now, as in the original method, the patterns are applied, and the results will be shared with the domain experts. Depending on the outcome, we propose an iterative cycle, where the to-be processes and the data sources will be refined. This shall be repeated until maturity is reached when no additional workarounds can be identified. This procedure is comparable to data science methods such as CRISP-DM [30]. The original SWORD method ends after presenting and confirming the found workarounds with the domain experts. However, we suggest continuing by documenting the found candidates, such that they are known and findable within the workaround stack of the organization [3]. Subsequently, the analysts should refine the causal DAG further to estimate the potential effects of the workaround. This iscrucial for deciding whether to leverage, prohibit, or tolerate the workaround. The approach ends here, in case no action is needed, or it will be continued by transforming the process and verifying the effect estimated in the DAG. Transforming the process is, of course, no trivial task. Early research on this topic, such as a workaround-to-innovation process, exists, which shall help structure such efforts [6].

5   Discussion

At Safety Solutions Inc., 17 out of 24 patterns were applicable, and seven successfully led to verified workarounds, denoted as true positives in reftab:Patterns. Further, a significant amount of eight false positives were found. Only two were applicable, yet no findings were yielded. At Fuel Logistics Inc., our analysis resulted in one true positive, five false positives, and eleven true or false negatives.

However, it should be noted that the respective ERP systems and, therefore, the data and its format limit the analysis. Overall, without the iterative approach regarding the process data as well as the to-be process and organizational routine, the number of findings would have been smaller as, in general, fewer patterns would have been applicable. Nonetheless, it was shown that not all patterns are always applicable in each use case, which does not have to be a weakness of the method, but is a limitation induced by the focal case. In retrospect, the approach to searching predefined patterns in a defined data structure could not be executed as efficiently as planned. Investigating data structures, understanding the inner workings of IT systems, and uncovering known edge cases not yet represented in available process models required considerable effort, especially as an external data analyst. Moreover, data integration challenges and data quality issues proved to be time and resource-consuming. The level of detail regarding process documentation and data documentation, especially regarding customized IT systems, appear to be major determinants for the effort required for workaround detection using LongSWORD. Furthermore, the efficiency of these cyclic activities is primarily constrained by the availability of domain experts and their willingness and openness to discuss workarounds.

So far, the effort of this approach to workaround detection can only be justified since the application of the method yielded other qualitative effects that have their own return on investment. That means, since applying the LongSWORD approach generated an improved understanding of the processes at hand, the interpretation of process trace data, and the causal structure of the KPIs involved, both companies appear to be better prepared regarding the following challenges. First of all, the cross-functional knowledge elicited and documented is valuable in itself from a knowledge management and business continuity point of view, i.e., as a reduction of process debt [15]. Numerous known special cases require special treatment not hitherto covered in normative process documentation.

We started our development process with the observation that the SWORD method [27] cannot fully be applied to an industrial manufacturing context. In its original form, it shows several limitations, for which we provide specific enhancements. First, the method does not provide insights into which patterns should be used for which process. Therefore, we added a loop in the mid-section of the method regarding both the data set and the to-be process. Further, we introduced causal DAGs to the method [20], supporting organizations to choose the relevant patterns and data snippets for the case at hand.

In contrast to the original SWORD method [27], applying LongSWORD requires more time from the domain experts but less time from data analysts. The diverging resource intensity of both methods enables organizations in diverse contexts to employ one or the other method based on the availability of specific resources. However, the application of both methods is relatively resource-intensive, but at the same time, it results in beneficial side effects regarding improved process and data understanding within the organization. Thus, it may be beneficial to approach the idea of systematic workaround detection for process innovation together with a data quality initiative, a process mining or redesign initiative, and culture-focused initiatives for continuous process improvements to benefit from the synergetic effects [10].

Beyond efficiency, the original SWORD method is limited to only providing a single case view while neglecting cross-case effects [24]. In contrast, the LongSWORD method comprises two additional patterns to investigate potential workaround chains and process drift [23]. In our case organizations, we noticed that workarounds may trigger other workarounds as a direct effect. Additionally, because patterns that rely on data variance struggle to identify the frequency with which workarounds are already enacted in a process, a drift analysis can counteract this shortcoming.

One major issue we encountered was the dependency of the SWORD method on available to-be process models. In both of our case organizations, process models solely existed on an abstract level. At the same time, the established routines of the employees were more precise and represented the ground truth of process activities—an issue already observed for SMEs [29]. This aligns with the fifth definition of workarounds found by Ejnefall et al. (2019) [9]. There, workarounds are deviations from the ostensive aspects described by Alter (2014) [2], which includes formal to-be processes and routines. This perspective should be kept in mind if using either the SWORD or the LongSWORD method.

Table 1. Overview of patterns and findings

Perspective	Pattern	Fuel Logistics Inc.	Safety Solutions Inc.
Control flow	1. Occurrence of an activity	-	-
	2. Occurrence of recurrent activity sequence	*TN/FN*	*FP*
	3. Activity frequency out of bounds	*TN/FN*	*FP*
	4. Occurrence of activities in an order different from process model	**TP**	**TP**
	5. Occurrence of mutually exclusive activities	-	-
	6. Occurrence of unusual neighboring activities	*TN/FN*	**TP**
	7. Occurrence of directly repeating activity	*TN/FN*	*FP*
	8. Missing occurrence of activity	*TN/FN*	*TN/FN*
Data	9. Data object with value outside boundary	*TN/FN*	*TN/FN*
	10. Change in value between events	*FP*	-
	11. Specific information in free-text fields	*TN/FN*	**TP**
Resource	12. Activity executed by unauthorized resource	-	-
	13. Activities executed by multiple resources	*TN/FN*	**TP**
	14. Activities executed by a single resource	*TN/FN*	*FP*
	15. Activity frequency out of bounds for a resource	*TN/FN*	-
	16. Value frequency out of bounds for a resource	-	-
Time	17. Occurrence of activity outside of time period	-	*FP*
	18. Delay between start of trace and activity is out of bounds	*FP*	**TP**
	19. Time between activities out of bounds	*FP*	*FP*
	20. Duration of activity out of bounds	-	*FP*
	21. Duration of trace out of bounds	*FP*	*FP*
	22. Delay between event and logging is out of bounds	-	-
Cross-Case	23. Workaround Chains	*TN/FN*	**TP**
	24. Drift of process KPI	*FP*	**TP**

-: Not applicable
TN/FN - True Negative / False Negative: No candidate(s) detected
TP - True Positive: Workaround candidate(s) detected
FP - False Positive: Candidate detected, but not identified as workaround by domain expert

6   Conclusion, Limitations, and Outlook

In this paper, we applied the SWORD framework [27] to two industrial case organizations within the manufacturing context and adapted it to the domain. This includes enhancing the framework by taking additional steps before, during, and after the core pattern application, including an iterative segment and two additional patterns enabling cross-case comparison. The resulting method, LongSWORD, provides guidance for both practitioners and researchers.

Naturally, our results are subject to limitations. First, the artifact was developed with solely two case organizations. Applying LongSWORD to more organizations from diverse contexts would strengthen its robustness. For example, in the Safety Solutions Inc. case, the flexibility and variance of the process posed the main challenge for its analysis. In contrast, in the Fuel Logistics Inc. case, the (statistically speaking) lack of variance was challenging for its analysis and the detection of workarounds, thus leading to a relatively small amount of findings. We expect to find more diverse challenges in other organizations, for example, referring to external dependencies of the process, sufficient freedom for process participants, and data coverage for identifying interesting workarounds as potential process innovations.

Acknowledgements. As part of the Change.WorkAROUND project (promotion sign 02J21C166), this research was funded by the German Federal Ministry of Education and Research.

References

1. van der Aalst, W.M.P.: Process Mining: A 360 Degree Overview, pp. 3–34. Springer International Publishing, Cham (2022)
2. Alter, S.: Theory of workarounds. Commun. Assoc. Inf. Syst. **34**(1), 1041–1066 (2014)
3. Bartelheimer, C., Wolf, V., Beverungen, D.: Workarounds as generative mechanisms for bottom-up process innovation-insights from a multiple case study. Inf. Syst. J. **33**(5), 1085–1150 (2023)
4. Beerepoot, I., Van de Weerd, I.: Prevent, redesign, adopt or ignore: improving healthcare using knowledge of workarounds (2018)
5. Beerepoot, I., van de Weerd, I., Reijers, H.A.: The potential of workarounds for improving processes. In: Di Francescomarino, C., Dijkman, R., Zdun, U. (eds.) BPM 2019. LNBIP, vol. 362, pp. 338–350. Springer, Cham (2019). https://doi. org/10.1007/978-3-030-37453-2_28
6. Beverungen, D., Bartelheimer, C., Assbrock, A., Löhr, B.: Workaround-to-innovation-exploring bottom-up process re-design. In: European Conference on Informations Systems (ECIS) (2024)
7. Carmona, J., van Dongen, B., Solti, A., Weidlich, M.: Conformance checking
8. Di Ciccio, C., Marrella, A., Russo, A.: Knowledge-intensive processes: characteristics, requirements and analysis of contemporary approaches. J. Data Semant. **4**(1), 29–57 (2015)

9. Ejnefjäll, T., Ågerfalk, P.J.: Conceptualizing workarounds: meanings and manifestations in information systems research. Commun. Assoc. Inf. Syst. **45**(1), 20 (2019)

10. Galic, G., Wolf, M.: Global process mining survey 2021: delivering value with process analytics-adoption and success factors of process mining. Deloitte (2021)

11. Goldratt, E.M.: Theory of constraints. North River Croton-on-Hudson (1990)

12. Gregor, S., Hevner, A.R.: Positioning and presenting design science research for maximum impact. MIS Q. **37**(2), 337–355 (2013)

13. Hosseinpour, M., Jans, M.: Auditors' categorization of process deviations. J. Inf. Syst., 1–23 (2024)

14. Janssenswillen, G., Depaire, B., Swennen, M., Jans, M., Vanhoof, K.: bupaR: enabling reproducible business process analysis. Knowl.-Based Syst. **163**, 927–930 (2019)

15. Martini, A., Besker, T., Bosch, J.: Process Debt: a first exploration. In: 2020 27th Asia-Pacific Software Engineering Conference (APSEC), pp. 316–325. IEEE (2020)

16. Martini, A., Bosch, J.: The danger of architectural technical Debt: contagious debt and vicious circles. In: 2015 12th Working IEEE/IFIP Conference on Software Architecture, pp. 1–10. IEEE (2015)

17. Mörike, F.: Inverted hierarchies on the shop floor: the organisational layer of workarounds for collaboration in the metal industry. Comput. Support. Coop. Work (CSCW) **31**(1), 111–147 (2022)

18. Outmazgin, N., Soffer, P.: Business process workarounds: what can and cannot be detected by process mining. In: International Workshop on Business Process Modeling, Development and Support, pp. 48–62. Springer (2013)

19. Outmazgin, N., van der Waal, W., Beerepoot, I., Hadar, I., van de Weerd, I., Soffer, P.: From automatic workaround detection to process improvement: a case study. In: International Conference on Business Process Management, pp. 372–390. Springer (2023)

20. Pearl, J.: Causal diagrams for empirical research. Biometrika **82**(4), 669–688 (1995)

21. Peffers, K., Tuunanen, T., Rothenberger, M.A., Chatterjee, S.: A design science research methodology for information systems research. J. Manag. Inf. Syst. **24**(3), 45–77 (2007)

22. Röder, N., Wiesche, M., Schermann, M., Krcmar, H.: Toward an ontology of workarounds: a literature review on existing concepts. In: 2016 49th Hawaii International Conference on System Sciences (HICSS), pp. 5177–5186. IEEE (2016)

23. Sato, D.M.V., De Freitas, S.C., Barddal, J.P., Scalabrin, E.E.: A survey on concept drift in process mining. ACM Comput. Surv. **54**(9) (2021)

24. Senderovich, A., Di Francescomarino, C., Ghidini, C., Jorbina, K., Maggi, F.M.: Intra and inter-case features in predictive process monitoring: a tale of two dimensions. In: Carmona, J., Engels, G., Kumar, A. (eds.) Business Process Management, pp. 306–323. Springer International Publishing, Cham (2017)

25. van der Waal, W., Beerepoot, I., van de Weerd, I., Reijers, H.A.: The sword is mightier than the interview: a framework for semi-automatic workaround detection. In: Proceedings of the ICPM, pp. 91–106. Springer (2022)

26. van der Waal, W., Kappen, T., van de Weerd, I., Reijers, H.A., Haitjema, S.: Whetting the sword: detecting workarounds by using active learning and logistic regression. In: Proceedings of the 57th Hawaii International Conference on System Sciences, pp. 3687–3696 (2024)

27. van der Waal, W., et al.: Putting the sword to the test: finding workarounds with process mining. Bus. Inf. Syst. Eng., 1–20 (2024)

28. Weinzierl, S., Wolf, V., Pauli, T., Beverungen, D., Matzner, M.: Detecting temporal workarounds in business processes - a deep-learning-based method for analysing event log data. J. Bus. Analyt. **5**(1), 76–100 (2022)
29. Wijnhoven, F., Hoffmann, P., Bemthuis, R., Boksebeld, J.: Using process mining for workarounds analysis in context: learning from a small and medium-sized company case. Int. J. Inf. Manage. Data Insights **3**(1), 100163 (2023)
30. Wirth, R., Hipp, J.: CRISP-DM: towards a standard process model for data mining. In: Proceedings of the 4th International Conference on the Practical Applications of Knowledge Discovery and Data Mining, vol. 1, pp. 29–39. Manchester (2000)

First Insights into the Impact of Concept Drift on Process Complexity

Maxim Vidgof[(✉)] [iD]

Vienna University of Economics and Business, Welthandelsplatz 1, 1020 Vienna,
Austria
maxim.vidgof@wu.ac.at
https://complex.wu.ac.at/nm/vidgof

Abstract. Process mining is concerned with extracting knowledge
about business processes based on digital traces left during their execution. Most existing process mining methods are focused on stable processes. However, processes tend to change over time in response to changing regulations or market conditions but also due to internal dynamics.
Concept drift is a phenomenon in process mining when a process changes
during the timeframe when it is analyzed. Following the change in how a
process is executed, process complexity also changes over time. Despite
this apparent connection, the relationship between these concepts have
not been explicitly studied. In this paper, some theoretical considerations on how concept drift can be reflected in process complexity are
presented. These considerations are then evaluated using artificial and
real-life event logs. The results support the considerations and invite a
multitude of questions for future research.

Keywords: Process complexity · Concept drift · Process mining

1 Introduction

Business processes change over time. This change can be caused by a number of reasons, both internal to the process and external. New regulations or
change in the market can cause the processes to add activities or variants. Process optimization and streamlining initiatives, on the other hand, can eliminate
redundant activities and shrink the variability of the process. Even without any
top-down incentive, processes can still change as process participants optimize
their work, find workarounds, or change preferences from one path to the other.
All these changes in the way a process is executed fall under the category of
concept drift.

Another important aspect of any business process is its complexity. Complexity of business processes has multi-faceted influence on their execution and
ultimately process performance [11]. As processes change over time, so does process complexity. However, the relationship between these two concepts have not
yet been made explicit and studied rigorously. In this paper, the first considerations about the impact of concept drift are presented and evaluated with a
simple artificial scenario as well as a real-life event log.

© The Author(s), under exclusive license to Springer Nature Switzerland AG 2025
K. Gdowska et al. (Eds.): BPM 2024 Workshops, LNBIP 534, pp. 332–337, 2025.
https://doi.org/10.1007/978-3-031-78666-2_25

2   Background

2.1   Concept Drift

Concept drift is a known problem in data mining. It describes the case where the relationship between the input and the target variable changes over time [9]. In process mining, it refers to a situation where the process changes over time.

Drift detection techniques analyze event logs or streams of events in search of concept drift. The simplest techniques merely report whether a drift was detected or not, mode advanced ones can also report when the drift happened, what type of drift it was and which part of the process was affected. Most techniques apply statistical hypothesis testing, i.e. they compute some characteristics of the process in different time windows and compare them using statistical tests. If the difference is significant, a drift is reported. Other techniques rely on trace clustering and detect drift by detecting the changes in cluster composition over time. Change point detection methods convert event logs into multivariate time series and identify changes in their values [14].

There are several aspects of concept drift. First, one can distinguish between momentary and permanent drift. The former is often considered an outlier and is filtered out, while most techniques focus on the latter. Another aspect is the type of drift. One can distinguish between *sudden, gradual, incremental* and *recurring* drift [3]. As process mining deals with multiple perspectives, also concept drift can be related to one (or more) of the following perspectives: *control-flow, time, resource* and *data*. While there are now emerging approaches to study e.g. drifts in resource perspective [5], most approaches focus on the control-flow perspective. [12] defines the following patterns of concept drift in imperative process models: adding/deleting fragments, moving/replacing fragments, adding/removing levels, adding control dependencies, changing transition conditions.

2.2   Process Complexity

Complexity of business processes manifests itself in multiple ways [8]. There have been several approaches to quantitatively assess process complexity, however, they either relied on perceptual measures or on complexity of process models. Recently, event log complexity has received increased attention as it uses evidence of how process was really executed to numerically assess process complexity. It can be distinguished between four categories of event log complexity [1,11]. Complexity measures related to *size* capture the number of traces, events, event classes and lengths of traces. *Variation*-related metrics capture the number of variants of how process can be executed. *Distance*-related metrics try to numerically assess the degree to which these variants differ. Finally, *entropy*-based metrics try to unify these three aspects. Most complexity metrics in these streams relate to control flow, however some approaches to also incorporate data complexity have emerged [10]. Some studies have already delved into change of processes over time and connected change of complexity [7,11,13]. However, none of these papers has explicitly connected process complexity with concept drift. Also, [7] used random drifts in the process, while the observations of business processes claim that the real drift patterns are more limited.

3   Theoretical Considerations

Concept drift implies that there are two versions of the process: *before* and *after* the change point. The claim of this paper is that a drift in control-flow perspective is connected to change in at least one of the complexity metrics. The following observations outline the expected connection in more detail.

Observation 1. If activities are added or removed from the process, this will be captured by size-related complexity metrics, i.e. size-related complexity values will differ in two states.

Observation 2. If process variant is added or removed, this will be captured by variation-related complexity metrics.

Observation 3. If process variants are added or removed, or some process variants are being changed while others remain intact, this will be captured by distance-related complexity metrics.

Observation 4. During the drift, when both process versions coexist, variation-related metrics will see bursts with higher values than both before and after the drift, since they consider both variants from the old process as well as variants from the new process.

Thus, for instance, especially gradual drifts increase the number of variants observable in transition period, which will be captured by variation-related metrics. After the old version is replaced, however, complexity will drop again. In case of recurring drift, these dynamics will be observed multiple times if the windows are granular enough. Otherwise, with coarse-grained windows, it may appear that variant-related complexity is permanently high.

4   Evaluation

In order to empirically evaluate the connection between process complexity and concept drift, both artificial and real-life event logs were used. This section describes the evaluation scenario and the observed results.

4.1   Artificial Logs

As a starting point, this work uses the first pattern described in Sect. 2.1, namely deleting the activities. The initial process model is a simple order-to-cash process described in [2]. It is a sequential process with single variant having 6 activities. Three of these activities are removed during the drift. The simulations were performed using CDLG tool [4]. For the sake of brevity, only a brief excerpt of the results is presented, however, full results together with the process model are available on GitHub[1] (Fig. 1).

---

[1] https://github.com/MaxVidgof/complexity-drift-data.

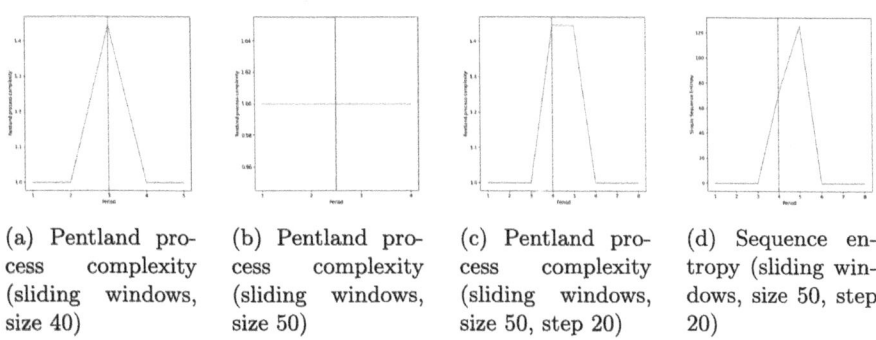

(a) Pentland process complexity (sliding windows, size 40)

(b) Pentland process complexity (sliding windows, size 50)

(c) Pentland process complexity (sliding windows, size 50, step 20)

(d) Sequence entropy (sliding windows, size 50, step 20)

Fig. 1. Complexity over multiple periods in presence of sudden drift.

As activities are removed, size-related complexity metrics report lower values after the drift. Bursts in variation-related complexity can indeed be seen. However, only if window size is such that some windows include both old and new version of the process. In case of gradual drift, the entire drift period has higher variation-related complexity. Entropy-based complexity can offer a more detailed view as is becomes more evident when and how many of the new versions are added and when old versions start to die out. In incremental drift, every drift point is associated with a burst in variation-related complexity.

4.2   Real-Life Logs

In the Italian helpdesk log[2], there are 2 known drifts: on July 25th 2011, and on September 11th 2012 [6, p. 45]. For the evaluation, this event log was split into non-overlapping monthly windows by trace start. The change in complexity observed in this event log is multi-faceted, however some complexity metrics seem to correlate with drifts. For instance, magnitude and support (size-related metrics) both drop around the drift points. Average trace length drops at the first drift and recovers at the second, while staying low in between. Both drifts are also characterized by abrupt falls in Pentland task complexity and Lempel-Ziv complexity (variation-related metrics) around the drift points, signalling a reduction in the variance of the process around these timestamps. This is also mirrored by the lower number of distinct traces.

4.3   Discussion

Even the first step of the evaluation allowed to confirm some of the observations and to make further ones. As in the artificial log the drift was related to number of activities, size-related metrics are different before and after the drift, confirming Observation 1. Bursts in complexity predicted by Observation 4 can indeed be observed, however, under some additional conditions. These bursts are only

---

[2] https://data.4tu.nl/articles/_/12675977/1.

seen when in some window both the old and the new versions of the process coexist. This, however, cannot be guaranteed by non-overlapping windows. Thus, in contrast to concept drift detection techniques that benefit from non-overlapping windows, observation of process complexity over time should be done with overlapping windows. This also provides the explanation to bursts in complexity observed in [7]: such bursts are present only for variation-related complexity and only if windowing strategy captures the coexistence of the process versions.

5   Conclusion

Concept drift is a phenomenon where a process changes while being analyzed. These changes have been classified, and algorithms have been developed for their detection. However, the relationship between concept drift and complexity of business processes has not been studied. In this paper, a first step has been made towards explaining this connection. Preliminary results support the expected relation, however, many open questions for further exploration remain.

5.1   Future Work

This paper presents several directions for future work. First, it is planned to extend the evaluation on artificial logs in multiple directions. This includes testing multiple drift types described in [3], multiple scenarios described in [12], employing further windowing strategies, including cumulative complexity measurements similar to [10], as well as evaluating on further existing artificial logs [9].

Second, more thorough evaluation using real-life logs is planned. This includes not only employing more event logs with known drifts but also more thorough data preparation (e.g. noise filtering) and more detailed analysis of the results.

Third, it is worth explicitly incorporating the model complexity into this discussion. In presence of concept drift, the models discovered from different observation periods may not only differ in their structure and content but also in their complexity, which would add another interesting angle to this phenomenon.

Fourth, given the observed connection, it may be reasonable to also develop a drift detection technique based on process complexity or at least incorporate process complexity as one of the variables into drift detection techniques based on statistical tests or multivariate time series analysis.

Finally, the opposite direction of this connection can be studied, as, for instance, increased complexity of the process may lead to its intentional or unintentional simplification, i.e. concept drift.

References

1. Augusto, A., Mendling, J., Vidgof, M., Wurm, B.: The connection between process complexity of event sequences and models discovered by process mining. Inf. Sci. **598**, 196–215 (2022). https://doi.org/10.1016/J.INS.2022.03.072

2. Dumas, M., Rosa, M.L., Mendling, J., Reijers, H.A.: Fundamentals of Business Process Management, 2nd edn. Springer (2018)
3. Gama, J., Zliobaite, I., Bifet, A., Pechenizkiy, M., Bouchachia, A.: A survey on concept drift adaptation. ACM Comput. Surv. **46**(4), 44:1–44:37 (2014). https://doi.org/10.1145/2523813
4. Grimm, J., Kraus, A., van der Aa, H.: CDLG: a tool for the generation of event logs with concept drifts. In: Janiesch, C., et al. (eds.) Proceedings of the Best Dissertation Award, Doctoral Consortium, and Demonstration & Resources Track at BPM 2022 Co-located with 20th International Conference on Business Process Management (BPM 2022), Münster, Germany, 11–16 September 2022. CEUR Workshop Proceedings, vol. 3216, pp. 92–96. CEUR-WS.org (2022). https://ceur-ws.org/Vol-3216/paper_241.pdf
5. Klijn, E.L., Mannhardt, F., Fahland, D.: Multi-perspective concept drift detection: including the actor perspective. In: International Conference on Advanced Information Systems Engineering, pp. 141–157. Springer (2024)
6. Ostovar, A., Leemans, S.J.J., Rosa, M.L.: Robust drift characterization from event streams of business processes. ACM Trans. Knowl. Discov. Data **14**(3), 30:1–30:57 (2020). https://doi.org/10.1145/3375398
7. Pentland, B.T., Liu, P., Kremser, W., Hærem, T.: The dynamics of drift in digitized processes. MIS Q. **44**(1) (2020). https://misq.org/the-dynamics-of-drift-in-digitized-processes.html
8. Revina, A., Aksu, Ü., Meister, V.G.: Method to address complexity in organizations based on a comprehensive overview. Information **12**(10), 423 (2021). https://doi.org/10.3390/info12100423
9. Sato, D.M.V., Freitas, S.C.D., Barddal, J.P., Scalabrin, E.E.: A survey on concept drift in process mining. ACM Comput. Surv. **54**(9), 189:1–189:38 (2022). https://doi.org/10.1145/3472752
10. Vidgof, M., Mendling, J.: Leveraging event data for measuring process complexity. In: Montali, M., Senderovich, A., Weidlich, M. (eds.) Process Mining Workshops - ICPM 2022 International Workshops, Bozen-Bolzano, Italy, 23–28 October 2022, Revised Selected Papers. LNBIP, vol. 468, pp. 84–95. Springer (2022). https://doi.org/10.1007/978-3-031-27815-0_7
11. Vidgof, M., Wurm, B., Mendling, J.: The impact of process complexity on process performance: a study using event log data. In: Francescomarino, C.D., Burattin, A., Janiesch, C., Sadiq, S. (eds.) Business Process Management - 21st International Conference, BPM 2023, Utrecht, The Netherlands, 11–15 September 2023, Proceedings. LNCS, vol. 14159, pp. 413–429. Springer (2023). https://doi.org/10.1007/978-3-031-41620-0_24
12. Weber, B., Reichert, M., Rinderle-Ma, S.: Change patterns and change support features - enhancing flexibility in process-aware information systems. Data Knowl. Eng. **66**(3), 438–466 (2008). https://doi.org/10.1016/J.DATAK.2008.05.001
13. Wurm, B., Grisold, T., Mendling, J., vom Brocke, J.: Measuring fluctuations of complexity in organizational routines. In: Academy of Management Proceedings, vol. 2021, p. 13388. Academy of Management Briarcliff Manor (2021)
14. Yeshchenko, A., Ciccio, C.D., Mendling, J., Polyvyanyy, A.: Visual drift detection for event sequence data of business processes. IEEE Trans. Vis. Comput. Graph. **28**(8), 3050–3068 (2022). https://doi.org/10.1109/TVCG.2021.3050071

Visualizing Routine Dynamics in Outpatient Medical Clinics with Topological Data Analysis

Li Zhang[1]([⊠]), Julie Ryan Wolf[2], Alice P. Pentland[2], and Brian T. Pentland[1]

[1] Eli Broad College of Business, Michigan State University, East Lansing, MI 48824, USA
zhan2070@msu.edu, pentland@broad.msu.edu
[2] School of Medicine and Dentistry, University of Rochester, Rochester, NY 14642, USA
{julie_ryan,Alice_Pentland}@urmc.rochester.edu

Abstract. We demonstrate the use of topological data analysis (TDA) to visualize the dynamics of organizational routines in outpatient medical clinics over the course of the COVID-19 pandemic. We use Electronic Health Record audit trail data from January 2020 to December 2021 to visualize changes in routines in outpatient clinics in four different medical specialties. By representing the pattern of action each day in each clinic as a weighted, directed graph, we see that some clinics bounced between several distinct patterns of action over the two year period. In contrast, other clinics never bounced back to the pre-COVID pattern of action. In all of the clinics, we see evidence of temporal auto-correlation: the pattern of action on any given day is similar to the pattern of action on days that are close in time. Because it can capture recurrence of action patterns that might be widely separated in time, TDA offers a substantial advance in the state of the art for visualizing the dynamics of organizational routines.

Keywords: Drift · Change · COVID-19 Pandemic · Electronic Health Records · Routine dynamics

1 Introduction

Routines can change on their own, but sometimes they are pushed. For example, the COVID-19 pandemic provided a strong external shock that disrupted nor-

This research was supported by the National Science Foundation under Grant No. BCS-2120530. Any opinions, findings, and conclusions or recommendations expressed in this material are those of the author(s) and do not necessarily reflect the views of the National Science Foundation. This research was also supported in part by University of Rochester CTSA (UL1 TR002001) from the National Center for Advancing Translational Sciences (NCATS) of the National Institutes of Health (NIH). The content is solely the responsibility of the author(s) and does not necessarily represent the official views of the National Institutes of Health.

K. Gdowska et al. (Eds.): BPM 2024 Workshops, LNBIP 534, pp. 338–349, 2025.
https://doi.org/10.1007/978-3-031-78666-2_26

mal processes and routines, especially in healthcare organizations [4]. Following guidance from the Centers for Medicare and Medicaid Services (CMS)[1] issued on March 18th, 2020, many clinics suspended their office visits entirely. Of course, many patients still had medical conditions that required attention, so healthcare providers scrambled to devise alternative ways to treat their patients. Standard procedures and routines needed to be reinvented to protect the patients and the clinical staff.

In this article, we use the example of the COVID-19 pandemic to demonstrate the use of topological data analysis (TDA) for visualizing recurrent patterns of action in organizational routines. We examine changes in routines in four outpatient clinics connected to a large U.S. medical center from January 2020 through December 2021. We build on the approach introduced by Pentland et al. [6] to visualize change in patterns of action over time. This approach is based on the idea of representing patterns of action as weighted, directed graphs and comparing the graphs over time. Here, we demonstrate the additional insights gained from topological data analysis (TDA) and discuss their implications for the theory of routine dynamics. Our findings show evidence of temporal auto-correlation in all the clinics: clinic-days that are close in time tend to have similar patterns of action. While changes from day to day are incremental, they are also systematic: the clinics move between distinctly different, recurrent patterns of action.

TDA describes a broad category of methods motivated by the intuition that the *shape* of data contains useful insights about the underlying phenomenon [1]. One particularly significant shape is the cycle (or circle or loop), which indicates recurrence (return to a prior state). Recurrence is fundamental to organizational routines [3], organizational path dependence [8], and institutions [5]. Here, we adapt a tool called Temporal Mapper [11] that uses TDA to visualize and analyze high-dimensional networks that change over time. Originally developed for the analysis of neurological networks, Temporal Mapper is a general purpose tool for identifying recurrent structures in temporal networks.[2] This perfectly describes the problem of analyzing routine dynamics using digital trace data.

2   Electronic Health Record Audit Trail Data

Audit trail data were extracted from the EPIC Electronic Health Record (EHR) system. Audit trail data is meta-data about the record keeping process, so it does not contain any clinical data. It provides an accurate view of who accessed or updated the clinic records, but it provides a limited view of the actual medical work.

The data we report here comes from four clinical areas: cardiology, dermatology, oncology and radiation oncology. The data cover a two year window from January 1, 2020 through December 31, 2021. This two year period spans

---

[1] https://www.cms.gov/files/document/covid-elective-surgery-recommendations.pdf.

[2] Code is available here: https://github.com/braindynamicslab/tmapper.

the most severe part of the COVID-19 pandemic, which hit the Northeastern United States in March 2020.

3 Visualizing Routine Dynamics with TDA

Audit trail data provides a rich source of information about patterns of action in the EHR record keeping process. The method we demonstrate here builds on the approach introduced by Pentland et al. [6] to visualize change over time by using a time-series of weighted, directed graphs (also known as directly follows graphs, or DFGs). This network time series is used in TDA to map the trajectory of clinical routines over time.

3.1 A Time Series of Weighted, Directed Graphs

We describe the pattern of action each day in each clinic as a weighted, directed graph. This graph provides a description of the "state" of the clinic on any given day and provides a way to compare any two clinic-days. The vertices in the graphs are the actions recorded in the audit trail, the edges are the sequential relations between actions within each patient encounter, and the weights denote the frequency of the sequential relations between actions.

We compute the graph for each day in each clinic from 7am to 7pm. This is a reasonable choice because the clinics only operate during the day and shut down at night. We compute the graph without filtering the data. This maximizes fitness without regard for any other metric of model quality. The resulting networks are very complex, but they provide a sensitive indicator of change.

As described by Pentland et al. [6], we use cosine distance based on the frequency of the edges in the graphs to compute the distance between clinic-days. The distance (or similarity) between clinic-days is an important input for topological data analysis (TDA). We use the graph for each clinic-day to represent the state of the clinic on that day and we analyze the trajectory of each clinic over time. This approach departs from methods where drift is detected or measured relative to a known process model. This approach is model-free; it is based on recurrence in the unfiltered DFG.

3.2 Simple Time Series

Pentland et al. [6] suggest two ways to plot the data based on the time-series of networks for each clinic-day. One way shows the change from prior day. This plot answers the question, "Is today the same as yesterday?" The second way shows the change from the pre-pandemic pattern of action. This plot answers the question, "How does today compare to a baseline?"

Figure 1 illustrates both of these visualization for one of the dermatology clinics. We smooth the plots and include a 95% confidence interval highlighted in gray. The plot on the left of the figure shows a comparison to the previous day, and the plot on the right of the figure shows a comparison to the pre-pandemic baseline. The pattern of action changes a lot and does not appear to bounce back. If anything, it appears to have stabilized into a "new normal" that is different from the original, pre-pandemic pattern of action.

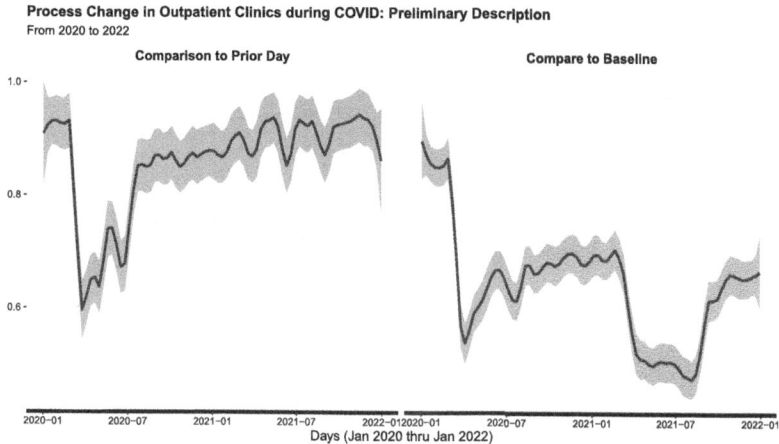

Fig. 1. Change over time in one dermatology clinic

3.3   Classic Recurrence Plot

Figure 1 is a helpful visualization of process change, but we can extract and visualize lot more information from the same data using a classic recurrence plot, as shown in Fig. 2. It is called a recurrence plot because it indicates recurrent states (similar patterns of action) over time. The plot in Fig. 2 shows the pairwise distance between *every* pair of clinic-days for the entire two-year time-series for one dermatology clinic. By comparison, the right-hand side of Fig. 1 shows only one row from Fig. 2 (comparison to a baseline). The left-hand side of Fig. 1 shows only the diagonal (offset by one, so it shows comparison to the previous day). Figure 2 is symmetric because the cosine distance measure is not directional.

The shading in this recurrence plot provides an intuitive picture of how the pattern of action changed over time. Dark areas show similar patterns of action; light areas show different patterns of action. For example, the small dark square in the upper-left corner is the pre-COVID routine, which was suddenly disrupted in March 2020. There is a significant period of moderate change, followed by two more rather sudden, substantial changes. The timing of these changes corresponds to the changes on the right-hand side of Fig. 1. Throughout the recurrence plot, there are bright lines that indicate days with exceptional patterns of action.

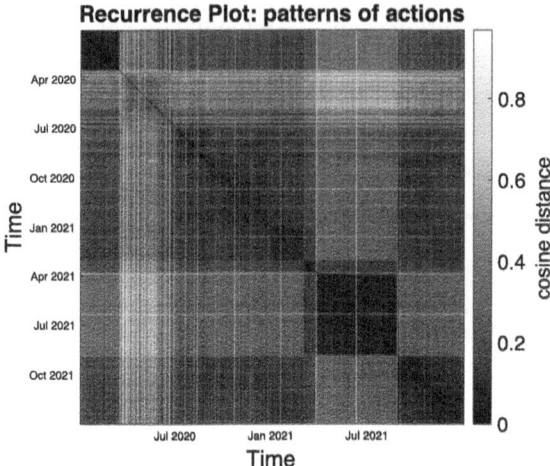

Fig. 2. Classic recurrence plot showing change over time in dermatology

3.4 Recurrence and Sequence of States Using TDA

TDA allows us to visualize the same data in a way that shows the temporal progression of the pattern of action. It allows us to see patterns of recurrence that are not evident in Figs. 1 or 2. In effect, it tells a visual story of how the patterns of action are related in terms of similarity and time.

3.5 Temporal Mapper Algorithm

To visualize the trajectory of the pattern of action, we use a TDA tool called Temporal Mapper [11]. The Temporal Mapper algorithm has four stages.

1. Cluster similar states using kNN. In the first stage, Temporal Mapper uses a k-nearest neighbor (kNN) algorithm to cluster clinic-days based on similarity (as measured by cosine distance). In principle, other clustering methods are possible, but KNN is ideally suited to the problem of identifying recurrence. For kNN clustering, the algorithm uses Fig. 2 as input. For the plots we show in this article, we set $k = 5$. This choice of k assumes that, on average, there are five non-adjacent work days that share the similar pattern of actions during the two year period. This assumption balances the inherent variability of organizational routines against their tendency to exhibit stability [2]. Lower values of k yield finer grained results (more clusters); higher values of k yield coarser grained results (fewer clusters).

The kNN algorithm results in a network of states with several clusters or network components. Each cluster contains a set of similar states. In this network, nodes represent patterns of action, and the edges represent similarity (as in Fig. 2). In this network, the edges are un-directed (or bi-directional). Even if clinic-days are widely separated in time, they can be clustered together as the same state if they are sufficiently similar.

2. Neighborhood graph. In the second stage, the algorithm introduces the temporal sequence back into the picture, forming a network with two types of edges: temporal edges and similarity edges. Temporal edges are directed and similarity edges are un-directed. In Temporal Mapper, this is called the neighborhood graph. Nodes can be neighbors in time, in similarity, or both.

3. State transition graph. In the third stage, the algorithm merges any circles formed by the directed edges that are shorter than a preset threshold. We set the threshold $d = 2$. This means that, for any two nodes a and b, we merge them together if the path lengths from a to b and from b to a are both less than or equal to 2. As the edges in stage 1 are all reciprocal, the distance between any two similarity neighbors is 1; therefore, the nodes in each cluster merge as a larger node. In this stage, the algorithm just continues to merge all circles under the threshold. After merging, the only visible edges are temporal edges: the "arrows of time". The resulting graph is called a state transition graph, showing the trajectory of the pattern of action as it changes over time. In this network, each node represents a "state", and each edge represents an "arrow of time."

4. Geodesic recurrence plot. Finally, using the state transition graphs, the algorithm calculates the pair-wise path lengths between each two clinic-days. In the state transition graphs, one node can contain multiple clinic-days, and the path lengths between two states are the shortest distances between clinic-days in the two states. The classic recurrence plot in Fig. 2 is based only on similarity (measured by cosine distance), but the geodesic recurrence plot is based on the distance between states in the state transition graph, which includes information about time and similarity.

3.6   Results of Temporal Mapper Visualization

Temporal Mapper produces plots that show the trajectory (or temporal progression) of the routine from one state to the next [7]. In this section, we show one example plot from each clinical specialty. In these plots, the size of the nodes indicates the number of times that state is observed in the data. In addition to the state transition graph, the geodesic recurrence plot shows distance between states on the path (or trajectory). Darker color indicates that the states are close together; light color indicates that the states are farther apart.

In all four clinics, the pre-pandemic "baseline" condition in January and February 2020 shows up as a small, dark box in the upper-left corner. When the pandemic hit in March 2020, we can see varying levels of reaction in the four clinics. Intuitively, one might expect more dramatic differences. In dermatology, office visits were reduced by 96% during the initial lockdown. Other clinical areas continued to see patients, but at a reduced rate. However, many differences in clinical practice (e.g., social distancing, use of masks and gloves, etc.) do not show up in the electronic health record because they did not change the clinical documentation process. In contrast, in all four clinics, changes in guidance from the Centers for Medicare and Medicaid Services (relating to the availability of COVID-19 vaccines, starting in March 2021) is readily visible. While each medicial specialty implemented this guidance in relation to their own caseload, it affected the documentation process in all four clinics. For example, clinical staff had to visit different screens and enter different information relevant to the vaccination status of the patient. As with any kind of digital trace data, we can only see what the EHR sees.

Cardiology (Fig. 3). In Cardiology, the trajectory has two main states. We can see that the pattern of action was affected by the initial lockdown in March 2020, but it bounced back fairly quickly. A much larger change occured later, in March 2021, when COVIC vaccines became available for different groups of patients. Eventually, the routines in the cardiology clinic have "bounced back" because the pattern of action is similar to the original, pre-pandemic pattern of action.

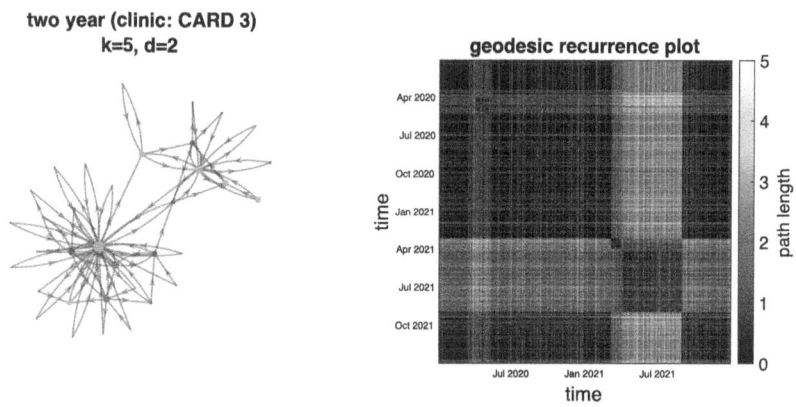

Fig. 3. Trajectory of Routine Cardiology Clinic

Dermatology (Fig. 4). In Dermatology, we can see a clear trajectory of distinct states. According to our co-authors (AP Pentland and J Ryan Wolf), Dermatology increased the use of telemedicine and implemented a novel "eConsult" service, so that primary care physicians could get quick consultations for patients with skin problems. As a result of these changes, the pattern of action continued to get more distant from the original, pre-pandemic pattern. The white (blank) area in the geodesic recurrence plot means that the there is no path back to the prior state (no recurrence). COVID appears to have provided the impetus for a permanent change to the routines in this Dermatology clinic.

Oncology (Fig. 5). Like Dermatology, the trajectory in Oncology has several distinct states. Unlike Dermatology, the trajectory bounces around between states and eventually returns to a state that is close to the original, pre-pandemic pattern of action. As in all of the other clinics, we can see a distinct change in the pattern of action when COVID-19 vaccines became available in March 2021.

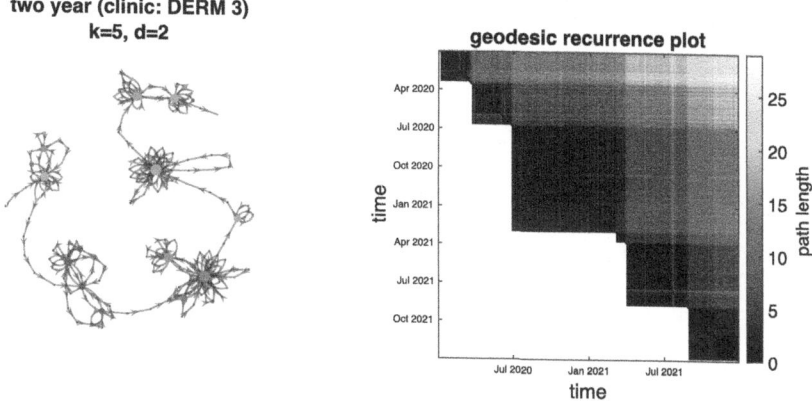

Fig. 4. Trajectory of Routine in Dermatology Clinic

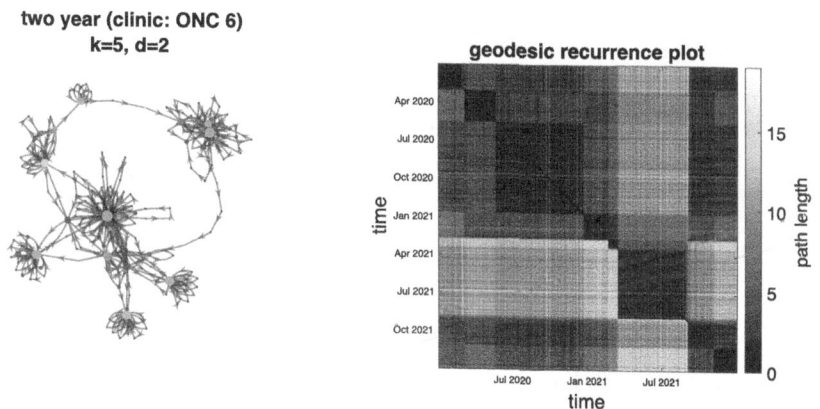

Fig. 5. Trajectory of Routine Oncology Clinic

Radiation-Oncology (Fig. 6). In this clinic, there are three distinct groups of states. Unlike the other clinics, the trajectory is not very much affected by the lockdown in March 2020. This may be because the patients in Radiation Oncology were receiving treatments that had to be provided on a strict schedule. Over the 2 year period, the clinic moved through three different states. By the end of 2021, the pattern of action in Radiation Oncology was similar to the pre-pandemic pattern, but not identical.

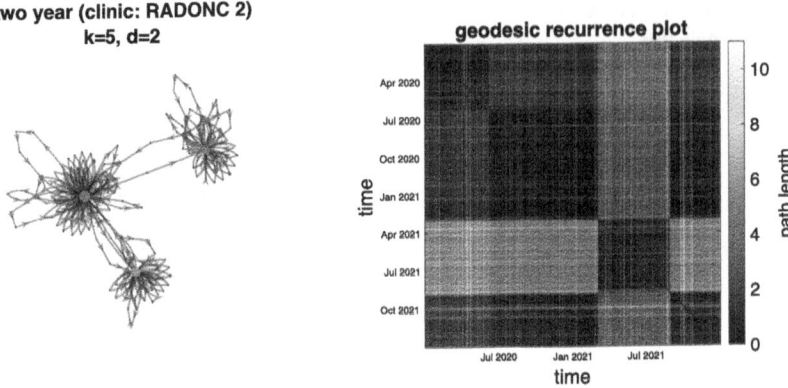

Fig. 6. Trajectory of Routine in Radiation Oncology Clinic

3.7 Visualizing the Pattern of Action in Different States

Visualizing the trajectory can help facilitate explanations, as well. For example, we can drill down into each of the states in the trajectory of the routine by visualizing the pattern of action when the clinic is in that state. To illustrate differences between states, we used Celonis to generate process maps for the pattern of action in two distinct states in the Dermatology clinic (see Fig. 7). We filtered the data to create plots that are easier to interpret. The first state is during the time period from January to March, 2020 (i.e., pre-pandemic). The second state is during the time period from July 2020 to February, 2021 (i.e., first phase after reopening). From Fig. 7, we can see that the processes in the two states are quite different. For example, before pandemic, Clinical_Tech performed 100% of the check-in work; however, during the first phase of reopening, only 4% of the check-in work is performed by Clinical_Tech.

4 Discussion

Feldman et al. [3] stated that "Routine Dynamics focuses on tracing actions and associations between actions" (p. 506). Focusing on actions within a routine has opened up many avenues of inquiry and insights about how routines work. Here, we are showing how Routine Dynamics can expand its focus to trace the trajectory of a routine as it changes over time. TDA provides a way of zooming out, so we can trace *patterns* and associations between *patterns*.

Of course, tracing patterns with digital trace data has limitations. In clinical settings, EHR data provides a shadow of the actual medical work. Neurological research has a similar limitation. We can detect electrical signals in the brain, but it's impossible to know what a subject is really thinking. The audit trail data we analyzed does not contain any information about social distancing, personal protective equipment, or other changes in clinical practice. The audit trail data only shows actions in the health record, not the actual medical work.

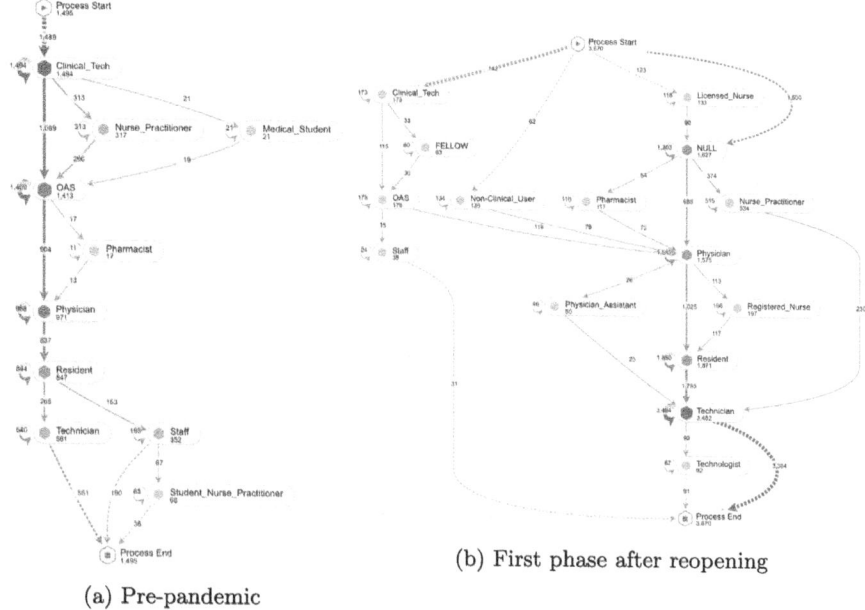

(a) Pre-pandemic

(b) First phase after reopening

Fig. 7. Comparing two states in the Dermatology clinic

The uncertainty (or noise) associated with trace data is evident in the visualizations we have presented here. In every clinic, there appear to be groups of closely related states – patterns of action that are similar but not similar enough to be grouped together. This may be an artifact of our choice to use "unfiltered" audit trail data to describe the patterns of action in each clinic-day. The data are noisy, and there is a lot of natural variability in clinical routines, so it is no surprise that the clinics appear to "bounce around" from day to day. At the same time, these examples illustrate the power of the Temporal Mapper algorithm to extract and display a clear temporal progression of action patterns.

4.1   Telling the Story of Change with TDA

This clarity of structure allows us to visualize the trajectory of change in a way that has never been possible before. We can detect drift, but we cannot understand its significance until we place in the context of a larger trajectory. The significance of any particular change is not evident until we compare it to other changes. By taking into account the pair-wise similarity of all clinic-days (as in Fig. 2), TDA uses all of the available information to construct a trajectory. Based on these trajectories, we see three important generalizations for routine dynamics.

Temporal Auto-correlation in Every Trajectory. While the trajectory of each clinic has unique features, they all illustrate some amount of temporal auto-correlation: States that are adjacent in time tend to be similar. In geography, Tobler's First Law of Geography [9] is based on spatial auto-correlation. It states that spatially adjacent locations tend to have similar properties. Of course, this "law" has many exceptions, such as shorelines. Here, we have visual evidence for a First Law of Routine Dynamics: temporally adjacent patterns of action tend to be similar. Temporal auto-correlation implies continuity in patterns of action over time.

More Than Just Incremental Drift. While there is evidence of continuity, every clinic also shows clear evidence that the pattern of action jumps from state to state in discrete, identifiable transitions. It does not look like incremental drift or a gradual accumulation of minor, situation-specific improvisations, workarounds, exceptions and so on. There are day-to-day variations in all of the clinics, but there are also discrete transitions as the pattern of action changes from state to state.

More Than Just Exogenous Shock. Some of the state changes we can see are clearly driven by external factors that affected all of the clinics at once. For example, guidance from the Centers for Medicare and Medicaid Services (in March 2021) appears to have a bigger effect than the onset of the virus itself (in March 2020). However, in spite of the fact that all of these clinics operated in the same hospital system, in the same regulatory context, and the same EHR software, the trajectory of every clinic is different. The Cardiology clinic has two distinct groups of states, but Radiation Oncology has three, and the other clinics have more. Three of the clinics bounce around between states, but Dermatology just keeps changing to new states. We clearly see the change introduced by vaccine availability in every clinic, but other aspects of the trajectory appear to be unique to each clinic.

4.2   Beyond Stability and Change: Visualizing Recurrence

Tsoukas and Chia [10] note that theorists tend to assume stability and then try to explain change. However, the patterns of action in medical clinics are highly variable. Every moment, in every clinic, patterns of action form and dissolve. There is continuous change. At the same time, as we can clearly see, the patterns of action are highly recurrent. There is continuous change, but also periods of stability.

A major theoretical advantage of the method presented here is that it does not assume stability or change; the only assumption is temporal progression, which is logically antecedent to both stability and change. Given a temporal progression of states, as suggested by Pentland et al. [6], Temporal Mapper helps us identify and visualize recurrence, which is an indicator of both stability and change.

5   Conclusion

Topological Data Analysis (TDA) provides a new way to visualize and analyze recurrence in high-dimensional, temporal data. This kind of data is increasingly available for organizational routines. Because it provides a way to visualize recurrence, TDA can provide new avenues for inquiry and insight into the dynamics of organizational routines and other processual phenomena in organizations.

References

1. Epstein, C., Carlsson, G., Edelsbrunner, H.: Topological data analysis. Inverse Prob. **27**(12), 120201 (2011)
2. Feldman, M.S., Pentland, B.T.: Reconceptualizing organizational routines as a source of flexibility and change. Adm. Sci. Q. **48**(1), 94–118 (2003)
3. Feldman, M.S., Pentland, B.T., D'Adderio, L., Lazaric, N.: Beyond routines as things. Organ. Sci. **27**(3), 505–513 (2016)
4. Herrera, C.A., et al.: COVID-19 disruption to routine health care services: how 8 Latin American and Caribbean countries responded: the study examines disruption to routine health care services brought about by the COVID-19 pandemic in eight Latin American and Caribbean countries. Health Aff. **42**(12), 1667–1674 (2023)
5. Ocasio, W.: Institutions and their social construction: a cross-level perspective. Organ. Theory **4**(3), 26317877231194370 (2023)
6. Pentland, B., Vaast, E., Wolf, J.R.: Theorizing process dynamics with directed graphs: a diachronic analysis of digital trace data. MIS Q. **45**(2) (2021)
7. Pentland, B.T., Yoo, Y., Recker, J., Kim, I.: From lock-in to transformation: a path-centric theory of emerging technology and organizing. Organ. Sci. **33**(1), 194–211 (2022)
8. Sydow, J., Schreyögg, G., Koch, J.: On the theory of organizational path dependence: clarifications, replies to objections, and extensions. Acad. Manag. Rev. **45**(4), 717–734 (2020)
9. Tobler, W.: A computer movie simulating urban growth in the Detroit region. Econ. Geogr. **46**(sup1), 234–240 (1970)
10. Tsoukas, H., Chia, R.: On organizational becoming: rethinking organizational change. Organ. Sci. **13**(5), 567–582 (2002)
11. Zhang, M., Chowdhury, S., Saggar, M.: Temporal mapper: transition networks in simulated and real neural dynamics. Netw. Neurosci. **7**(2), 431–460 (2023)

1st International Workshop
on Managing Process Innovation
and Value Creation in the Era of Digital
Transformation (Innov8BPM 2024)

1st International Workshop on Managing Process Innovation and Value Creation in the Era of Digital Transformation Innov8BPM 2024)

In the era of digital transformation and innovation, business process management (BPM) as a field needs to adapt to new topics and themes. This workshop addressed the necessary advancements in the BPM discipline for managing process innovation and creating business value in today's dynamic environment. Organizations are undergoing rapid transformation fueled by the pervasive diffusion of digital technologies. The opportunities brought by digital technologies push organizations to make fundamental changes in their business processes. This evolving digital landscape is challenging the role of BPM. The traditional role of BPM, centered around structuring and optimizing operational processes, may not fully capitalize on the opportunities presented by digitalization. Yet, BPM should facilitate the exploration of new possibilities in adapting to the dynamic nature of digital innovation and transformation. Researchers and practitioners must rethink the role of BPM in managing process innovations to leverage digital transformation with a broader perspective covering strategies, business models, and new offerings through digital technologies.

While most companies are going through digital transformation, few manage to reap the benefits from such transformations, while many struggle to create, communicate, capture, and demonstrate the business value of these initiatives. Recent technological innovations incorporated in BPM (e.g., AI, Big Data, Internet of Things), new organizational forms (e.g., business and digital ecosystems), as well as the new logics that underlie BPM in the context of digital transformation, together create new avenues for business value creation. BPM should facilitate business value creation through digital transformation and process innovations. On a theoretical level, a need exists to understand the conceptual underpinnings related to the business value of BPM to better delineate the affordances and explain the phenomena revolving around it. On a practical level, there is a need to understand the practical steps organizations may follow to create and capture value.

Based on those needs, this workshop aims to explore the dynamic relations between Digital Transformation and Business Process Management (BPM) for managing process innovation and conceptual and practical understanding of the business value of BPM, with a specific emphasis on the managerial aspects of business processes. This year, we set out to advance BPM based on this goal through three sessions and presentations of five accepted papers.

In the first session, Michael Rosemann delivered his keynote speech titled *"BPM 2025: New ways and new value?"*. He presented a framework structuring the three ways new technology BPM to be planned and conducted. These three ways constitute BPM drifts, namely transaction to conversation, automation to autonomization, and simplification to sophistication in BPM. Through these drifts, BPM no longer follows the traditional reductionist approach to process improvement and innovation but rather extends and enriches processes and experiences, thus enabling innovative business value.

The keynote was followed by a discussion of the paper entitled *"How Can We Create an Inclusive Business Process Management?"* authored by Mahendrawathi Er. The paper marks the first attempt to define diversity and inclusivity for BPM, conceptualizing the term and exemplifying the benefits of recognizing this dimension and the associated challenges. The paper complemented the keynote, addressing how BPM should open up to new concerns and trends.

The second session included the presentation of four papers. First, the second author presented the paper of Astria Hijriani and Marco Comuzzi entitled "Business Value of Process Mining: A Contingency Perspective". The paper suggests using the contingency theory to reveal how process mining can help organizations create business value. The contingency theory recognizes that the optimal way of actions depends on internal and external situations. The paper reveals negative contingencies for process mining business value. Based on a number of case studies, it identifies a preliminary list of negative contingencies, such as multi-site and multi-national operations, as well as the availability of industry success stories.

The following presentation was made by the last author of the paper by Andrea Burattin, Ekkart Kindler, Nicholas Dyhre, Sebastian Vestrup, Francesca Zerbato, and Barbara Weber, entitled "Process Mining Pipelines with Controlled Sharing of Data and Algorithms". The paper developed a set of requirements for a platform based on a distributed architecture to implement process mining pipelines. Supported by a prototype implementation, this platform can help organizations overcome the challenges of setting up process mining pipelines, including data collection, programming, privacy, and intellectual property issues.

The third paper was presented by the first two authors of the paper by Gerald Kremer, Luiz Ricardo Ribeiro, Till Blüher, Silvia Ines Dallavalle Padua, and Rainer Stark, entitled "Towards a Unified Approach: Developing a Reference Model for Digital Twins and Business Process Management in Clinical Trials". The paper enabled the workshop to observe the technology of digital twins implemented for clinical trial processes. The authors depicted clearly that drug and medical device development processes have unique challenges, such as protocols and regulatory requirements, resource and data management, and participant recruitment. A digital twin, typically having the core functions of monitoring, simulation, prediction, assessment, and control, can be developed for clinical trial processes to overcome such challenges.

The final presentation was delivered by the last author of the paper by Fabian Stiehle, Finn Klessascheck, Martin Kjäer, and Ingo Weber, entitled "Business in the Age of Platform Economics: Managing Decentralised Business Processes Beyond Blockchain". The paper raises a problem statement for the BPM community on deriving business value from platforms through managing decentralized business processes. Decentralization provides a potential solution to overcome the challenges of the platform economy. Blockchain-based systems constitute a way of decentralization but are not the only solution. The paper poses the question of creating profitable and sustainable business models on platform economics through decentralization solutions.

We concluded the workshop through an interactive discussion session on "Drifts in BPM: Challenges and Opportunities". The participants were divided into three groups to discuss each BPM drift defined during the keynote; each group was facilitated by a workshop co-chair. Each group identified challenges and opportunities for BPM and future research directions. A group moved to the next drift after 20 minutes. The participants followed a futuristic view, recognizing that new technologies can offer value and create challenges that we cannot directly envision currently. The findings were presented by the facilitators of each drift.

We believe that the workshop has taken a big step in questioning and investigating the evolving role of BPM in the digital era through the inspiring keynote, paper presentations exemplifying BPM innovations through various digital technologies and changes that BPM needs to embrace to realize new value, as well as a fruitful interactive session. We would like to thank all participants and the organizers of BPM 2024.

Organization

Organizing Committee

Banu Aysolmaz   Eindhoven University of Technology, The Netherlands

Amy Van Looy   Ghent University, Belgium

Oktay Turetken   Eindhoven University of Technology, The Netherlands

Marta Indulska   University of Queensland, Australia

Flavia Santoro   University of the State of Rio de Janeiro, Brazil

Panagiotis Keramidis   Copenhagen Business School, Denmark

Program Committee

Amine Abbad-Andaloussi   University of St. Gallen, Switzerland

Marco Comuzzi   Ulsan National Institute of Science and Technology, South Korea

Søren Debois   DCR Solutions, Denmark

Mahendrawathi Er   Institut Teknologi Sepuluh Nopember, Indonesia

Paul Grefen   Eindhoven University of Technology, The Netherlands

Thomas Grisold   University of St. Gallen, Switzerland

Sybren de Kinderen   Eindhoven University of Technology, The Netherlands

Thomas Hildebrandt   University of Copenhagen, Denmark

Rob Kusters   Open University, The Netherlands

Morten Marquard   DCR Solutions, Denmark

Jan Mendling   Humboldt-Universität zu Berlin, Germany

Fredrik Milani   University of Tartu, Estonia

Hajo Reijers   Utrecht University, The Netherlands

Maximilian Röglinger   University of Bayreuth, Germany

Arisa Shollo   Copenhagen Business School, Denmark

Mojca I. Stemberger   University of Ljubljana, Slovenia

Han van der Aa   University of Vienna, Austria

Inge van de Weerd   Utrecht University, The Netherlands

Iris Beerepoot   Utrecht University, The Netherlands

Peter Loos   Saarland University, Germany

Pascal Ravesteijn   HU University of Applied Sciences Utrecht, The Netherlands

Process Mining Pipelines with Controlled Sharing of Data and Algorithms

Andrea Burattin[1]([✉]), Ekkart Kindler[1], Nicholas Dyhre[1], Sebastian Vestrup[1], Francesca Zerbato[2], and Barbara Weber[3]

[1] Technical University of Denmark, DTU Compute, Lyngby, Denmark
`andbur@dtu.dk`
[2] Eindhoven University of Technology, Eindhoven, The Netherlands
[3] University of St. Gallen, St. Gallen, Switzerland

Abstract. Process mining leverages execution traces within an organisation's IT systems to gain insights into its processes. Despite being a mature discipline in academia and industry, setting up process mining pipelines is still a complex task and involves programming, manual steps, and considerations of privacy and intellectual property. This paper introduces a platform based on a distributed architecture that helps define, deploy, and execute process mining pipelines across organisations. The requirements for this distributed architecture and platform are derived from a set of process mining scenarios, whose relevance is validated through a survey. Furthermore, this paper introduces a prototype for an initial version of the platform, demonstrating feasibility and supporting the specified requirements.

1 Introduction

Process mining is a data-driven discipline aiming to extract insights from process execution data typically stored in event logs and combining it with other artefacts, such as process models or business rules [2]. Over the years, a multitude of algorithms and techniques have been developed, many of which are available as open-source tools (e.g., ProM [6], bupaR [10], or PM4Py [5]), or as commercial products offered as services or on-premise solutions. Consequently, when creating a process mining pipeline, process analysts are faced with a wide range of choices, including many techniques and their different implementations [17].

The decision of which specific tools to employ hinges on several factors, including context-specific requirements and constraints on data and algorithm sharing. Additionally, the availability of algorithms, whether locally within an organisation or remotely as a service, and their open or closed-source nature further influence these choices. Obtaining the best possible results for a specific purpose might require the combination of diverse data sources and algorithms across different organisations. Setting up such process mining pipelines remains cumbersome: It involves programming for adjusting interfaces and manual steps for adjusting and transferring data and for invoking algorithms on them; and it

K. Gdowska et al. (Eds.): BPM 2024 Workshops, LNBIP 534, pp. 357–369, 2025.
https://doi.org/10.1007/978-3-031-78666-2_27

needs to take care of the privacy of some data and intellectual property rights when running third-party mining algorithms.

In this paper, we investigate the issues and problems when setting up process mining pipelines across different organisations. Then, we propose a Distributed Architecture for Process Mining (DAPM). This architecture and the supporting platform make it possible to easily define, deploy, and execute process mining pipelines across different organisations. This is driven by the research question: "How can we assist process analysts in sharing data and algorithms for process mining in a controlled manner?" This allows analysts to share and re-use data, algorithms and processing pipelines, while simultaneously affording them the ability to maintain the privacy of these resources when the context demands it.

In Sect. 2, we articulate the problem at hand and delineate a set of representative scenarios. In Sect. 3, we derive functional and non-functional requirements. In Sect. 4, we introduce a first architectural design. A comparison with existing systems in Sect. 5 and the Conclusion, Sect. 6, complete the paper.

2   Problem Description and Validation

In this section, we delineate the research problem by presenting a concrete example in Sect. 2.1, and by defining a set of representative scenarios in Sect. 2.2 that capture key aspects of the problem. Finally, in Sect. 2.3, we present the results of a survey conducted to validate the representativness of the scenarios.

2.1   Motivating Example

Let's consider the following illustrative example:

> Hospital FirstAid would like to improve its Delivery of Care process in the Emergency Department. Process mining has been identified as a viable technology. Alice is a data scientist and researcher at the hospital in charge of the project. Among other aspects, Alice identified the needed data sources, which are distributed across different systems and providers. Alice is particularly interested in predictive process mining and plans to develop her own algorithm and benchmark it against the state-of-the-art.

In order to make the situation more specific, let us elaborate on some aspects of this example in more detail:

Data sources and repositories: To test different algorithms, Alice uses two repositories of event logs that have different levels of confidentiality: *(i)* a private database from the hospital FirstAid, containing sensitive data about patient histories, medical procedures and outcomes, and *(ii)* a public repository of health records, including data about health interventions across all the hospitals in the country.

Algorithm benchmarking: Alice wants to compare her new technique with state-of-the-art process mining algorithms. One of these algorithms is made available by an external provider, who does not share the source code of it. Alice is not allowed to share sensitive data outside the hospital and wants to maintain the confidentiality of her technique during the benchmarking phase.

Algorithm sharing: Upon successful benchmarking and validation, Alice wants to share her technique and make it available for use by other researchers. Then, she also shares the source code to be compliant with open science principles.

Algorithm reuse and extension: Bob is a researcher working for a different hospital who wants to integrate Alice's technique into his predictive analytics pipeline. Since he works with anonymised data, he can use Alice's technique and outsource the computational task. However, after a while, he notices that the interoperability of his and Alice's algorithms can be improved. Thus, he gets the source code from Alice and extends it for this purpose.

This example highlights some of the challenges associated with creating and configuring process mining pipelines: *(i)* work with and integrate data sources having different levels of privacy and confidentiality, *(ii)* orchestrate the execution of algorithms with different requirements of source code visibility, deployment, and execution, and *(iii)* allow for the interoperability of data and techniques across organisations while maintaining organisational boundaries.

2.2   Distilling the Problem into Representative Scenarios

The problem presented in Sect. 2.1 captures a complex setting with several subproblems that we broke down into its basic features, devising *scenarios*.

Scenario Elicitation Process. To identify the scenarios, we followed a systematic approach where we considered the main phases involved in process mining projects [7]: defining research questions, data collection, data pre-processing, mining and analysis, stakeholder evaluation, and implementation. We observed that the relevant entities/actors across these phases are: *(i)* stakeholders and service providers (i.e. organisations), *(ii)* data, *(iii)* algorithms, and *(iv)* analysts/developers. As reported by Zimmermann et al. [20], in areas like healthcare, banking, or insurance, the data cannot leave the organisation due to data protection and security regulations—presenting serious challenges for process mining projects. Having an explicit way of contextualising data, algorithms, and analysts/developers with respect to organisations represents an important step towards acknowledging the problem and thus devising a solution. With these entities and actors identified, the possible *conceptual* interactions among them can be modelled. The elements for defining scenarios are shown in Table 1:

– Stakeholders and service providers are *organisations* that, in the context of this paper, own data and are capable of executing mining algorithms.

- The concept of data is represented in our scenarios as two separate elements: *repositories* and *objects*, where repositories are data containers and objects are data produced or consumed by mining algorithms.
- Algorithms are represented as *miners* with corresponding interfaces for specifying the input and the output objects. A miner is not necessarily a control-flow discovery algorithm but can be any process mining algorithm.
- The role of developers and analysts consists of creating new miners or, respectively, constructing an analysis pipeline by *wiring* objects and miners using *connectors*. Data and algorithms can reside with different *organisations*, yet analysts can decide to connect them across organisation borders.

Table 1. Description of the elements used in the presented scenarios.

Element	Description
O1	**Organisation.** An organisation, called O1, has a collection of repositories and miners. In our scenarios, we assume that no information leaves O1 unless a connector connects it to a miner outside of O1.
R1	**Repository.** A repository, called R1, is a place where objects (e.g., data, event logs, models) can be created, read, updated, and deleted. This can be as simple as a file system or a database.
out1	**Object.** An object, called *out1*, which can be created, read, updated, or deleted by a Miner in a Repository: they can be the input or output of a Miner.
Miner 1	**Miner.** An executable miner "Miner 1". Miners expect inputs and create output objects. These are represented by » and ▶, respectively. A miner can be shadowed from source code; indicated by stereotypes «*source*» and «*shadow*».
	Connector. Connectors wire specific objects as input and output of a miner.

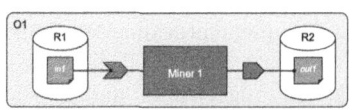

Fig. 1. S1 - Single organisation and single miner.

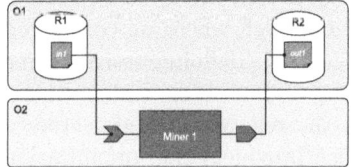

Fig. 2. S2 - Two organisations and single miner.

Scenario Description. With these graphical elements, we can define five scenarios that, taken together, capture the most relevant interactions among the entities that appear in a process mining project.

In the first scenario, S1 (Fig. 1), we can imagine a person working for a given organisation, O1, with access to two repositories: R1 and R2. In R1, an event log, called *in1*, is stored, whereas R2 is the container for the results of the analysis. The analysis itself will be performed using *Miner 1*, which is an algorithm installed within organisation O1. Figure 2 presents scenario S2, where a second stakeholder is involved as a provider of the used mining algorithm. In S3 (Fig. 3), the analysis requires two different operations (i.e., two different miners), where the results of the first are used as inputs of the second. In addition, the two miners are running at two different organisations. In scenario S4 (Fig. 5) the miner is *shadowed* (i.e., shared as executable without sharing the source code) to the organisation where the data resides (as opposed to moving the data). In this way, the data does not leave the premise of the organisation that owns it. Finally, S5 (Fig. 5) emphasises the fact that one of the "organisations" is actually a personal machine, and the owner is planning to perform some comparisons (e.g., benchmarking) with other miners located at other organisations.

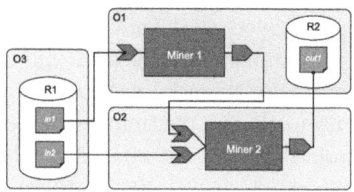

Fig. 3. S3 - Multiple organisations and multiple miners.

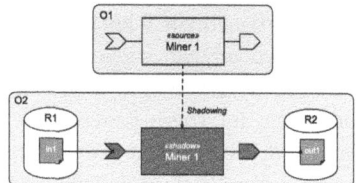

Fig. 4. S4- Shadowing of a miner from a different organisation.

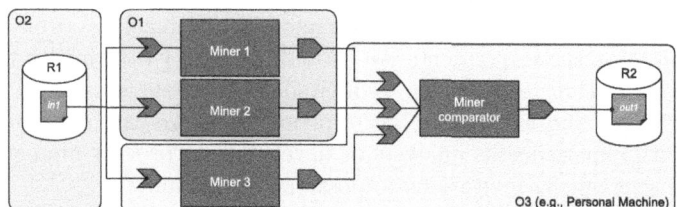

Fig. 5. S5- Multiple organisations and miners (with a personal machine).

Please note that the illustrative example described in Sect. 2.1, can be fully captured by a combination of the concepts mentioned in the scenarios above and that these scenarios are conceptual models (i.e., they miss technical information for actually executing the pipeline, which we will discuss later in Sect. 3).

2.3   Scenario Validation

To validate the representativeness of the described scenarios, we conducted a qualitative survey with process mining users. *Representativeness* can be split up into two parts: how *relevant* are these scenarios and how *complete* are they.

Survey Design and Execution. The goal of the survey was to validate whether the presented scenarios are understandable and relevant for process mining practice and whether they are complete with respect to the possible conceptual interactions among the stakeholders, data and algorithms involved in a process mining project. Moreover, we aimed to gather insights into situations in which such scenarios might manifest themselves and identify scenarios we might have missed from our elicitation process. To this end, we designed a qualitative survey where we provided for each scenario: *(i)* a textual description of the scenario; *(ii)* a concrete example instantiating the scenario; *(iii)* a graphical representation of the scenario, *(iv)* a set of closed and open questions about the scenario and *(v)* an open question about scenarios we might have overlooked.

More specifically, the questions in *(iv)* asked whether the respondent had understood the scenario, experienced it and in which situations, and about "envisioned situations" in which the scenario could occur. Here, we followed an adaptive design based on whether the respondents indicated that they understood and experienced the scenario. If the respondents reported that they did not understand the scenario, we asked them to explain why they did not understand it and, then, let them proceed to the next scenario. If the participant had not experienced the scenario, we skipped the question about the situations in which they encountered it and asked only about envisioned situations. In this way, respondents had to answer at most 5 questions per scenario and could avoid wasting time on scenarios they did not understand or experience firsthand.

The questions about the scenarios were preceded by some demographic questions asking respondents about their job role, process mining expertise, and experience. The latter included experience as "analysts" and "developers" of process mining. Additionally, participants were provided with a sample scenario to get familiar with the scenario description and the graphical notation used.

Before inviting the respondents, we piloted the survey design with two people external to the author team. Then, we invited respondents following a purposive sampling strategy and we reached out to people in our research and professional networks with experience as analysts or developers in process mining projects. Overall, we sent out 32 invitations to possible participants.

Respondents. 16 people responded and fully completed our survey. All but one respondent, who works as a process mining R&D lead, work in academia as researchers or professors. The respondents reported an average to advanced process mining expertise, with the majority indicating *good* process mining expertise. All participants have 3+ years of process mining experience, and a quarter of them have 10+. All but one have experience as analysts in process mining projects involving customers, and all have experience with developing process mining algorithms or tools, with half having developed five or more algorithms.

Results. The results of the survey confirmed that the scenarios are relevant depictions of process mining practice, i.e., they can be understood by process mining users, and are complete wrt. situations that our respondents have either encountered or can envision. In Fig. 6, we report the results of the survey for the two closed questions related to relevance. The figure shows that scenarios S1, S2, and S4 have been encountered by at least half of the respondents. When referring to S1, participants reported that *"any process mining project matches this scenario"* and that it is very common when *"clients provide data for the analysis"* or *"when developing our own algorithms"*. Similarly, S2 was a good example of using third-party software, either *"developed by colleagues or other institutions"* or *"commercial tools"*, in case *"the data is not confidential or protected by a non-disclosure agreement"*. S4 is another example where data confidentiality matters, which was often encountered when dealing with sensitive or private data as one respondent describes: *"We typically install our system on the hospital's servers"*. This scenario was also considered relevant for settings where federated learning is applied. Instead, scenarios S3 and S5 turned out to be encountered by only a minority of the respondents. However, when looking at the answers to "Could you envision situations in your work where such a scenario might occur?" it turns out that these are encountered less often because existing systems do not support them, and not because these scenarios are less relevant. Indeed, participants indicated S3 to be an ideal scenario for software-as-a-service, *"where providing access to several algorithms as a service would ease the need of having to search, download and set the environment to run algorithms"* while S5 seemed relevant for experiments and benchmarking in research, where researchers compare *"algorithms from established frameworks with proprietary ones"*. When looking at the answers to the last question "Are there other scenarios in your work with process mining that you think are relevant to be supported?" the majority of the respondents indicated that the five scenarios cover all possible situations they might think of. However, two respondents hinted at situations in which the *"data is distributed between organisations"* and *"the objective is to discover a comprehensive process model."* We agree that this is a shortcoming of the presented scenarios, but what the respondent asked for (i.e. having repositories in different organisations serving as input to the same miner, similarly to S3) can be modelled using the presented concepts. Another important limitation we are aware of is that of the 16 respondents, 15 came from academia, which limits the generalizability towards industry.

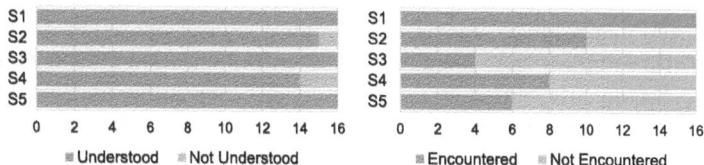

Fig. 6. Results of the questions "Have you understood the scenarios?" and "Have you encountered this scenario before in your work or research experience?".

3 Requirements for a DAPM

Following a Scenario Based Requirement Elicitation [9] process, this section derives more detailed requirements for a platform and a Distributed Architecture for Process Mining. At the highest level of abstraction, the functional requirements for DAPM can be captured in the following use cases:

1. Process mining scenarios:
   - DAPM must allow a user to define some specific process mining scenario.
   - DAPM must allow a user to deploy and run (enact) a defined process mining scenario.
2. Access control: DAPM must support user management and access control.

Section 3.1 explains the process mining scenarios, Sect. 3.2 explains access control and Sect. 3.3 outlines the non-functional requirements.

3.1 Functional Requirements Process Mining Scenarios

For defining scenarios, we distinguish the *conceptual* (or logical) view and the *physical* view, which defines some more technical details. In Sect. 2.2, we focus mainly on the conceptual view; with the details added here for the *physical* view, they can be enacted, which is when we also call them *pipelines*.

Defining Scenarios: Conceptual View. To define scenarios, we need *repositories*, *algorithms*, and *organisations* (cf. Sect. 2.2). In our scenarios, the algorithms are mostly miners; but there can be other kinds of algorithms, like algorithms for computing alignments [2]. Each algorithm has a set of *inputs* from which it produces some *outputs*. Most importantly, repositories and the inputs and outputs of algorithms can be *wired* with each other through *connectors*, defining the data flow of the processing pipeline represented by the scenario. Typically, the wiring involves connecting a repository object as an algorithm's input or linking the output of an algorithm to a repository object. So far, the scenarios could actually be any data processing pipeline. However, there are two aspects that are specific to distributed and cross-organisational process mining. First of all, each repository and algorithm is associated with an organisation. When the scenario is *enacted*, this implies that no data leaves an organisation unless defined by the wiring. Second, the different repositories and each input and output of an algorithm are actually associated with a *type*. This type defines which kind of data it represents. In the context of process mining, the predominant types are event logs employed as inputs and process models generated as outputs. However, other types, such as alignments, might come into play in scenarios involving conformance monitoring. The spectrum of potential types extends beyond these examples, warranting further exploration. In addition, there could be a *macro* feature for defining some scenarios with inputs and outputs that are not connected to a repository. Such a macro can then be *instantiated* and used as an *algorithm* within other scenarios. This way, mining scenarios can be built up in a modular way with reusable components.

As discussed before, the wiring of a scenario makes explicit which data is exchanged among different organisations by a wiring that crosses organisational boundaries. In some situations, organisations also want to protect the intellectual property of their algorithm and, therefore, do not want to share the source code of their algorithm. However, they are willing to share a binary version or some *shadow*, which can be executed by a third party without sharing the source code. This *shadowed* execution of an algorithm by a third party can also be defined in a mining scenario (as, for example, in the scenario S4 from Fig. 4).

While defining the conceptual view of a scenario, DAPM should provide features for validating the scenarios, so that they can actually be deployed and enacted once the information from the physical view is provided.

Defining Scenarios: Physical View. The physical view of a scenario defines how the data is actually stored and accessed in a repository, where the algorithms are running and how they can be initiated and provided with the necessary data, and which protocols are used for that. We do not go into the details, but the prototype of Sect. 4, will give some idea.

3.2 Functional Requirements – Access Control

For actually deploying and running a scenario as a process mining *pipeline*, some access information and credentials to external services and repositories might be required. In order to increase reusability, however, these credentials are kept separate from the conceptual view and the physical view of a scenario.

Since these credentials are personal and sensitive, DAPM needs a user management system. In this system, a users will be able to manage and maintain their credentials and assign them for algorithms (services) and repositories used in certain scenarios. Once this information is provided, DAPM should be able to validate the complete scenario and to deploy it as a pipeline. Hereby the deployment of a pipeline should only be possible for valid scenarios and for a user with the necessary credentials for this scenario.

3.3 Non-functional Requirements

This section explains the most important non-functional requirements for DAPM. Most importantly, there are *security* and *interoperability*. For security, DAPM must specifically guarantee *confidentiality* and *integrity* ensuring that data and algorithms can only be accessed and changed through DAPM and the scenarios when explicitly granted access by the defined scenarios and access credentials. This includes the protection of the intellectual property (IP) of the source code of certain algorithms. In turn, interoperability guarantees that the data from the repositories and the algorithms work seamlessly together across different organisations once a scenario is validated. DAPM must be *extensible*, so that existing data sources and implementations of algorithms can be easily added, with the needed input and output types. This might require implementing wrappers for existing algorithms or transformation of data. However, once these are added to DAPM, these data types and algorithms should be available for defining scenarios and be deployable and runnable without further implementation work. The features of the scenarios introduced above guarantee *configurability* and *flexibility*, i.e. scenarios can be defined, deployed, and run without any programming and result in the same outcome for the same data. The separation of the definition of scenarios and their enactment as well as the separation of the conceptual, the physical, and the access information will provide *reusability*, in particular when combined with macros for defining new algorithms. With the above features of the scenarios, DAPM can provide *data provenance* [19], i.e. provide information when and how certain data was produced by a reference to the scenario. To some extent, the features of the scenarios also help achieving *reproducibility*. However, this will depend on strong assumptions on algorithms provided via the cloud (being available and not changed) and explicit versioning of the algorithms.

Moreover, when using a micro-service architecture for implementing the distributed architecture, *scalability* and *elasticity* can be achieved by applying vertical and horizontal scaling.

4   Prototype Implementation

In order to demonstrate some of the feasibility of such a platform, we have implemented a prototype application that employs a micro-service architecture (MSA) [12]. Note that this prototype focuses on interoperability, extensibility, and flexibility and not on security and usability. A general overview of the services is reported in Fig. 7. The first type of service is the *"Repository"*, which stores resources and manages their physical location. The second type of service is the *"Miner"*, providing implementations of algorithms. The *wiring* and the configuration of the algorithm are managed through a configuration file that describes the input and output resources (i.e. the corresponding metadata objects, with information about the respective type and location) and parameters of each mining algorithm. A Miner service can be shared as a Docker container for *shadowing* algorithms from foreign Miner nodes into themselves. This is done through a REST API request containing the necessary information for a Miner to fetch the algorithm file and its configuration from the source. Finally, to support the discovery and the handling of the nodes, the Service Registry design pattern [16] is implemented as another type of service: the *Service Registry*. These nodes use a database to maintain knowledge of registered miners and repository nodes. We call all the components discussed up to now *backend*. The *Frontend* is a Javascript application that allows the addition of Miner and Repository nodes. All the source code of the application is available as open-source and is distributed via GitHub[1] with Apache-2.0 license. Our implementation addresses most of the requirements identified in Sect. 3. Specifically, from the conceptual point of view, *shadowing* is already available. Data confidentiality cannot be guaranteed for repositories as the current implementation does not have user

---

[1] See https://github.com/DistributedArchitectureProcessMining.

management and access control. The *wiring* is possible in the backend but not in the current frontend. The current implementation ensures interoperability and extensibility via RESTful protocol and micro-service architecture, which also permits flexible scaling and elasticity of service instances. The backend supports reproducibility and metadata files are used to record provenance information.

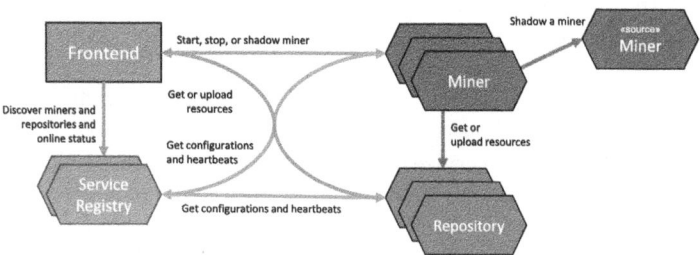

Fig. 7. Software architecture of the components of DAPM.

5 Related Work

For analysing the state-of-the-art, we look at it from two different perspectives: the *scientific* and the *platform/tools* perspective. When examining the scientific literature, to the best of our knowledge, there is no approach tackling the identified requirements comprehensively. However, isolated aspects, such as the wiring (a.k.a., data processing pipelines, data flow, scientific workflow), have been extensively discussed [4,14]. Moreover, multiple approaches for privacy in process mining exist, tackling both the anonymisation of data [8] and the techniques themselves [15]. These aspects, however, have been studied in isolation from the deployed platform. Taking the platform/tool perspective, we conducted a thorough analysis of how existing tools support the non-functional requirements[2]. The main tools identified are: ProM [6] is an open-source desktop application used to prototype new process mining algorithms via a plugin architecture. RapidProM [1] ports some ProM plugins into the RapidMiner platform. This platform allows the definition of process mining pipelines for processing an event log. Similarly to RapidProM, PM4KNIME [11] embeds ProM capabilities into the KNIME platform. Process mining pipelines can be defined and run locally or, in principle, in remote servers too. RuM [3] is an open-source desktop application that focuses on Declare models for performing discovery and conformance. Cortado [18] is an open-source desktop application built on top of the open-source library PM4Py [5], which focuses on interactive process discovery. As PM4Py is a library that requires programming knowledge to be used and is not designed as an end user tool, we did not include it directly in our comparison. Smyrida [13] is a web application supporting basic mining algorithms. Fluxicon Disco is a commercial desktop application that focuses on

---

[2] See https://doi.org/10.5281/zenodo.10298566.

data exploration by providing interactive ways of filtering and tuning the mining parameters. Apromore is a commercial web-based application that models and mines processes in the cloud. SAP Signavio Process Intelligence and Celonis offer functionalities comparable to Apromore, while emphasising the capabilities of easily ingesting data from other systems. When comparing the tools, desktop applications permit data confidentiality (as data does not have to leave the premises of the data owner, *on-premise* deployments could be viable options too for web applications), while IP protection is possible for systems that allow all or some components to be closed-source. Interoperability is generally possible but requires the manual exchange of files (though with slightly different meaning compared to what was discussed). Extensibility is possible either via open-source code or micro-services/plug-in-based architectures (in the former case, it requires a development team, while the latter is much more tailored to end users). Scalability and elasticity are easily achievable only for web applications. Regarding the requirements focusing on the execution of the scenarios, only partial and very sparse support is seen, with no obvious patterns emerging.

All in all, no platform fully meets the requirements. Despite this, our prototype is the only tool with at least partial fulfilment for all requirements.

6   Conclusion and Future Work

In this paper, we focused on integrating data sources and algorithms across organisations in order to enable process mining. We elicited and validated scenarios, and identified corresponding requirements. We also presented a prototype capable of addressing most of the requirements. In the future, we plan to advance the prototype by focusing on end users both from an implementation as well as a conceptual point of view. For example, we will implement and evaluate a type-checking mechanism that would allow to specify only "correct" wirings or define a proper Domain Specific Language for specifying scenarios tailored to end users.

References

1. van der Aalst, W.M.P., Bolt, A., van Zelst, S.J.: RapidProM: Mine Your Processes and Not Just Your Data. CoRR **abs/1703.03740** (March 2017)
2. Process Mining. Springer, Heidelberg (2016). https://doi.org/10.1007/978-3-662-49851-4_16
3. Alman, A., Di Ciccio, C., Maggi, F.M., Montali, M., van der Aa, H.: RuM: declarative process mining, distilled, pp. 23–29 (2021)
4. Barker, A., van Hemert, J.: Scientific workflow: a survey and research directions. In: Parallel Processing and Applied Mathematics (PPAM), pp. 746–753 (2008)
5. Berti, A., van Zelst, S.J., van der Aalst, W.M.: Process mining for python (PM4Py). In: Proceedings of the ICPM Demo Track (2019)
6. van Dongen, B., de Medeiros, A.K.A., Verbeek, E.H.M.W., Weijters, T.A.J.M.M., van der Aalst, W.M.: The ProM framework: a new era in process mining tool support. Appl. Theory Petri Nets **3536**, 444–454 (2005)

7. Emamjome, F., Andrews, R., ter Hofstede, A.H.M.: A case study lens on process mining in practice. In: Panetto, H., Debruyne, C., Hepp, M., Lewis, D., Ardagna, C.A., Meersman, R. (eds.) OTM 2019. LNCS, vol. 11877, pp. 127–145. Springer, Cham (2019)
8. Fahrenkrog-Petersen, S.A., van der Aa, H., Weidlich, M.: PRETSA: event log sanitization for privacy-aware process discovery. In: ICPM. IEEE (2019)
9. Holbrook, H.: A scenario-based methodology for conducting requirements elicitation. ACM SIGSOFT Softw. Eng. Notes **15**(1), 95–104 (1990)
10. Janssenswillen, G., Depaire, B., Swennen, M., Jans, M., Vanhoof, K.: BUPAR: enabling reproducible business process analysis. Knowl. Based Syst. **163** (2019)
11. Kourani, H., van Zelst, S.J., Lehmann, B.D., Einsdorf, G., Helfrich, S., Liße, F.: PM4KNIME: process mining meets the KNIME analytics platform. In: ICPM Doctoral Consortium/Demo, pp. 65–69 (2022)
12. Lewis, J., Fowler, M.: Microservices: a definition of this new architectural term (2014). https://martinfowler.com/articles/microservices.html
13. Merkoureas, I., Kaouni, A., Theodoropoulou, G., Bousdekis, A., Voulodimos, A., Miaoulis, G.: Smyrida: a web application for process mining and interactive visualization. SoftwareX **22**, 101327 (2023)
14. Munappy, A.R., Bosch, J., Olsson, H.H.: Data pipeline management in practice: challenges and opportunities. In: PROFES, pp. 168–184 (2020)
15. Rafiei, M., van der Aalst, W.M.P.: Mining roles from event logs while preserving privacy. In: Business Process Management Workshops, pp. 676–689 (2019)
16. Richardson, C.: Microservices Patterns. Manning Publications, New York (2019)
17. dos Santos Garcia, C., Meincheim, A., Junior, E.R.F., Dallagassa, M.R., Sato, D.M.V., Carvalho, D.R., et al.: Process mining techniques and applications-a systematic mapping study. Expert Syst. Appl. **133**, 260–295 (2019)
18. Schuster, D., van Zelst, S.J., van der Aalst, W.M.: Cortado: a dedicated process mining tool for interactive process discovery. SoftwareX **22**, 101373 (5 2023)
19. Zerbato, F., Burattin, A., Völzer, H., Becker, P.N., Boscaini, E., Weber, B.: Supporting provenance and data awareness in exploratory process mining. In: Proceedings of CAiSE, pp. 454–470. Springer Nature, Switzerland, Cham (2023)
20. Zimmermann, L., Zerbato, F., Weber, B.: What makes life for process mining analysts difficult? A reflection of challenges. Softw. Syst. Model. (2023)

How Can We Develop Inclusive Business Process Management?

E. R. Mahendrawathi[✉] [iD]

Institut Teknologi Sepuluh Nopember, Surabaya, Indonesia
mahendra_w@is.its.ac.id

Abstract. Diversity and inclusion are fundamental principles within the Sustainable Development Goals (SDGs). While the BPM discipline has grown to tackle more contemporary issues, diversity and inclusion remain understudied. This problem statement paper aims to trigger the BPM community to think about developing more inclusive BPM. This paper first introduced a definition of inclusive BPM. Then, areas for further studies on developing inclusive BPM are proposed with several examples in the context of Indonesia.

Keywords: Inclusive · Inclusive BPM · Sustainable Development Goals

1 Introduction

Diversity and inclusion are fundamental principles within the Sustainable Development Goals (SDGs), a set of global objectives to tackle a range of social, economic, and environmental issues. Diversity and inclusion are important for individual SDGs and the overall framework, highlighting the importance of equitable and inclusive approaches in all aspects of development [18].

Business Process Management (BPM) as a discipline has grown tremendously since its global initiation in the early 90s by Hammer & Champy [9] and Davenport [6]. A literature review on the state-of-the-art of BPM research has revealed that many papers highlight ways BPM research could be more diverse, inclusive, or innovative [15]. The call for innovativeness has been answered as the discipline has expanded to include contemporary topics such as agile [2], green [5], and benevolence BPM [16].

There has been a movement to broaden the scope of the BPM discipline to contribute to achieving SDGs in 2030. The advancement of technology and the change in the environment, emphasized even more by the COVID-19 global pandemic, has driven the emergence of Process Science with several key characteristics: a focus on processes, the scientific investigation of these processes, an interdisciplinary lens, and the intention to influence and change processes to create an impact [3]. Despite this effort, a quick search of the literature for inclusive and BPM does not find any literature that specifically deals with business processes or BPM.

This fact highlights the problem that, despite the strong push for inclusivity, it has not received enough attention in the BPM discipline. Several research questions emerge:

K. Gdowska et al. (Eds.): BPM 2024 Workshops, LNBIP 534, pp. 370–375, 2025.
https://doi.org/10.1007/978-3-031-78666-2_28

What does inclusivity mean for BPM? How can the BPM discipline contribute to the attainment of inclusivity?

The remainder of this paper is structured as follows. First, related works that help define inclusive BPM are presented in Sect. 2. Then, in Sect. 3, a definition of inclusive BPM is provided. Section 4 proposes areas for further studies. A brief concluding remark is given in the final section.

2 Related Work

The United Nations Development Programme (UNDP) is a key source of inclusivity. As a UN agency, it helps countries eradicate poverty and achieve sustainable growth. UNDP defines inclusive business models as including the poor as clients, customers, employees, producers, and business owners in the value chain [19].

A literature search is also conducted with the terms "inclusive" AND "process" OR "business process" AND "management," resulting in studies on inclusive business, business models, and inclusive education. A quick review selects relevant studies to extract key points on inclusiveness. An inclusive business model engages income-constrained groups in the value chain by addressing neglected problems. An inclusive business is any self-sustaining entity that adopts this model to generate net value for these groups [17].

Other work offers similar definitions, emphasizing that it is a private sector strategy involving people at the base of the economic pyramid by integrating them into a company's core business value chain as suppliers, distributors, retailers, or customers [14]. Inclusive education guarantees access for all, including those with disabilities, and aims to eliminate discrimination [1]. These definitions refer to the target of inclusivity as the poor, income-constrained groups or people at the base of the economic pyramid and people with disability.

The UNDP recently recognized digital transformation (DT) as a significant driving force, but DT alone will not guarantee the achievement of the SDGs [12]. DT must be intentionally inclusive, carefully designed, and implemented, focusing on people and human rights. Inclusive DT addresses the needs of the poorest and most vulnerable, including women and people with disabilities, empowering underrepresented groups and promoting gender equality. This approach prevents worsening existing inequalities and aligns with the vision of leaving no one behind. It also protects individuals from the negative impacts of digital technologies and promotes open, responsible, and rights-based digital technology. While governments typically lead these efforts, true inclusivity requires a whole-of-society approach, engaging the government, private sector, civil society, academia, and residents.

3 Defining Inclusive BPM: What, Who and Where?

Based on the review of related works, several key aspects of inclusivity can be extracted: 1) it should be intentionally and thoughtfully designed, 2) focusing on people and human rights, 3) addressing the needs of the poor (people at the bottom of the economic pyramids or income-constrained groups) and most vulnerable groups, including women and people with disabilities.

These key aspects are mapped to the definition and stakeholders of BPM provided in [7] to come up with a working definition of Inclusive BPM: a discipline of managing work processes that must be **intentionally inclusive**, thoughtfully **designed** and **implemented** with **people and human rights** at the center of all aspects. This broad definition can be elaborated further into internally and externally oriented inclusivity for different stakeholders of BPM: individuals, organizations, value chains or countries, as shown in Fig. 1.

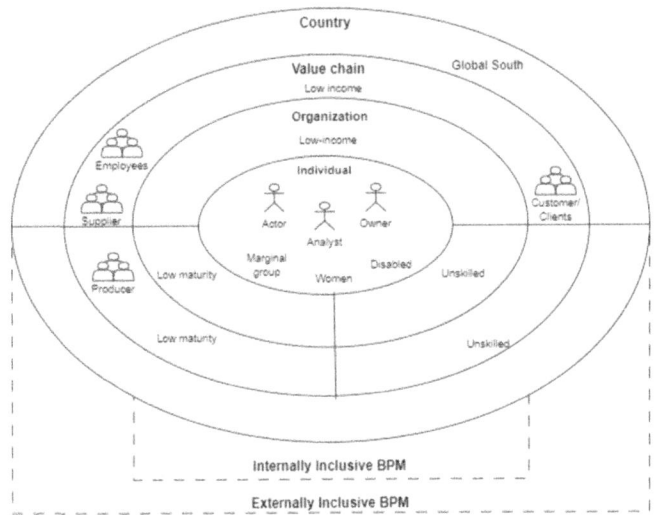

Fig. 1. Conceptualization of Inclusive BPM

Internally inclusive BPM ensures that **no individuals as actors, process analysts, process owners, or organizations that apply BPM** are left behind. Key inclusivity at the individual level is people from marginal groups, women and disabled people. At the organizational level, the key inclusivity target is an organization owned by low-income people with low maturity processes and unskilled workers. **Externally inclusive BPM** means the discipline **actively involved in ensuring inclusivity at the society/country level**. This means expanding the scope of the discipline to consider the issues facing poor/income-constrained groups/people at the base of the economic pyramid on the demand side as clients and customers and on the supply side as employees, producers and business owners at various points in the value chain. Inclusivity at the value chain level should also consider organizations with unskilled labor and low maturity. Finally, inclusivity at the country level should target countries in the global south.

Inclusive BPM aligns with the work of context-aware BPM as it provides an additional goal to consider in BPM adoption, which will influence the BPM planned action [4]. It also fits with benevolent BPM [16], as both focus on aligning BPM initiatives with the well-being of all stakeholders, including employees, customers, and society. Inclusive BPM adds extra consideration of income-constrained groups, women and people with disabilities.

4 Developing Inclusive BPM: Opportunities for Further Studies

Based on the definition, an ideation process is done to identify areas that need to be addressed in developing Inclusive BPM. The BPM lifecycle, with special highlights on modeling, is used to structure the key areas. Two main approaches can be considered in developing inclusive BPM: mainstream or targeted, inspired by [11]. Mainstream research is when the BPM projects consider inclusivity in all aspects of its development and application. Targeted research creates BPM-specific interventions to address the specific needs, gaps and challenges of the inclusivity target group.

Table 1 summarizes areas for internally inclusive BPM. It provides examples rather than a comprehensive list. A mainstream or targeted approach can be chosen for each stage in the BPM lifecycle. For example, in the identification stage, a mainstream research area will be to specifically identify barriers and challenges to inclusivity in the current business processes. A more targeted area is specifically identifying a set of processes targeted to inclusivity. Other opportunities arise in other stages of the BPM life cycle. Several works on women and BPM competency [8, 22] provide a good example of a targeted approach to inclusive BPM, but further studies are needed in other areas.

Table 1. Opportunities for Further Studies in Internally Inclusive BPM

	Mainstream	Targeted
Identification	Identify barriers and challenges to inclusivity in an organization's business process	Identifying a set of processes specifically for inclusivity target groups (e.g., people with disability)
Modeling	Identify how the current way of modeling business processes may challenge certain stakeholders (e.g., using patterns instead of color in modeling)	Develop a way to visualize inclusive sensitizing points in the business processes (e.g., modeling standards for visually impaired users)
Analyze	Analyzing inclusivity in the organization's business process	Analyzing inclusive processes (e.g., onboarding and HR process)
Re-design	Inclusive as redesign goals of organizations business processes	Redesigning inclusive processes with specific inclusivity goals
Implement	Implement inclusive sensitive solutions to current processes	Implementation of inclusive processes and inclusive BPM solutions
Monitor	Monitor inclusivity attainment	Monitor inclusive processes

The areas presented in Table 1 can be expanded into Externally Inclusive BPM. The main difference will be the scope and inclusivity of target groups. For Externally Inclusive BPM, the scope should be the value chain and country, focusing on low-income, low maturity and unskilled groups who act as employees, suppliers, producers, and customers. For example, mainstream research can focus on ensuring inclusivity issues in the processes, tools and methods of BPM across the entire value chain.

Several examples of inclusive BPM issues in the context of Indonesia are as follows.

Individual level. As developing country and part of the Global South, Indonesia is characterized by a significant number of small businesses, with 99% of its organizations being Small and Medium Enterprises (SMEs) or Microbusinesses (MBs) employing fewer than ten people [10]. In Indonesia, a specific type of microbusiness is a warung. These small-scale general stores are typically owned and operated by individuals, often within a single family, and they sell goods and services that are relevant and valuable to the local community. Recently, several startups in Indonesia have sought to digitally transform the traditional business processes of warungs. With the help of mobile applications provided by the startups, warung is now connected to the retail value chain of big corporations. Warung owners can conduct procurement, sales and accounting with the application. However, many warung owners are challenged in the DT [13]. For these owners, simple tasks of ordering products are difficult. The question is: how can we simplify processes for such individuals with low skills?

Organization level. Many organizations in Indonesia may not be aware of the BPM as a holistic concept because they have very low process maturity and are constrained by human resources knowledge and skills related to BPM [8]. These tend to be SMEs with low process maturity that opted for low-cost solutions. This is a challenge for BPM provider to sensitize their offering to include the needs of such organizations.

Organization and Country level. Some of Indonesia's philosophies and cultural practices align with BPM. Pancasila, the five principles of the Indonesian state philosophy, Gotong Royong (mutual cooperation), musyawarah mufakat (deliberation and consensus), rukun (harmony and social order) principles emphasize community, harmony, and collective effort, which are valuable in enhancing organizational processes and management practices. However, many organizations face difficulty breaking the functional silos. The questions for inclusive BPM are: why does the national culture that aligns with BPM not transpire at the organizational level? how can we address this issue?

5   Concluding Remarks

This paper aims to trigger the BPM community to think about developing more inclusive BPM. In arguing so, this paper introduced a definition of Inclusive BPM. Then, the paper proposes areas for further studies and gives examples in the context of Indonesia.

References

1. Ari, R., Altinay, Z., Altinay, F., Dagli, G., Ari, E.: Sustainable management and policies: the roles of stakeholders in the practice of inclusive education in digital transformation. Electronics (Switzerland) **11**(4) (2022)
2. Badakhshan, P., Conboy, K., Grisold, T., vom Brocke, J.: Agile business process management: a systematic literature review and an integrated framework. Bus. Process. Manag. J. **26**(6), 1505–1523 (2020)
3. Brocke, J.v., van der Aalst, W.M.P., Berente, N., et al.: Process science: the interdisciplinary study of socio-technical change. Process Sci. **1**, 1 (2024). https://doi.org/10.1007/s44311-024-00001-5

4. vom Brocke, J., Zelt, S., Schmiedel, T.: On the role of context in business process management. Int. J. Inf. Manag. **36**(3), 486–495 (2016)
5. Couckuyt, D., Van Looy, A.: A systematic review of green business process management. Bus. Process. Manag. J. **26**(2), 421–446 (2020)
6. Davenport, T.: Process Innovation: Reengineering Work Through Information Technology. Harvard Business School Press, Boston (1993)
7. Dumas, M., La Rosa, M., Mendling, J., Reijers, H.A.: Fundamentals of Business Process Management. Springer, Berlin, Heidelberg, Berlin, Heidelberg (2018)
8. ER, M., Nurkasanah, I., Pratama, A.R.: A configuration taxonomy based on business process orientation: evidence from organizations in Indonesia. Benchmarking **30**(10), 3837–3859 (2023)
9. Hammer, M., Champy, J.: Reengineering the Corporation. Harper (1993)
10. Indonesia Investment: Micro, Small & Medium Enterprises in Indonesia: Backbone of the Indonesian Economy (2022). https://www.indonesia-investments.com/finance/financial-col umns/micro-small-medium-enterprises-in-indonesia-backbone-of-the-indonesian-economy/ item9532
11. Koneksi: Grant Guideline of KONEKSI Call for Proposal Digital Transformation (2024)
12. Lister, S., et al.: Inclusive by Design: Accelerating Digital Transformation for the Global Goals (2022)
13. Mahendrawathi, E.R., Muhammad, G., Ramadhany, E.D., Nurkasanah, I.: Digital transformation of warungs in Indonesia: the interrelation of socio-technical and indigenous factors. Electron. J. Inf. Syst. Dev. ctries. (2024)
14. Mangnus, E.: An assessment of "inclusive" business models: vehicles for development, or neo-colonial practices?: Case studies from Ghana and Kenya. J Agric. Environ. Ethics **36**(3) (2023)
15. Recker, J., Mendling, J.: The state of the art of business process management research as published in the BPM conference: recommendations for progressing the field. Bus. Inf. Syst. Eng. **58**(1), 55–72 (2016)
16. Rosemann, M., Ostern, N., Voss, M., Bandara, W.: Benevolent business processes - design guidelines beyond transactional value. In: LNCS, pp. 447–464. Springer (2023)
17. Schoneveld, G.C.: Sustainable business models for inclusive growth: towards a conceptual foundation of inclusive business (2020)
18. SDG Resource Centre: Diversity and inclusion. https://sdgresources.relx.com/diversity-and-inclusion. Accessed 06 July 2024
19. United Nations Development Program: Brokering Inclusive Business Models (2010)

Business Value of Process Mining: A Contingency Perspective

Astria Hijriani[1,2] and Marco Comuzzi[1(✉)]

[1] Ulsan National Institute of Science and Technology, Ulsan, South Korea
{astria.hijriani,mcomuzzi}@unist.ac.kr
[2] University of Lampung, Bandar Lampung, Indonesia

Abstract. This short paper introduces our ongoing research on the value conversion contingencies of process mining. We believe that this novel perspective on process mining business value complements the existing literature, which is mostly focused on critical success factors and affordances of process mining. We delineate the theoretical background of our proposed framework and briefly discuss the methods that we have chosen to achieve our intended goals.

Keywords: business value · process mining · contingency theory

1 Introduction

As a crucial part of business process management, process mining provides a means for organizations to gain insights into their operational processes by allowing the extraction of valuable information from event logs, laying the groundwork for data-driven analysis and improvement of business processes [3]. According to industry research, process mining technology has grown significantly during the last five years, with market analysts predicting a remarkable increase in the coming years. Despite the widespread adoption of process mining across various industries, comprehensive research still needs to be conducted to explain the value organizations derive from this technology.

The ability to recognize the business value of process mining is crucial to demonstrating its strategic value for an organization. Currently, the value of process mining is demonstrated based on its impact on organizational performance, i.e., its measurable outcomes [5] or, more generally, by highlighting the opportunities that it provides for process improvement [1]. However, implementing a process mining initiative can also be seen as an investment in information technology to develop an information capability [7] focused on analyzing business operations extensively. Such a perspective on process mining opens up various challenges and opportunities for researchers. Specifically, we focus on the contingencies that could arise in a process mining initiative. These are the different events or circumstances that may influence, in either a positive or negative way, the creation of business value through process mining. Contingency theory has

K. Gdowska et al. (Eds.): BPM 2024 Workshops, LNBIP 534, pp. 376–381, 2025.
https://doi.org/10.1007/978-3-031-78666-2_29

already been considered in the context of technology implementation and process improvement [8] and critical success factors of BPM [12], but it has not been applied specifically to the study of the value created by process mining initiatives. In this context, we aim to answer the following research question: "What are the value conversion contingencies of process mining, and how do they influence the creation of business value?"

We argue that a contingency theory-based framework can effectively bridge the gap between process mining and business value, offering practical implications for achieving organizational goals and increasing business process performance. The theoretical foundation for this approach emphasizes the importance of aligning environmental variables with the organizational structure for optimal performance [2]. As an analytic and information processing capability, process mining can be strategically leveraged for tangible organizational benefit.

This short paper delineates our effort to discover value conversion contingencies in process mining and the research method that we adopt for identifying and validating such contingencies. The paper is structured as follows: First, we present the theoretical background of our work in Sect. 2. Then, we discuss the core of our contribution in Sect. 3. Finally, in Sect. 4, we introduce our study's research design, providing examples of preliminary results supporting our core contribution.

2 Theoretical Background

The success of a process mining project is often measured by analyzing process improvements, such as decreasing process durations or more effective dashboards to support process-related decision-making. A critical yet frequently overlooked step is value recognition, that is, recognizing the impact for an organization of process mining adoption [13]. Value recognition involves more than understanding the potential value for business process improvement. Value stems from implementing careful planning and execution to avoid frustration due to the failures [2], developing correct estimations of project efforts, measuring and monitoring changes, and evaluating project outcomes [5]. Additionally, ensuring that the executives and process owners perceive and appreciate the actual changes is crucial, especially if we consider process mining as an IT investment [2].

Research on process mining implementation success has focused mainly on understanding its critical factors. Mamudu et al. [6] present a set of process mining critical success factors extracted from case studies and interviews. Stein Dani et al. [11] identify certain factors as challenges in process mining, deriving recommendations. The approach is consistent with the concept of creating business value with process mining, which explains how organizational and individual traits interact to create effective systems [13]. Badakhsan et al. [1] address value more explicitly, identifying the affordances of process mining that are likely to create value from analyzing a set of case studies. We argue that these research works focus on the actual (realized) value of process mining, but fail to conceptualize the potential value of process mining that may (or not) be reached by a process mining initiative.

To address this crucial gap in the literature, we propose to conceptualize process mining as an information (analytic) capability and to consider contingency theory [2,7], which helps situate process mining outcomes within organizational objectives and contextual factors. The contingency theory states that a capability's effectiveness level depends on how well it fits the organization's context. This theory assumes that there is no universal approach to capability development, and the best approach will depend on factors such as the organization's resources, existing processes, capabilities, and external environment [2,10]. Contingency theory emphasizes the importance of understanding where potential value lies and how best to relate it contextually to measuring the firm's realized value across multiple levels of analysis. This approach introduces the idea that complementary assets, especially business process design and human capital, influence the firm's value realization. The theory utilizes concepts such as locus of value and value conversion contingencies [2].

3   Contingency Perspective on Process Mining Value

Figure 1 illustrates how an approach centered on value conversion contingencies can be useful for recognizing the business value of process mining.

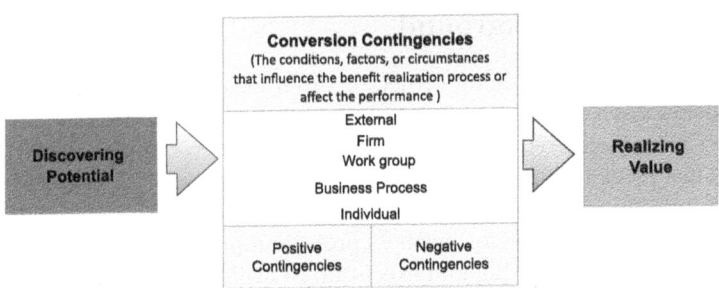

Fig. 1. Illustration of contingency approach.

As shown in Fig. 1, value conversion contingencies refer to the conditions, factors, or circumstances that influence the transformation of potential value (before an IT investment, or pre-implementation) into realized value (the actual outcomes achieved after implementing the IT solution, i.e., post-implementation) [2]. These conversion contingencies could be positive or negative. The conversion contingencies can be defined and bear an impact at different levels: individual, business process, that is, routines and procedures, workgroup, firm, and external. External contingencies, specifically, are related to market conditions, such as competitor strategy or job market characteristics, and they are not usually under an organization's direct control.

For instance, organizations with substantial financial resources and innovative capabilities (i.e., positive, firm-level value conversion contingencies) can pursue more aggressive growth strategies, improving their chances of converting the potential value of IT into realized (actual) value. Conversely, organizations struggling to acquire technical skills on the job market (i.e., a negative, external-level conversion contingency) may have to scale down and delay their effort in developing advanced technology, e.g., AI.

Our approach is centered on conceptualizing process mining as an information capability. While an organization's resources are its stable, functional properties, capabilities are the action possibilities that emerge from using them. As an analytic capability, process mining can be linked to different maturity levels in an organisation [4]. For instance, process mining may be only at an initial, exploratory stage, or it may have already led to the development of advanced predictive dashboards of process execution. The journey along the maturity of process mining may depend on the level of fit of process mining with organizational characteristics, which we conceptualize as value conversion contingencies.

For process mining, a practical illustration of the process depicted in Fig. 1 can be as follows. Suppose an organization has just begun to look into process mining. The value they pursue will differ from company B, which has already run a successful process mining initiative for three or five years. At the same time, the value conversion contingencies for the two organizations may differ, and the strategies to cope with them may be different. Company A may have to deal with negative contingencies related to acquiring skilled process mining experts (external-level) and resistance from staff (individual-/workgroup-level) related to increased scrutiny over human resources behaviour entailed by process mining. While for company B these contingencies may be no longer relevant, company B may have to address an efficiency gap related to more effective average process mining adoption from its competitors (external-level). An analysis of conversion contingencies is necessary to understand the constraints and to be able to prioritize an organization's journey to higher process mining maturity.

4 Research Method

Our research aims to establish a framework for defining the business value of process mining using a contingency approach. We apply in-depth qualitative analysis to build, re-specify, and validate the framework. The research unfolds in three phases: (i) identifying the value conversion contingencies of process mining, (ii) integrating the identified contingencies into the levels of process mining maturity, and (iii) validating the overall framework with first-hand data.

Identifying Conversion Contingencies from Case Studies. After conducting a thorough review of the related literature, our first objective is to identify a list of positive and negative value conversion contingencies in process mining. We are particularly interested in the negative contingencies since they represent a viewpoint that is often under-investigated in the literature. We begin to identify the contingencies inductively analyzing the case studies in process mining made available by the Task Force for Process Mining[1] and other bibliographic

---

[1] https://www.tf-pm.org/resources/casestudy.

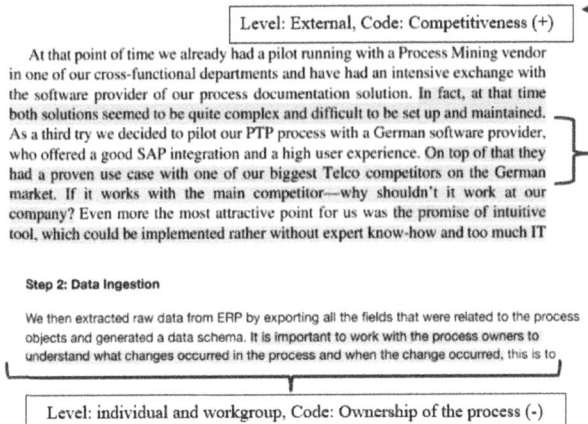

Fig. 2. Code examples for identifying conversion contingencies

resources, such as [9]. These cases present PM implementation projects from the viewpoints of PM users, vendors, and practitioners, emphasizing successful outcomes and the immediate advantages of PM. Figure 2 shows examples of the coding procedure of passages from case studies. The first passage suggests that the choice of process mining technology provider could be influenced by the competitor's choices (external-level contingency); this may result in a positive or negative contingency depending on the complexity of the implementation of the competitor's choice (firm-level contingency). The second passage highlights the significance of involving the process ownership to understand the insights obtained from process mining (i.e., a negative contingency at the individual-/workgroup-level). Since the case studies focus mainly on positive project outcomes, we also plan to rely on interviews with industry experts to expand on the negative contingencies.

Integrate the Contingencies into Process Mining Maturity. After identifying the conversion contingencies, we aim to integrate these with the P3M process mining maturity model [4], which we developed in previous research, determining how the positive contingencies help improve maturity and how negative ones could hinder the maturity improvement journey. To do so, we will match contingencies levels to the dimensions of P3M, revising the P3M's maturity model descriptors if necessary. Matching contingencies and maturity levels will help develop more effective guidelines to navigate the maturity levels.

Validating the Framework with Experts. The comprehensive framework of value conversion contingencies and their integration with P3M will be validated by a diverse panel of experts through structured interviews.

5   Conclusions

This paper has introduced our plan to develop a framework for the business process value of process mining based on the identification of value conversion contingencies and the integration of those into our P3M maturity model. We argue that such a perspective complements the existing research on PM business value, helping to look beyond PM success factors.

References

1. Badakhshan, P., Wurm, B., Grisold, T., Geyer-Klingeberg, J., Mendling, J., vom Brocke, J.: Creating business value with process mining. J. Strateg. Inf. Syst. **31**(4), 101745 (2022)
2. Davern, M.J., Kauffman, R.J.: Discovering potential and realizing value from information technology investments. J. Manag. Inf. Syst. **16**(4), 121–143 (2000)
3. Eggers, J., Hein, A., Böhm, M., Krcmar, H.: No longer out of sight, no longer out of mind? How organizations engage with process mining-induced transparency to achieve increased process awareness. Bus. Inf. Syst. Eng. **63**(5), 491–510 (2021)
4. Hijriani, A., Comuzzi, M.: Towards a maturity model of process mining as an analytic capability. In: Hawaii International Conference on System Science (2024)
5. Kubrak, K., Milani, F., Nolte, A.: Process mining for process improvement-an evaluation of analysis practices. In: International Conference on Research Challenges in Information Science, pp. 214–230. Springer (2022)
6. Mamudu, A., Bandara, W., Leemans, S.J., Wynn, M.T.: A process mining impacts framework. Bus. Process. Manag. J. **29**(3), 690–709 (2023)
7. Marchand, D.A., Kettinger, W.J., Rollins, J.D.: Information orientation: people, technology and the bottom line. MIT Sloan Manag. Rev. **41**(4), 69 (2000)
8. Pacheco-Cubillos, D.B., Boria-Reverter, J., Gil-Lafuente, J.: Transitioning to agile organizational structures: a contingency theory approach in the financial sector. Systems **12**(4), 142 (2024)
9. Reinkemeyer, L.: Process mining in action. Process mining in action principles, use cases and outloook (2020)
10. Schweikl, S., Obermaier, R.: Lost in translation: it business value research and resource complementarity–an integrative framework, shortcomings and future research directions. Manag. Rev. Q. **73**(4), 1713–1749 (2023)
11. Stein Dani, V., Leopold, H., van der Werf, J.M.E., Reijers, H.A.: Progressing from process mining insights to process improvement: challenges and recommendations. In: EDOC 2023, pp. 152–168 (2023)
12. Trkman, P.: The critical success factors of business process management. Int. J. Inf. Manag. **30**(2), 125–134 (2010)
13. Vom Brocke, J., Jans, M., Mendling, J., Reijers, H.A.: A five-level framework for research on process mining. Bus. Inf. Syst. Eng. 1–8 (2021)

Towards a Unified Approach: Developing a Reference Model for Digital Twins and Business Process Management in Clinical Trials

Gerald Kremer[1]([envelope]), Luiz Ricardo Brito Ribeiro[2], Till Blüher[1], Silvia Dallavalle[2], and Rainer Stark[1]

[1] Technische Universität Berlin, Pascalstraße 8-9, 10587 Berlin, Germany
kremer@tu-berlin.de

[2] University of São Paulo, Av. Bandeirantes, Ribeirão Preto 3900, Brazil
luizrbr@usp.br

Abstract. The present study aims to develop a reference model that integrates Digital Twins (DTs) and Business Process Management (BPM) to enhance the efficiency and effectiveness of clinical trials. To achieve this, a five-step method was employed: (1) a collaborative workshop with Clinical Research Center (CRC) managers, scientists, and principal investigators (PIs); (2) validation of stakeholder needs; (3) conceptualization of a DT use case; (4) development of a synergetic reference model; and (5) evaluation of this model with a clinical research specialist. The results identified key business needs from various stakeholders, proposed a DT solution to address these needs, and created a reference model that integrates BPM and DT perspectives. This integrated approach addresses challenges such as ensuring continuous and interoperable data capture, motivating systematic data collection, and facilitating data transfer between CRCs, thus improving stakeholder satisfaction in clinical trials. The developed reference model and the identified research challenges provide a foundation for future research and practical implementation of DTs and BPM in clinical trials. This study advances the field by bridging DT and BPM frameworks, offering a comprehensive approach to managing clinical trial processes, and demonstrating the significant potential of these technologies to improve clinical research efficiency and effectiveness.

Keywords: digital twin · business process management · clinical trial · reference model

1 Introduction

One of the key Sustainable Development Goals set by the United Nations is to ensure "good health and well-being" [1]. Achieving this goal requires the provision of affordable and effective pharmaceuticals and medical devices, which depend on robust industry capabilities that leverage digital advancements. Clinical Research Centers (CRCs) play a crucial role in this, evaluating medical products globally to verify their efficacy and

© The Author(s), under exclusive license to Springer Nature Switzerland AG 2025
K. Gdowska et al. (Eds.): BPM 2024 Workshops, LNBIP 534, pp. 382–393, 2025.
https://doi.org/10.1007/978-3-031-78666-2_30

safety, thereby supporting the SDG on "Global partnership and sustainable development" [1].

Digital twins (DTs) are increasingly significant in traditional industrial product development. These virtual replicas combine operational data with executable models to derive insights or predict future behavior [2]. DTs require high-quality data from various lifecycle stages, enabling product analysis and process management. Given their industrial success, there's growing interest in exploring DTs' potential in clinical trials. However, the view on DT in CT is mostly human centered and involves AI-based data analysis [2] whereas the industrial DT approach is rather model-based.

Healthcare organizations' management practices significantly impact the development of new drugs or medical equipment. The sector's dynamic environment necessitates adaptable processes. Business Process Management (BPM) focuses on delivering value through end-to-end processes [3], encompassing all work performed by an organization in delivering products and services.

However, key components in the healthcare chain, such as CRCs, face significant challenges in managing operations efficiently to provide access to innovative treatments [4]. The range of techniques and tools offered by BPM is crucial for facilitating structured data collection and process management, both of which are essential for the effective operation of DTs [5, 6]. DTs rely on structured data, accurate models, connected IT infrastructure, and an operational approach that adapts as business processes evolve [7]. Identifying synergies between DT and BPM concepts is vital for developing a DT framework that supports BPM in clinical trials.

CRCs are motivated to manage operations efficiently, with BPM methods playing a key role. BPM facilitates structured process management and data collection, which is crucial for developing DTs. The effective operation of DTs relies on structured data, accurate models, connected IT infrastructure, and an operational approach that adapts as business processes change. DTs can provide insights useful to various stakeholders, including principal investigators, sponsors (pharmaceutical and medical device companies), and CRCs. Identifying synergies between DT and BPM concepts is essential for a DT framework supporting BPM in clinical trials.

This context leads to the research question: What are the application potentials and challenges for integrating BPM and DT frameworks to address the complexities of clinical trials more efficiently? Which DT and BPM use cases are most promising? This article aims to develop a reference model integrating DTs and BPM to enhance clinical trials' efficiency and effectiveness. The research identifies challenges and gaps in implementing these technologies and proposes digital solutions to meet stakeholder needs. The article includes a literature review, a five-step research method, results, discussion, conclusion, and next steps.

2 Literature Review

The literature review examines the current state of research and practice in the integration of DTs and BPM within clinical trials, highlighting key advancements, challenges, and gaps that this study aims to address.

2.1 Clinical Trials

Clinical trials are a pivotal aspect of clinical research conducted primarily in clinical research centers (CRCs) or units (CRUs), where new pharmacological treatments, devices, and technologies are tested on humans to verify their effects and safety before being approved for public use. This process is critical for developing new drugs and healthcare products, assessing their efficacy, and ensuring they meet safety standards before they can be registered or modified with relevant authorities. The types of clinical research range from epidemiological and prevention studies to translational and health service evaluation research, including cost-effectiveness analyses [8, 9].

The development of these medical products and treatments relies heavily on the proper execution and management of clinical trials, which are handled by the CRCs. These centers are not only responsible for the operational management of the trials but also contribute socially by providing participants with access to new treatments. Despite protocols being consistent across different centers, each CRC autonomously makes decisions regarding task organization, recruitment methods, and resource allocation [10, 11].

Clinical Trials are usually segmented into four primary stages: Study Start Up, Study Conduction, Study Close Out, and providing administrative assistance [12].

During the initiation phase of the study, essential tasks consist of drafting the protocol, identifying sites and investigators, conducting feasibility assessments, organizing qualification visits, developing case report forms, selecting investigators, creating budgets, managing contracts, and submitting for approval to the Institutional Review Board or Regulatory Authority. This phase is chiefly overseen by principal investigators (PIs), and managers, establishing a solid groundwork for the clinical trial.

The study conduction phase concentrates on implementing the trial, conducting site inspections, developing data, managing investigational products (IP), reporting safety concerns, and overseeing site monitoring, which entails the participation of PIs, data specialists, health authorities, and clinical research associates. The study closure phase deals with data organization, statistical evaluations, reporting findings from the clinical study, and archiving documents, ensuring that all trial data is appropriately handled and analyzed. Administrative support guarantees the continuous oversight of contracts, risks, and financial transactions, amalgamating all essential administrative tasks for seamless trial procedures [12].

However, the clinical trial process faces numerous challenges that can impede efficient execution. These include protocol amendments, unengaged participants, complex procedures that complicate recruitment, and the high resource demands for data management [13, 14]. Furthermore, the recent pandemic has underscored the necessity for well-coordinated multicenter trials, which suffer from complex and often inefficient processes due to a lack of formal process representation and fragmented process views among staff [15].

2.2 Business Process Management

To address these challenges, there is a growing emphasis on the importance of improving process management within these research environments. Implementing continuous process improvements in clinical research is vital for enhancing efficiency and effectiveness. Such improvements require cultural changes within organizations and can yield significant benefits, including streamlined operations, reduced costs, and enhanced quality of healthcare services. Business process modeling plays a crucial role in this context by providing a structured method of analyzing and improving clinical trial processes. It allows for a clearer understanding of existing processes and facilitates the identification and implementation of necessary improvements to support the complex logistics of multicenter studies [16–18]. This concerted effort towards process innovation helps Clinical Research Centers orchestrate functions across various specialized medical departments and disciplines, ultimately leading to better health outcomes and more efficient use of resources.

Process modelling is a typical tool used in the diagnostic phase BPM lifecycle, offering a systematic approach to enhancing the efficiency and effectiveness of clinical trials and healthcare services. In clinical research, it excels as a method for targeting and ameliorating data quality issues. By providing a detailed visualization of data collection processes, process modelling enables researchers and healthcare professionals to pinpoint the root causes of poor data quality, thus paving the way for substantial improvements in the collection and management of scientific data [19]. This enhancement of data integrity is crucial not only for the accuracy of clinical trials but also for ensuring the reliability of the findings, which ultimately support medical advancements and patient safety.

The adoption of process modelling in healthcare not only streamlines data collection but also fosters better collaboration across different disciplines involved in clinical research. For instance, the work of Kononowech et al. [20] illustrates how visual process models can support implementation efforts by clarifying roles and responsibilities within interdisciplinary teams. This clarity is essential for the efficient execution of clinical trials, where timely and accurate data collection is paramount. Additionally, by documenting and assessing consistent behaviors and activities, process models help in refining these processes, thus reducing inefficiencies and enhancing the overall effectiveness of the research being conducted.

Process mining is an algorithmic approach used within BPM, that involves analyzing event data from processes collected and maintained in organizational information systems (e.g. ERP systems) to extract insights about the processes, i.e. to analyze the operational state of a process (e.g. it's progress), or to generate process models that depict the actual steps conducted by humans and systems within the process [21]. The approach of process mining therefore can create digital representations of operational processes in the field based on event data and underlying models of these processes creating a DT [22]. Nevertheless, the resulting views and models on processes are just a partial aspect and cannot cover all views that might be relevant for stakeholders in the context of clinical trials, because they are reliant on existing event data.

2.3 Digital Twins

DTs, originally rooted in systems engineering, have evolved beyond physical product replication and extended into the realm of services and processes [23–25]. By creating digital replicas based on underlying causal models, DTs provide insights into operational systems, enabling predictions and prescriptions as well as interactions through remote control of actuators [26]. This way they can also be used to monitor and forecast relevant aspects of processes and provide recommendations to manage processes more effectively [27]. DTs can be implemented on different levels of complexity and through diverse technologies leading to a heterogeneous landscape of DT maturity in industry [28]. DTs consist conceptually of so called "Master Models", representing the underlying system of interest (e.g. through simulation models, graph-based models or data-based models), coupled with an incoming stream of operational data from the asset or process, to make a digital assessments of that real object [29].

The concept of DT therefore in principle is applicable to a wide range of physical and intangible objects (products, processes and hybrids). However, the development of a DT requires a deep domain understanding to be able to develop a feasible digital abstraction. Feasible in that sense includes not only satisfying information and control needs but also the ability of an organization to provide the required master models and operational data flows in an integrated manner with the necessary digital infrastructure [5]. Besides the conceptual part, implementing DTs requires a wide skill set of understanding digital technology to provide the necessary infrastructure for DT operations and IoT-Infrastructure for data acquisition, connectivity and actuator control in the physical world.

3 Research Method

As outlined above DTs are highly reliant on the quality of data, they are processing [29]. BPM on the other hand can serve as the missing auxiliary element which enables this data provision. The present study's objective is to identify and highlight the potential for improved stakeholder satisfaction in clinical trials through the synergetic application of BPM and DT. To enable future system developers a reference model is to be developed including the system's main solution elements. More than that associated challenges and research questions need to be identified.

To meet these objectives the study followed a five-step method. The first step was to organize and conduct a workshop with 13 presentations and interviews with CRC managers, scientists, PIs and to conduct two CRC site visits. Afterwards protocols of the interviews were screened and needs of the stakeholders were identified and clustered. The second step was a validation of the identified needs by a clinical research specialist. The third step was to conceptualize three DT use cases for clinical trials incl. The support through BPM, from which one is presented in this paper. Based on this use case a reference model for the synergetic approach was to be conceptualized in the fourth step. In the final fifth step the envisioned solution and the reference model was to be evaluated with a clinical research specialist to identify challenges and derive open research questions.

4 Results

The first step of the method was realized through a workshop week with presentations and interviews with CRC managers, PIs, sponsors and scientists. During the workshop week, protocols of discussions with various stakeholders were maintained. From these protocols, the business needs mentioned were extracted. These identified needs were subsequently thematically clustered to organize them into coherent groups. Additionally, an analysis was conducted to determine which business needs were commonly identified across different stakeholders, to facilitate a structured understanding of individual and collective requirements but also highlight areas of consensus and divergence among the participants. The outcome was the venn diagram in Fig. 1. From the management of a CRC a total of ten business needs were identified. Only mentioned by the CRCs were the need for support of finding the best fitting participants and the minimal resource consumption (incl. costs for staff) and to keep a good relationship to the participants due to high acquisition costs. Together with the PIs they share the need for the lean but digitalized support for data capturing processes and automation of the documentation. Together with the sponsors they share the need to fulfill the sponsor's requirements in time and the exact and not misunderstanding communication of CT protocols. The sponsor's needs were to have positive CT outcomes and get new insights for the development of new products and protect their intellectual property rights. Together with the PIs sponsors shared the need for more context information of participants for the ex-post analysis of CTs and the understanding about the mode of action of the medical products with individual participants. The PIs alone formulated the need for the collaborative knowledge exchange with other PIs and the early information of possible side effects of products and intervention methods. All three stakeholders shared the need for the safety of participants, knowledge about the effects of drugs for participants with different preconditions and the efficient monitoring and controlling of manual processes.

In the second step of the method the diagram was evaluated by the clinical research specialist. The specialist added four more needs to the diagram, which are depicted in the white boxes in Fig. 1. The added needs include: appropriate compensation for conducting research, access to innovative treatment, low dropout rates of participants, increase automation level of CRC processes.

In the third research step based on the needs a possible use case for a DT solution was conceptualized by researchers from the domains of Industrial IT and BPM. The goal of the envisioned solution was to best address the identified business needs from each stakeholder. Nonetheless in the conceptualization phase it was not possible to develop a "one-fits-all" solution to address all stakeholder needs. Instead, multiple solutions of DT were identified to best address each of the needs. This would also go along with the usage of different master models, shadow data, IT infrastructure and methodological and technological support of BPM frameworks. Still, one DT solution was conceptualized to meet the research objective.

The developed use case can be found in Fig. 2. The use case focuses on the business needs coming from the sponsor's side. Here a solution was conceptualized so that insights through clinical trial data are generated to be used in future developed products and better predict outcomes of clinical trials before the trial start. From a DT perspective a solution was developed, in which toxicological models combined with patient and trial data should be used to predict the outcome of new medical products. More than that sponsors could use models to control the mode of action in case of severe adverse events (SAE). This way employees in CRC could directly get information/alerts about their need of intervention based on patient and trial data. The CT data was to be captured by both automatic data capturing approaches through process mining technologies and through reviewing existing models of CRC by sponsors and including appropriate process models in CT protocols.

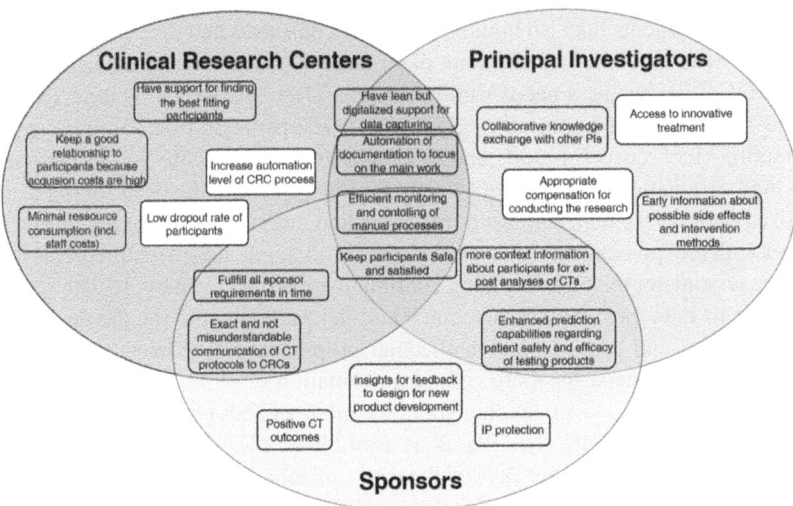

Fig. 1. Identified Business Needs of the Relevant Stakeholders in Clinical Trials

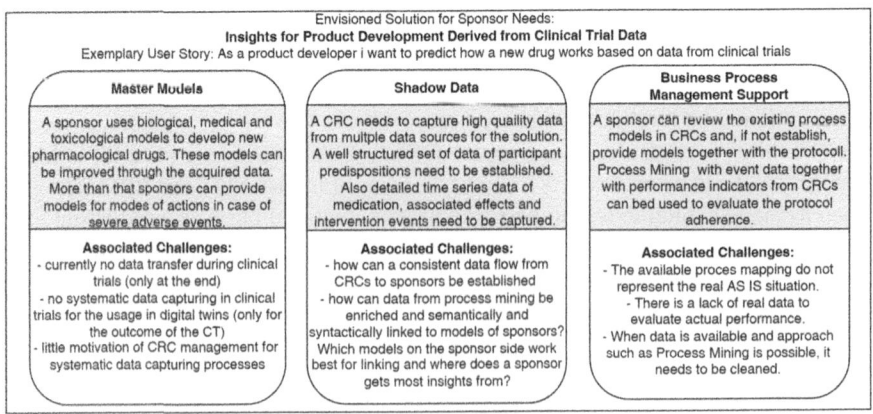

Fig. 2. Envisioned Concept of a DT Solution Including the Support through BPM

In the fourth step of the method, based on the identified business needs and the envisioned solution the synergetic reference model was developed to guide the creation of new DT solutions in clinical trials targeting these needs. The goal of the reference model was to help future solution developers identify the users and comprehend the interrelations of solution elements within future use cases. More than that contribution areas of BPM should be highlighted for identifying new research areas. The reference model is depicted in Fig. 3, which illustrates the various stakeholders within the system alongside the artifacts and resources available for solution development. The DT solution is depicted in the middle and consists of different models that are provided by both CRCs, sponsors and PIs. These models are linked via a twinning engine with the aggregated shadow data.

The data can derive from the real-world processes in CRCs, from patient data but also from proprietary software solutions including their data bases that are used in CRCs. The DT solution itself is to be used by different stakeholders in the process by user and use case specific user interfaces. Red highlighted in the reference model are the key elements in which the BPM perspective is needed to build a feasible DT solution.

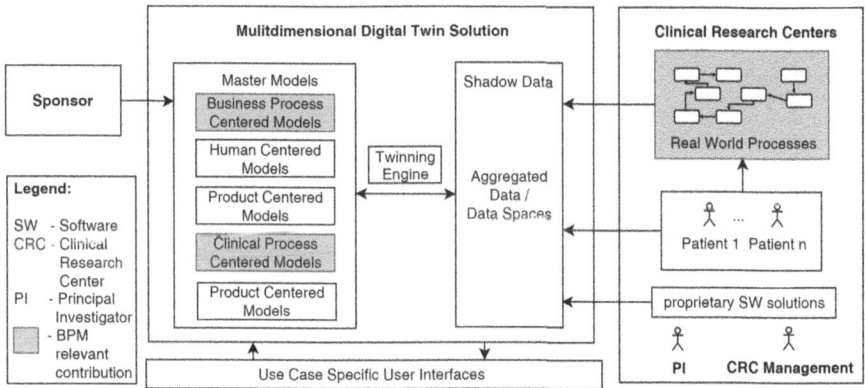

Fig. 3. Reference Model of Elements for DT Solutions for Clinical Trials incl. Elements of BPM Contribution

In the fifth step of the method the envisioned use case and the developed reference model was evaluated by the clinical research specialist. Identified possible additional use cases with high potential of the DT solution were the simulations of CT progress to effectively modify protocols to shorten CT execution times. It was suggested to also include data from pharmacological studies that are executed before the CT. The most prominent remark about the model is about the ownership of patient data. Data captured during the CT is, by contract, owned by the sponsors. The usage of collected data in digital twin solutions therefore must be authorized by sponsors beforehand.

Based on the reference model and the envisioned use case associated challenges were analyzed together with the clinical research specialist. Based on the challenges research needs were derived. They can be found in Table 1.

Table 1. Challenges and Research Needs for DT and BPM in Clinical Trials.

Challenge	Potential Research Question
Data availability only after, not during, clinical trials	How can continuous and interoperable data capture be ensured during clinical trials?
Lack of systematic data capturing processes	How can CRCs be motivated to systematical-ly capture a wide variety of data?
Absent process, product, and human-centered models	What models need to be developed for the creation of DTs in the context of clinical trials?
No data transfer between CRCs	How can interconnected and efficient data transfer between CRCs be established?
Insufficient knowledge for model creation in CRCs	How can CRCs be empowered in the development and utilization of models for DTs?
Available process maps do not represent the "AS IS" situation	How can process maps accurately reflect the current situation in CRCs?
Lack of data to evaluate process models	What types of data are necessary for an accurate evaluation of process models?
High effort for data preprocessing for process mining	How can the effort required for data preprocessing in process mining be reduced?
Dependence on humans to share their expertise	How can reliance on human experts in process modeling and optimization be minimized?
Variations in standard procedures	How can variations in standard procedures during clinical trials be managed?
Need to semantically and syntactically enrich data from process mining for use in DTs	How can data from process mining be semantically and syntactically enriched for DT creation?
Identification of models to link process mining data to DTs	What are the most suitable models to link process mining data to DTs?
Integration of multimodal data representing patient conditions	How can multimodal data representing patient conditions be dynamically integrated?
Automatic derivation and sharing of knowledge through DTs	How can knowledge be automatically derived and shared through DTs?
Protection of individual patient data	How can the protection of patient data be ensured during the use of DTs and BPM?

5   Discussion

As evidenced by previous studies, the integration of DTs and BPM frameworks holds significant potential for enhancing the efficiency and effectiveness of clinical trials. Stark et al. [5] emphasized the necessity of deep domain understanding and the availability of structured data for the successful implementation of DTs. Our findings align with this, showing that the structured data captured through BPM significantly supports the creation and application of DTs.

Process mining, as discussed by Park and van der Aalst [21] plays a critical role in generating the master models required for digital process twins. However, our study revealed that while BPM facilitates the structured recording and management of processes, the motivation for extensive data aggregation remains low among CRCs. This finding echoes the challenges highlighted by Dagliati et al. [15] regarding the fragmented process views and lack of formal process representation in multicenter trials.

The simplicity and ease of integration of BPM into CRC workflows, as noted by Rebuge & Ferreira & others [16], were particularly favored in our study. This facilitated smoother adoption and highlighted the importance of continuous process improvements in clinical research environments, a point also emphasized by Zong et al. [19].

An integrated approach, combining data aggregation through BPM and utilization for DTs, proved to be the most beneficial. However, significant research challenges remain, particularly in identifying and developing suitable models for DTs and ensuring the semantic and syntactic linkage of data from BPM to DTs.

6   Conclusion and Next Steps

The primary objective of this study was to develop a reference model integrating DTs and BPM to enhance the efficiency and effectiveness of clinical trials. This research successfully identified the challenges and gaps in implementing these technologies and proposed digital solutions tailored to various stakeholders' needs. The reference model developed, the use cases, and the integrated framework provide a robust foundation for addressing these challenges, demonstrating the potential of DTs and BPM to improve clinical trial processes significantly.

This study advances the state-of-the-art by bridging the gap between DTs and BPM within the context of clinical trials. This research contributes by integrating these frameworks, providing a comprehensive approach to managing clinical trial processes. The findings underscore the necessity of structured data and deep domain understanding and offer a practical pathway for implementing these technologies in real-world scenarios.

Despite the many derived research gaps and challenges the idea of developing digital twins with the help of BPM in the clinical field has had positive outcomes. Nonetheless a standardized approach and framework to enable all stakeholders to develop those solution is yet missing. In the next steps the reference model needs to be further enriched with information about the system's infrastructure, the classification of relevant data and identified models in the domain of CTs.

Acknowledgments. 2023/08335-3 Fundação de Amparo à Pesquisa do Estado de São Paulo (FAPESP) 406168/2021-0 National Council for Scientific and Technological Development (CNPq) and the internal SPRINT funding of TU Berlin for mobility of researchers.

References

1. Universal sustainable development goals, 2015. http://sustainabledevelopment.un.org/content/documents/1684sf_-_sdg_universality_report_-_may_2015.pdf
2. Liu, H., Xia, M., Williams, D., Sun, J., Yan, H.: Digital twin-driven machine condition monitoring: a literature review. J. Sens. **2022**(1), 1–13 (2022). https://doi.org/10.1155/2022/6129995
3. Maddern, H., Smart, P.A., Maull, R.S., Childe, S.: End-to-end process management: implications for theory and practice. Prod. Plan. Control **25**(16), 1303–1321 (2014). https://doi.org/10.1080/09537287.2013.832821
4. Roehrich, J.: Healthcare Operations Management. In: Cooper, C.L. (ed.) Wiley Encyclopedia of Management, pp. 1–3. Wiley, Hoboken (2015)
5. Stark, R., Fresemann, C., Lindow, K.: Development and operation of digital twins for technical systems and services. CIRP Ann. **68**(1), 129–132 (2019). https://doi.org/10.1016/j.cirp.2019.04.024
6. Antonacci, G., Calabrese, A., D'Ambrogio, A., Giglio, A., Intrigila, B., Ghiron, N.L.: A BPMN-based automated approach for the analysis of healthcare processes. In: 2016 IEEE 25th International Conference on Enabling Technologies: Infrastructure for Collaborative Enterprises, Paris, France, 2016
7. Dang, J., Hedayati, A., Hampel, K., Toklu, C.: Personalized medical workflow through semantic business process management. In: Proceedings of the 11th International Conference on Enterprise Information, Milan, Italy, 2009
8. United States Of America. National Institute of Health, 2018. https://ncats.nih.gov/ctsa/about
9. Brasil. Ministério da Saúde. https://bvsms.saude.gov.br/bvs/saudelegis/gm/2018/prt0559_14_03_2018.html
10. Zucchetti, C., Morrone, F.B.: Perfil da pesquisa clínica no Brasil. Revista HCPA, Porto Alegre **32**(3), 340–347 (2012)
11. Huckman, R.S., Zinner, D.E.: Does focus improve operational performance? Lessons from the management of clinical trials. Strateg. Manag. J. **29**(2), 173–193 (2008). https://doi.org/10.1002/smj.650
12. Park, Y.R., et al.: Utilization of a clinical trial management system for the whole clinical trial process as an integrated database: system development. J. Med. Internet Res. **20** (2018). https://doi.org/10.2196/jmir.9312
13. Karara, A.: Clinical trial–end to end optimization. Imp. J. Interdiscip. Res. India **2**(11) (2016)
14. Lee, C., et al.: Clinical trial metrics: the complexity of conducting clinical trials in North American cancer centers. JCO Oncol. Pract. **17**(1), e77–e93 (2021). https://doi.org/10.1200/OP.20.00501
15. Dagliati, A., Malovini, A., Tibollo, V., Bellazzi, R.: Health informatics and EHR to support clinical research in the COVID-19 pandemic: an overview. Brief. Bioinform. **22**(2), 812–822 (2021). https://doi.org/10.1093/bib/bbaa418
16. Rebuge, Á., Ferreira, D.R.: Business process analysis in healthcare environments: a methodology based on process mining. Inf. Syst. **37**(2), 99–116 (2012). https://doi.org/10.1016/j.is.2011.01.003

17. Yarmohammadian, M.H., Ebrahimipour, H., Doosty, F.: Improvement of hospital processes through business process management in Qaem teaching hospital: a work in progress. J. Educ. Health Promot. **3**, 111 (2014). https://doi.org/10.4103/2277-9531.145902

18. Daudelin, D.H., Selker, H.P., Leslie, L.K.: Applying process improvement methods to clinical and translational research: conceptual framework and case examples. Clin. Transl. Sci. **8**(6), 779–786 (2015). https://doi.org/10.1111/cts.12326

19. Zong, W., Lin, S., Gao, Y., Yan, Y.: Process-driven quality improvement for scientific data based on information product map, EL, vol. 40, no. 3, pp. 177–195, 2022. https://doi.org/10.1108/EL-08-2021-0157

20. Kononowech, J., et al.: Visual process maps to support implementation efforts: a case example. Implement. Sci. Commun. **1**(1), 105 (2020). https://doi.org/10.1186/s43058-020-00094-6

21. van der Aalst, W.M.P., Berti, A.: Discovering Object-Centric Petri Nets, October 2020. http://arxiv.org/pdf/2010.02047

22. Rojas, E., Munoz-Gama, J., Sepúlveda, M., Capurro, D.: Process mining in healthcare: a literature review. J. Biomed. Inform. **61**, 224–236 (2016). https://doi.org/10.1016/j.jbi.2016.04.007

23. Hänel, A., et al.: The development of a digital twin for machining processes for the application in aerospace industry. Procedia CIRP **93**, 1399–1404 (2020). https://doi.org/10.1016/j.procir.2020.04.017

24. Zhang, H., Ma, L., Sun, J., Lin, H., Thürer, M.: Digital twin in services and industrial product service systems. Procedia CIRP **83**, 57–60 (2019). https://doi.org/10.1016/j.procir.2019.02.131

25. Tao, F., Cheng, J., Qi, Q., Zhang, M., Zhang, H., Sui, F.: Digital twin-driven product design, manufacturing and service with big data. Int. J. Adv. Manuf. Technol. **94**(9–12), 3563–3576 (2018). https://doi.org/10.1007/s00170-017-0233-1

26. Michael Grieves, "Origins of the Digital Twin Concept," 2016

27. Ebel, H., Riedelsheimer, T., Stark, R.: Enabling automated engineering's project progress measurement by using data flow models and digital twins. Int. J. Eng. Bus. Manag. **13**, 184797902110336 (2021). https://doi.org/10.1177/18479790211033697

28. Digital twin readiness assessment, 2020. https://publica.fraunhofer.de/bitstreams/eb53ea9f-7aba-4afd-a4cf-b1434c4fff99/download

29. Stark, R.: Virtual product creation in industry: the difficult transformation from IT enabler technology to core engineering competence. Springer, Berlin 2022

Business in the Age of Platform Economics: Managing Decentralised Business Processes Beyond Blockchain

Fabian Stiehle[1]([✉]), Finn Klessascheck[1,2], Martin Kjäer[3], and Ingo Weber[1,4]

[1] Technical University of Munich, School of CIT, Munich, Germany
{fabian.stiehle,finn.klessascheck,ingo.weber}@tum.de
[2] Weizenbaum Institute for the Networked Society, Berlin, Germany
[3] TU Wien, Vienna, Austria
[4] Fraunhofer Gesellschaft, Munich, Germany

Abstract. The promise of an internet-based 'sharing economy' stands corrected by the reality of a 'platform economy' under the control of a few central proprietary markets. With increasing awareness of the many downsides of these oligopolistic markets, regulatory policies like the EU *Digital Markets Act* (DMA) have been enacted, while in research and practice, the term 'decentralisation' gained renewed interest. However, as we argue in this *problem statement*, decentralisation is often confused as an aim, while in reality it encompasses many different (often opposing) drivers and principles to achieve certain aims. In order to stimulate discussion, we discuss decentralisation referencing a multi-disciplinary perspective and exemplary projects in practice. We ask how profitable and sustainable business models can be created in the age of platform economics. We claim that, to reap the many proclaimed benefits of platforms, while managing and preventing its many centralistic downsides, an approach integrating technical and organisational capabilities is needed. We see Business Process Management (BPM) ideally positioned to be at the forefront of such a movement. With this contribution, we hope to stimulate discussion within the BPM community. Further, we see opportunity even beyond BPM. We draw comparisons to privacy engineering, which emerged as a discipline to operationalise the abstract notion of privacy.

Keywords: Platform Economy · Decentralisation · Blockchain · Business Process Management

1 Introduction

Early theorizing on the internet contemplates the many promises of a decentralised network, where information flows freely (see e.g., [12]). Some authors proclaimed the dawn of a new *sharing economy* [20]. Taking stock of the actual state of the web paints a different picture: the dominant business models of the

K. Gdowska et al. (Eds.): BPM 2024 Workshops, LNBIP 534, pp. 394–400, 2025.
https://doi.org/10.1007/978-3-031-78666-2_31

internet are proprietary markets under the control of few. More aptly, this model has become known as *platform economy* or *digital capitalism* [21]. Platforms serve as intermediaries in so-called two-sided markets [17], where the attractiveness of one side (e.g., number of users) has a direct influence on the attractiveness of the other side (e.g., advertisers). Through network effects, price suppression, or large sums of investment capital, some platforms have established quasi-monopolies [1,4]. Yet, even these platforms are arranged in a hierarchy, where the offerings of giants like Alphabet, Amazon, Apple, or Meta are regarded as *superstructures* [1], *meta-platforms* [21], or, in the EU *Digital Markets Act* (DMA), as *gatekeepers*. Platforms of a lower hierarchy cannot reach relevance without entering the ecosystems of these few gatekeepers [1]. These control market access, command large amounts of data, and have created strong lock-in effects [4].

Even in the case of technologies like blockchain, tightly associated with decentralisation,[1] early "success" stories often were centralised platforms, offering e.g., proprietary markets for non-fungible tokens.[2]

Awareness of the downsides of these platforms has risen recently, fuelled by major scandals like Cambridge Analytica, where large amounts of data was exploited.[3] Gatekeepers are criticised for creating strong lock-in effects and stifling competition [4]. Or more drastically, for exploiting large amounts of personal data [26], creating precarious work environments,[4] and even eroding the fabric of liberal democracies [25].

To tackle these issues from a policy side, the European Union has enacted a range of new regulations. First, the *General Data Protection Regulation* was introduced, which gave rise to an entire new discipline: *privacy engineering*, an interdisciplinary field of research with the aim to translate the abstract notion of privacy into engineering knowledge and applicable practice [7]. Second, and more recently, the *Digital Services Act* and the DMA were introduced, with the latter specifically addressing the gatekeeper platforms, and obligating them to, a.o., interoperability and greater transparency. Simultaneously, the idea of decentralisation and decentralised systems has gained renewed interest in research and practice as a possible remedy [23].

We postulate that, in order to reap the many proclaimed benefits of platforms while managing and preventing their many centralistic downsides, an interdisciplinary approach that integrates technical and organisational aspects of decentralisation is needed—following the example of privacy engineering.

We see the discipline of *Business Process Management* (BPM) ideally positioned to be at the forefront of a line of research to explore this notion. BPM

---

[1] A more nuanced analysis [24] characterizes blockchain as "logically centralized [...] organizationally decentralized and physically distributed" system.

[2] https://www.theverge.com/2022/2/2/22914081/open-sea-nft-marketplace-web3-fundraising-finzer-a16z [Accessed: 13/06/2024].

[3] https://www.theguardian.com/news/2018/mar/17/cambridge-analytica-facebook-influence-us-election [Accessed: 14/06/2024].

[4] https://www.thenation.com/article/archive/how-crowdworkers-became-ghosts-digital-machine [Accessed: 13/06/2024].

integrates organisational aspects with socio-technical systems, and focuses on business value and business improvement and change [5]. Most importantly, a process view allows a detailed analysis of information flow through a business.[5] Furthermore, BPM has a track record of integrating different perspectives into the management of processes (e.g., [3]). Although the notion of decentralised processes is nothing new *per se*, the focus so far has been largely on technical capabilities. One could say that our proposal follows a BPM tradition, in that the primary focus is on technical capabilities followed by a call for better integration of the managerial capabilities later (cf. [3]).

Indeed, as we claim, the fallacy of many previous hopes and promises was the postulate that organisational decentralisation would follow technical decentralisation (cf. also [2]). In this problem statement, we want to pose the overarching challenge: *"How can we create profitable and sustainable business models in the age of platform economics?"* Viewing current market dynamics through the lens of platform economics allows us to re-contextualize the need for decentralised process management and chart new opportunities for business and research— while avoiding the fallacy to treat decentralization as an unquestioned aim in itself. We believe this topic to be central to today's societal issues and see immense opportunity. Our epistemological viewpoint hereby follows Rosemann et al.: *doing good pays off* [18].

In the following, we will discuss the challenges and pitfalls, but also the opportunities of decentralised process management, referencing an inter-disciplinary view. We will exemplify the need for this new sub-discipline by introducing examples from organisational practice. We conclude by presenting exemplary problem statements for different aspects of BPM, in order to stimulate discussion.

2 Decentralisation: Drivers, Challenges and Opportunities

For a long time, the term decentralisation in the context of information systems mostly relied on a topology-based view (arrangement of computing nodes), with early works investigating how such topologies map to organisational structure (e.g., [11]).

The term rose to more significance with the popularity of blockchain and its many promises to revolutionise business [13]. Today, decentralisation is mostly defined with reference to this novel technology. (see e.g. [22,23]). We claim that this sole focus hampers many more practical and useful instantiations of decentralised systems.

As we will exemplify, many related, practical challenges exist at present— while proven use-cases of blockchain are still rare [14]. By referencing an inter-disciplinary view, we propose to refocus on the actual *drivers* of decentralisation. This allows to approach decentralisation as an umbrella term for different design principles to achieve certain aims. While we argue that the sole focus on

---

[5] Compare this to the coarse-grained notion of the decentralised organisation [6].

blockchain is problematic, we also envision how such a viewpoint can help shape blockchain research in information systems—shifting focus, with blockchain as one option of many, not an aim in itself.

2.1 Drivers, Challenges and Opportunities

Bodó et al. discuss four drivers for decentralisation that emerge from the viewpoint of different disciplines: *economics, power* (or distribution thereof), *politics* (or disintermediation thereof), and *information security* [2]. For the purpose of this problem statement, we forego the discussion on information security, as it is well covered by the distributed systems field. We reference these drivers, but discuss it in the light of platform economics and exemplary projects to highlight the current challenges, as well as opportunities, of decentralisation. From this, we formulate exemplary research challenges.

Economics. Applying a platform economy lens uncovers the—presently strong—incentives to form closed markets. Platforms often seek to acquire massive market dominance, following a loss-leader strategy that is projected to eventually pay off [1]. Examples of such business models are numerous, with companies like Uber reporting losses for fifteen years before turning a profit.[6] Consequently, such business models are reliant on large capital investment or high stock market valuations [1].

New business models for emerging players can be incentivised through decentralisation. A present example is the movement towards open banking [15], where financial institutions are obligated to allow account holders to permit data access to other services. As the delicate nature of financial transactions exemplifies, such new obligations must be accompanied by technical and organisational *safeguards* to prevent, e.g., abuse of data. Moreover, any new unregulated decentralised market can lead, in principle, to strong *re-centralisation* effects (cf. [2]).

Distribution of Power. Dolata ascribes to closed platforms economic power, power over data, and infrastructural and rule-setting power. Indeed, the design of new features on such platforms are described to be "rule-setting, action-oriented, and opinion forming" [4]. An often fielded example of such rule-setting power is the case of Twitter (now 'X'), where the recent acquisition by Elon Musk led to major changes in its policies, directly affecting a place of public discourse (cf. [25]). Many controversial actions were taken, such as the suspension of accounts belonging to journalists critical of the platform.[7] Subsequently, many users declared their intent to move to alternative platforms (which X actively tried to undermine),[8] often with claims of a more decentralised structure. However, such *decentralised social media* still faces many challenges, often facing

---

[6] https://www.businessinsider.com/uber-ceo-dara-khosrowshahi-helped-turn-billion-dollar-profit-2024-2 [Accessed: 13/06/2024].

[7] https://www.bbc.com/news/world-us-canada-63996061 [Accessed: 13/06/2024].

[8] https://www.bbc.com/news/technology-63999452 [Accessed: 13/06/2024].

incentives toward re-centralisation [16]. However, we also warn of condemning all forms of centralisation: Many centralised institutions fulfil important societal functions. We see *accountable centralisation* (cf. [19]) as a major opportunity, where central control is made accountable and a certain degree of openness is preserved. A first reference frame for what such accountable centralisation might look like could be recent regulation such as the DMA.

Disintermediation. Platforms can be described as intermediaries in two-sided markets [17], bringing different actors together on their proprietary markets. Disintermediation promises to connect parties directly without the need of a central platform. Examples are initiatives like GAIA-X and other data spaces, which create decentralized data infrastructures [9]. They aim to facilitate the standardized exchange of *sovereign, locally stored* data and provide services such as identity management, data discovery, and access control. The data exchange itself occurs solely between data owners and consumers, thereby reducing the aggregation of data at centralized platforms. Similarly, the emerging idea of local-first software advocates data ownership and promotes software usage within local networks. Users can then choose to connect to centralised cloud offerings for additional features [10].

Disintermediation is often accompanied by the notion of reducing certain trust assumptions; however, in more precise terms, trust assumptions just shift (cf. [19]), e.g., from a trusted third party to a distributed network of actors.

3   Outlook: Managing Decentralised Business Processes

Based on our investigation in the previous section, we formulate exemplary research questions and challenges in Table 1, in relation to the BPM discipline. A strong viewpoint that emerges is that decentralisation cannot be thought of as a *purely technical* concept. Only if it is accompanied by organisational structures can it achieve a positive outcome (cf. [2,19]). With this problem statement paper, we want to stimulate discourse on how the presented drivers for decentralisation may shape a decentralised BPM framework to create new but sustainable business opportunities in the age of platform economics. A cornerstone of BPM is that it fuses organisational capabilities with engineering knowledge—precisely what is needed to operationalise the potential benefits. Towards this end, we propose to treat decentralisation not as the mere question of who operates a server, but as an umbrella term encompassing many design principles. Such design principles aim to improve qualities including interoperability, accountability, or equal-access, with an awareness that possible downsides must be controlled for. The success of privacy engineering can serve as a reference. Notably, privacy engineering successfully translated the abstract notion of privacy to organisational and engineering design theory.[9] Consequently, while we think that this

---

[9] See, for example, the *privacy design strategies* [8], now the 'little blue book', containing organisational and technical strategies to implement privacy.

Table 1. Exemplary challenges that emerge from different drivers for decentralisation.

Driver	Exemplary Challenges
Economics	• What innovative business models could emerge through decentralisation; how could such business be sustainably funded (without exploiting re-centralisation effects)?
Power	• What does the notion of accountable centralisation imply for process-aware systems? • How can such systems support accountable decisions and dispute resolution? • How could process re-design heuristics look like to achieve positive effects (for client, worker, or business) through accountable centralisation or decentralisation?
Disinter-mediation	• How can techniques such as conformance checking not just technically but also be organisationally distributed, without a centrally determined prescriptive model or centralized determination of conformance? • How can the effects of disintermediation be quantified and compared?

journey should begin within the BPM community, it must ultimately eclipse it and draw from a larger inter-disciplinary context.

References

1. Andersson Schwarz, J.: Platform logic: an interdisciplinary approach to the platform-based economy. Policy Internet **9**(4), 374–394 (2017)
2. Bodó, B., Brekke, J.K., Hoepman, J.H.: Decentralisation: a multidisciplinary perspective. Internet Policy Rev. **10**(2) (2021)
3. Couckuyt, D., Van Looy, A.: Green BPM as a business-oriented discipline: a systematic mapping study and research agenda. Sustainability **11**(15) (2019)
4. Dolata, U.: Apple, amazon, google, facebook, microsoft: market concentration-competition-innovation strategies. Technical report, SOI Discussion Paper (2017)
5. Dumas, M., Rosa, L.M., Mendling, J., Reijers, A.H.: Fundamentals of business process management. Springer (2018)

6. Ellinger, E.W., Gregory, R.W., Mini, T., Widjaja, T., Henfridsson, O.: Skin in the game: the transformational potential of decentralized autonomous organizations. MIS Q. (2023)

7. Gürses, S., Del Alamo, J.M.: Privacy engineering: shaping an emerging field of research and practice. IEEE Secur. Priv. **14**(2), 40–46 (2016)

8. Hoepman, J.H.: Privacy design strategies. In: IFIP International Information Security Conference, pp. 446–459. Springer (2014)

9. Kari, A., Schurig, T., Gersch, M.: The emergence of a new European data economy: a systematic research agenda for health data spaces. SMR-J. Serv. Manag. Res. **7**(4), 176–198 (2024)

10. Kleppmann, M., Wiggins, A., Van Hardenberg, P., McGranaghan, M.: Local-first software: you own your data, in spite of the cloud. In: ACM SIGPLAN International Symposium on New Ideas, New Paradigms, and Reflections on Programming and Software, pp. 154–178 (2019)

11. Leifer, R.: Matching computer-based information systems with organizational structures. MIS Q. 63–73 (1988)

12. Mason, P.: Postcapitalism: A Guide to Our Future. Macmillan, New York (2016)

13. Mendling, J., Weber, I., Aalst, W.V.D., Brocke, J.V., Cabanillas, C., Daniel, F., et al.: Blockchains for business process management-challenges and opportunities. ACM Trans. Manag. Inf. Syst. (TMIS) **9**(1), 1–16 (2018)

14. O'Leary, D.E.: Blockchain: trouble in the enterprise? ICIS Proceedings, no. 2 (2023)

15. O'Leary, K., O'Reilly, P., Nagle, T., Filelis-Papadopoulos, C., Dehghani, M.: The sustainable value of open banking: insights from an open data lens. In: HICSS, pp. 5891–5901 (2021)

16. Raman, A., Joglekar, S., Cristofaro, E.D., Sastry, N., Tyson, G.: Challenges in the decentralised web: the mastodon case. In: IMC, pp. 217–229. IMC '19, ACM (2019)

17. Rochet, J.C., Tirole, J.: Platform competition in two-sided markets. J. Eur. Econ. Assoc. **1**(4), 990–1029 (2003)

18. Rosemann, M., Ostern, N., Voss, M., Bandara, W.: Benevolent business processes - design guidelines beyond transactional value. In: BPM, pp. 447–464. Springer (2023)

19. Schneider, N.: Decentralization: an incomplete ambition. J. Cult. Econ. **12**(4), 265–285 (2019)

20. Schor, J.: Debating the sharing economy. J. Self-Gov. Manag. Econ. **4**(3), 7–22 (2016)

21. Staab, P.: Markets and Power in Digital Capitalism. Manchester University Press, Manchester (2024)

22. Sunyaev, A., et al.: Token economy. Bus. Inf. Syst. Eng. **63**(4), 457–478 (2021)

23. Sunyaev, A., Lacity, M., Avital, M.: Call for papers for a special issue of the journal of information technology: decentralised information systems: new thinking and new paradigms (2024)

24. Weber, I., Staples, M.: Programmable money: next-generation blockchain-based conditional payments. Digit. Financ. (DFIN) **4**(2), 109–125 (2022)

25. Weinhardt, C., Fegert, J., Hinz, O., van der Aalst, W.M.P.: Digital democracy: a wake-up call: how is research can contribute to strengthening the resilience of modern democracies. Bus. Inf. Syst. Eng. s12599–024–00862-x (2024)

26. Zuboff, S.: The age of surveillance capitalism. In: Social Theory Re-wired, pp. 203–213. Routledge (2023)

Author Index

A

Acitelli, Giacomo 5
Agostinelli, Simone 5
Ahmadi, Zahra 138
Alile, Daniel 319
Al-Zamkan, Amgad 61
Andersen, Julia 101
Armas-Cervantes, Abel 113
Arnold, Lisa 287

B

Bartelheimer, Christian 319
Baumann, Lisa 155
Beerepoot, Iris 180
Bein, Leon 18
Berti, Alessandro 84
Białas, Sylwia 168
Blüher, Till 382
Braakman, Mari A. J. 180
Breitmayer, Marius 271
Brennig, Katharina 197
Brzychczy, Edyta 125, 209
Burattin, Andrea 357
Burger, Mara 247
Busch, Kiran 61
Buss, Alina 221

C

Callisto De Donato, Massimo 113
Casciani, Angelo 5
Comuzzi, Marco 376

D

Dallavalle, Silvia 382
Dees, Marcus 44
Delwaulle, Maxime 247
Di Francescomarino, Chiara 31
Dijkman, Remco 61
Dissegna, Sebastiano 31
Dyhre, Nicholas 357
Dyong, Julian 247

E

Ehrendorfer, Matthias 61
El Kari, Jana 61

F

Fornari, Fabrizio 113
Fournier, Fabiana 233
Franzoi, Sandro 247

G

Goel, Asvin 73

H

Hadian, Raheleh 271
Hartig, Olaf 300
Hijriani, Astria 376

J

Jalali, Amin 300

K

Kaltenpoth, Sascha 197
Kampik, Timotheus 271
Kęsek, Marek 125
Khayatbashi, Shahrzad 300
Kindler, Ekkart 357
Kjäer, Martin 394
Klessascheck, Finn 394
Kluza, Krzysztof 209
Knies, Eva 180
Köhne, Frank 319
Köpke, Julius 259
Kourani, Humam 44
Kratsch, Wolfgang 221
Kremer, Gerald 382
Kunkler, Michel 61

L

Landsiedel, Olaf 101
Lanz, Stefanie 61

K. Gdowska et al. (Eds.): BPM 2024 Workshops, LNBIP 534, pp. 401–402, 2025.
https://doi.org/10.1007/978-3-031-78666-2

Latten, Andrees   319
Leopold, Henrik   61
Limonad, Lior   233
Löhr, Bernd   319
Lugtigheid, Sven   180

M

Mahendrawathi, E. R.   370
Marrella, Andrea   5
Martens, Thomas   180
Müller, Oliver   197

N

Nordlohne, Sina   319
Norouzifar, Ali   44

P

Peeters, Maria   180
Pegoraro, Marco   84
Pentland, Alice P.   338
Pentland, Brian T.   338
Podobińska-Staniec, Marta   125
Pufahl, Luise   18

R

Rathje, Patrick   101
Reichert, Manfred   271, 287
Reijers, Hajo A.   180
Ribeiro, Luiz Ricardo Brito   382
Rodzik, Olga   61

S

Sadeghibogar, Zahra   84
Safan, Aya   259
Sani, Mohammadreza Fani   138
Schaffner, Jan   247

Schmauch, Martin   61
Schmid, Sebastian Johannes   221
Schumann, Felix   61
Seidel, Anjo   155
Serral, Estefanía   138
Shirali, Mohsen   138
Skarbovsky, Inna   233
Stark, Rainer   382
Stiehle, Fabian   394
Szała, Leszek   209

T

Tıraş, Efe   61

U

Urny, Leon   61

V

van der Aalst, Wil M. P.   44, 84
Vestrup, Sebastian   357
Vidgof, Maxim   332
Voelter, Marvin   271
vom Brocke, Jan   247

W

Wang, Hongyang   221
Weber, Barbara   357
Weber, Ingo   394
Weske, Mathias   155
Wolf, Julie Ryan   338
Wróbel, Piotr   168

Z

Zerbato, Francesca   357
Zhang, Li   338
Zuijderwijk, Jos   180

The manufacturer's authorised representative in the EU is Springer
Nature Customer Service Centre GmbH, Europaplatz 3, 69115 Heidelberg,
Germany. If you have any concerns regarding our products, please
contact ProductSafety@springernature.com

Printed and bound by CPI Group (UK) Ltd, Croydon, CR0 4YY
24/04/2026
02096367-0014